Discrete Mathematical Structures for Computer Science

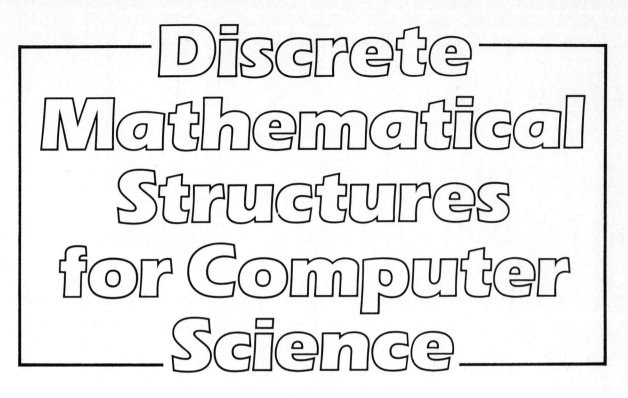

Bernard Kolman

Robert C. Busby
Drexel University

D0068795

Prentice-Hall, Inc., Englewood Cliffs, New Jersey 07632

Library of Congress Cataloging in Publication Data

Kolman, Bernard, (date)
  Discrete mathematical structures for computer
science.

  Includes bibliographies and index.
  1. Electronic data processing—Mathematics.
I. Busby, Robert C.   II. Title.
QA76.9.M35K64 1984   511.3   83-17154
ISBN 0-13-215418-8

To the memory of Lillie
B.K.

**To my wife, Patricia,
and our sons, Robert and Scott**
R.C.B.

Editorial/production supervision and interior design: Maria McKinnon
Chapter openings and cover design: Christine Gehring-Wolf
Cover art: *TriVega*, by Vasarely, courtesy of Vasarely Center, New York.
Manufacturing buyer: John Hall

Printed in the United States of America
10   9   8   7   6   5   4   3   2

ISBN 0-13-215418-8

Prentice-Hall International, Inc., *London*
Prentice-Hall of Australia Pty. Limited, *Sydney*
Editora Prentice-Hall do Brasil, Ltda., *Rio de Janeiro*
Prentice-Hall Canada Inc., *Toronto*
Prentice-Hall of India Private Limited, *New Delhi*
Prentice-Hall of Japan, Inc., *Tokyo*
Prentice-Hall of Southeast Asia Pte. Ltd., *Singapore*
Whitehall Books Limited, *Wellington, New Zealand*

$[a]$	the equivalence class of a, p. 105
A/R	a partition of a set A determined by the equivalence relation R on A, p. 106
$R \circ S$	the composition of R and S, p. 125
1_A	the identity function on A, p. 145
$a \vee b$, LUB(a, b)	the least upper bound of a and b, p. 182
$a \wedge b$, GLB(a, b)	the greatest lower bound of a and b, p. 182
a'	the complement of a, p. 191
$\mapsto$	a finite relation, p. 233
$\Rightarrow$	a substitution relation, p. 233
$\langle \ \rangle ::=$	BNF specification of a grammar, p. 242
S^S	the set of all function from S to S, p. 285
Z_n	the quotient set $Z/\equiv$ (mod n), p. 297
aH	a left coset of H in G, p. 317
$\mathcal{F}$	the set of transition functions of a finite state machine, p. 324
$\mathcal{M}$	the monoid of the machine M, p. 333
$l(w)$	the length of a string w, p. 334
e	an (m, n) encoding function, p. 353
$\delta(x, y)$	the distance between the words x and y, p. 355
$\mathbf{A} \oplus \mathbf{B}$	the mod 2 sum of $\mathbf{A}$ and $\mathbf{B}$, p. 359
$\mathbf{A} * \mathbf{B}$	the mod 2 Boolean product of $\mathbf{A}$ and $\mathbf{B}$, p. 359
d	an (n, m) decoding function, p. 365
ε_j	a coset leader, p. 368
$x * H$	the syndrome of x, p. 372
$\sim p$	not p, p. 377
$p \wedge q$	p and q, p. 378
$p \vee q$	p or q, p. 379
$p \rightarrow q$	p implies q, p. 379
$p \leftrightarrow q$	p is equivalent with q, p. 381

Kolman, Bernard,
1932-

Discrete
mathematical
structures for
computer science

DATE			
DE 1 '86			
MR 9 '?			
OC 17 '88			

Contents

3 Functions 141

4 Order Relations and Structures 164

5 Trees and Languages 220

Preface

Discrete mathematics for computer science is a difficult course to teach to freshmen or sophomores for several reasons. It is a hybrid course. Its content is mathematics, but many of its applications, and most of its students, are from computer science. Thus careful motivation of topics and previews of applications are important. Although the course covers a wide variety of diverse topics, we have integrated the material in various ways.

First, we have limited both the areas covered and the depth of coverage to what we deemed prudent in a *first* course taught at the freshman or sophomore level. We have identified a set of topics that we feel are of genuine use in computer science and that can be presented in a logically coherent fashion. We have presented an introduction to these topics, along with indications how they may be pursued in greater depth. For example, we cover the simpler finite state machines, not Turing machines. Many of the mathematically interesting aspects of graph theory, that we feel are not central to computer science, have been omitted. We have limited the coverage of abstract algebra to a discussion of semigroups and groups and have given applications of these to the important topics of finite state machines and error correcting and detecting codes, respectively. Error correcting codes, in turn, have been primarily restricted to simple linear codes. We have also omitted probability theory, which we view as a topic that needs more time for development than it could receive in this course, but we have included the rudiments of combinatorics.

Second, the material has been organized and interrelated to minimize the mass of definitions and the abstraction of some of the theory. Relations and directed graphs are treated as two aspects of the same mathematical idea, with a directed graph being a pictorial representation of a relation. The material on partially ordered sets has been organized into one chapter that includes lattices and Boolean algebras. Whenever possible each new idea uses previously encountered material and, in turn, is developed in such a way that it simplifies the more complex ideas that follow.

We have tried to produce a text that is neither mathematics nor computer science, but is, instead, a genuine blending of the two. Difficult topics are explained with great care, and many examples and diagrams are provided. This text has been repeatedly tested at Drexel and several other colleges and universities over a period of years. Every effort has been made, by the publisher and authors, to incorporate improvements gained from these classroom experiences into the text.

Chapter 1 contains a miscellany of basic material required in the course. These include sets and sequences, combinatorics, introduction to pseudocode, induction and recursion, and matrices. We have included a simplified pseudocode for the description of algorithms, thereby eliminating a programming language prerequisite. Chapter 2 covers basic types and properties of relations, along with their representation as directed graphs. Connections with matrices and other data structures are also explored in this chapter. Chapter 3 presents the notion of a function, and gives several important examples of functions, including permutations. Chapter 4 deals with partially ordered sets, including lattices and Boolean algebras. Chapter 5 introduces directed and undirected trees, languages, and grammars. In Chapter 6 we give the basic theory of semigroups and groups. These ideas are applied in Chapters 7 and 8. Chapter 7 is devoted to finite state machines. It complements and makes effective use of ideas developed in previous chapters. Chapter 8 treats the subject of binary coding. Logic is briefly discussed in an appendix, which can be consulted as needed. The appendix contains exercises and can be used as Chapter 0 if desired.

The book has a wide variety of exercises at all levels. Every chapter contains a summary of Key Ideas for Review and references for further reading. Answers to all odd-numbered exercises appear in the back of the book. Solutions to all exercises appear in the Instructor's Manual which is available (to instructors only) gratis from the publisher. The Instructor's Manual also contains a collection of sample test questions with solutions.

This text can be used for either a one or two semester (quarter) course. The material in Chapters 1, 2, and 3 may be covered in a more rapid fashion, in order to devote increased time to the less familiar, applications-oriented chapters. Ordinarily, there is more material in the text than can be covered in one semester. If a more leisurely pace is desired, selected topics may be omitted. There are many alternatives, including the following. For a computer science orientation one may select Chapters 1, 2, 3, 5, Sections 6.1, 6.2, Chapter 7, or Chapters 1–6 omitting Sections 4.3, 5.4, 5.6, 6.3, 6.5, and Chapter 8. For an applied algebra orientation one may select Chapters 1, 2, 3, 4, 6, 8. If the text is to be used for two semesters (quarters), then the appendix can be treated as Chapter 0, and all chapters can be covered.

Acknowledgments

We are pleased to express our thanks to the following reviewers, whose suggestions, comments, and criticisms greatly improved the manuscript: Harold Fredricksen, Naval Postgraduate School; Thomas E. Gerasch, George Mason University; Samuel J. Wiley, La Salle College; Kenneth B. Reid, Louisiana State University; Ron Sandstrom, Fort Hays State University; Richard H. Austing, University of Maryland; Nina Edelman, Spring Garden College; Paul Gormley, Villanova University; and Herman Gollwitzer, and Loren N. Argabright, both at Drexel University. We wish to express our thanks to Todd Rimmer for proofreading galleys and pages and for checking the solutions to many of the exercises, to Debbie Hughes and Sandy Sees for cheerfully typing the several versions of the manuscript, and to Connie Mulhern for help with many clerical tasks. Finally, we thank Maria McKinnon, our production editor; Robert Sickles, mathematics editor; and the entire staff of Prentice-Hall, Inc. for their support, encouragement, creative imagination, and unfailing cooperation during the conception, design, production, and marketing phases of this book. We wish to express our sincere gratitude to all those mentioned above.

B.K.
R.C.B.

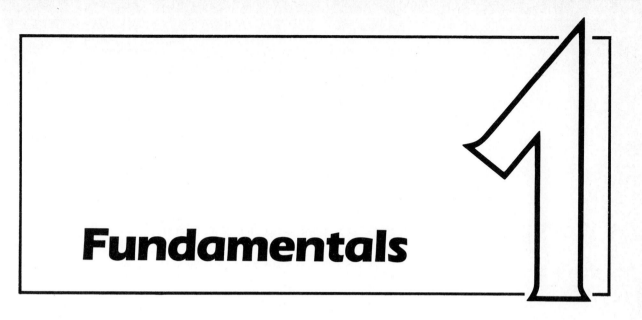

Fundamentals

Prerequisites: There are no formal prerequisites, but the reader is encouraged to consult the Appendix on logic as needed.

In this chapter we introduce the basic tools of discrete mathematics. We start with sets, subsets, and their operations, notions that many of you are already familiar with. Next we deal with ways of counting the elements in a set. We present pseudocode, a language for describing many of the algorithms in this book. Induction and recursion, central ideas in computer science, are introduced along with some of their applications. Then we review some of the basic divisibility properties of the integers. Finally, we introduce matrices and their operations.

1.1 Sets and Subsets

Sets

A **set** is any well-defined collection of objects called the **elements** or **members** of the set. For example, the set of all wooden chairs, the set of all one-legged black birds, or the set of real numbers between zero and one. Almost all mathematical objects are first of all sets, regardless of any additional properties they may possess. Thus set theory is, in a sense, the foundation on which virtually all of mathematics is constructed. In spite of this, set theory (at least the informal brand we need) is quite easy to learn and use.

One way of describing a set that has a finite number of elements is by listing the elements of the set between braces. Thus the set of all positive integers that are less than 4 can be written as

$$\{1, 2, 3\} \tag{1}$$

The order in which the elements of a set are listed is not important. Thus $\{1, 3, 2\}$, $\{3, 2, 1\}$, $\{3, 1, 2\}$, $\{2, 1, 3\}$, and $\{2, 3, 1\}$ are all representations of the set given in (1). Moreover, repeated elements in the *listing* of the elements of a set can be ignored. Thus $\{1, 3, 2, 3, 1\}$ is another representation of the set given in (1).

We use uppercase letters such as A, B, C, S, T, ..., to denote sets, and lowercase letters such as a, b, c, x, y, z, t, ..., to denote members (or elements) of sets.

We indicate the fact that x is an element of the set S by writing

$$x \in S$$

and we indicate the fact that x is not an element of S by writing

$$x \notin S$$

Example 1　Let

$$S = \{1, 3, 5, 7\}$$

Then $1 \in S$, $3 \in S$ but $2 \notin S$.

Sometimes it is inconvenient or impossible to describe a set by listing all its elements. Another useful way to define a set is by specifying a property that the elements of the set have in common. We use the notation $P(x)$ to denote a proposition (sentence or statement) P concerning the variable object x. The set defined by $P(x)$, written

$$\{x \mid P(x)\}$$

is just the collection of all objects x for which P is sensible and true. For example,

$$\{x \mid x \text{ is a positive integer less than } 4\}$$

is the set

$$\{1, 2, 3\}$$

described in (1) by listing its elements.

Example 2　The set consisting of all the letters in the word "byte" can be denoted by

$$\{b, y, t, e\}$$

or by

$$\{x \,|\, x \text{ is a letter in the word "byte"}\}$$

Example 3   We introduce here several sets (and their notations) that will be used throughout this book.

(a) $Z^+ = \{x \,|\, x \text{ is a positive integer}\}$

Thus Z^+ consists of the numbers used for counting: 1, 2, 3, ...

(b) $N = \{x \,|\, x \text{ is a positive integer or zero}\}$

Thus N consists of the positive integers and zero: 0, 1, 2, ...

(c) $Z = \{x \,|\, x \text{ is an integer}\}$

Thus Z consists of all the integers: ..., $-3, -2, -1, 0, 1, 2, 3, ...$.

(d) $\mathbb{R} = \{x \,|\, x \text{ is a real number}\}$

(e) The set that has no elements in it is denoted by the symbol $\varnothing$.

Example 4

$$\varnothing = \{x \,|\, x \text{ is a real number and } x^2 = -1\}$$

since the square of a real number x is always nonnegative.

Sets are completely known when their members are all known. Thus we say that two sets A and B are **equal** if they have the same elements and we write

$$A = B$$

Example 5   If

$$A = \{1, 2, 3\}$$

and

$$B = \{x \,|\, x \text{ is a positive integer and } x^2 < 12\}$$

then

$$A = B$$

Example 6   If

$$A = \{\text{ALGOL, FORTRAN, BASIC}\}$$

and

$$B = \{\text{BASIC, FORTRAN, ALGOL}\}$$

then

$$A = B$$

Subsets

If every element of A is also an element of B, that is, if whenever $x \in A$, then $x \in B$, we say that A is a **subset** of B, or that A is **contained** in B, and we write

$$A \subseteq B$$

If A is not a subset of B, we write

$$A \nsubseteq B$$

(see Fig. 1).

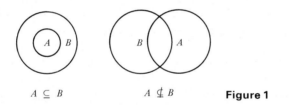

$$A \subseteq B \qquad\qquad A \nsubseteq B \qquad\qquad \textbf{Figure 1}$$

Diagrams, such as those in Fig. 1, which are used to show relationships between sets are called **Venn diagrams** after the British logician John Venn. Venn diagrams will be used extensively in the next section.

Example 7 We have

$$Z^+ \subseteq Z$$

Moreover, if Q denotes the set of all rational numbers, then

$$Z \subseteq Q$$

Example 8 Let

$$A = \{1, 2, 3, 4, 5, 6\}, \qquad B = \{2, 4, 5\}, \qquad C = \{1, 2, 3, 4, 5\}$$

Then

$$B \subseteq A, \qquad B \subseteq C, \qquad C \subseteq A$$

However,

$$A \nsubseteq B, \qquad A \nsubseteq C, \qquad C \nsubseteq B$$

Example 9   If A is any set, then $A \subseteq A$. That is, every set is a subset of itself.

Example 10   Let A be a set and let

$$S = \{A, \{A\}\}$$

Then, since A and $\{A\}$ are elements of S, we have $A \in S$ and $\{A\} \in S$. It then follows that $\{A\} \subseteq S$ and $\{\{A\}\} \subseteq S$. However, it is not true that $A \subseteq S$.

Since an implication is true if the hypothesis is false (see the Appendix at the end of the book), it follows that $\varnothing \subseteq A$.

It is easy to show (Exercise 17) that $A = B$ if and only if $A \subseteq B$ and $B \subseteq A$.

The set of everything, it turns out, cannot be considered defined without destroying the logical structure of mathematics. To avoid this and other problems, which need not concern us here, we will assume that for each discussion there is a "universal set" U (which may vary with the discussion) containing all objects for which our discussion is meaningful. Any other set S mentioned in the discussion will automatically be assumed to be a subset of U. Thus, if we are discussing real numbers and we mention sets A and B, then A and B must (we assume) be sets of real numbers, not matrices, electronic circuits, or rhesus monkeys. In most problems, the universal set will be apparent from the setting of the problem. In Venn diagrams, the universal set U will be denoted by a rectangle, while sets within U will be denoted by circles as shown in Fig. 2.

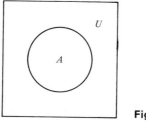

Figure 2

A set is finite if it has n distinct elements, where $n \in N$. In this case, n is called the **cardinality** of A and is denoted by $|A|$. Thus the sets of Examples 1, 2, 4, 5, and 6 are finite, but the sets introduced in Examples 3 and 7 are not finite.

If A is a set, then the set of all subsets of A is called the **power set** of A and is denoted by $P(A)$.

Example 11   Let $A = \{1, 2, 3\}$. Then $P(A)$ consists of the following subsets of A: $\varnothing$, $\{1\}$, $\{2\}$, $\{3\}$, $\{1, 2\}$, $\{1, 3\}$, $\{2, 3\}$, $\{1, 2, 3\} = A$.

It can be shown (Exercise 18, Section 1.6) that if A has k elements, then $P(A)$ has 2^k elements.

EXERCISE SET 1.1

1. Let $A = \{1, 2, 4, a, b, e\}$. Answer each of the following as true or false.
   - (a) $2 \in A$
   - (b) $3 \in A$
   - (c) $c \notin A$
   - (d) $\varnothing \in A$
   - (e) $\varnothing \notin A$
   - (f) $A \in A$

2. Let $A = \{x \mid x$ is a real number and $x < 6\}$. Answer each of the following as true or false.
   - (a) $3 \in A$
   - (b) $6 \in A$
   - (c) $5 \notin A$
   - (d) $8 \notin A$
   - (e) $-8 \in A$
   - (f) $3.4 \notin A$

3. In each part, write the set of letters in each word by listing the elements in the set.
   - (a) AARDVARK
   - (b) BOOK
   - (c) MISSISSIPPI

4. In each part, write the set by listing the elements.
   - (a) The set of all positive odd integers that are less than 10.
   - (b) $\{x \mid x \in Z$ and $x^2 < 12\}$

5. In each part, write the set in the form $\{x \mid P(x)\}$, where $P(x)$ is a property that the elements of the set have in common.
   - (a) $\{2, 4, 6, 8, 10\}$
   - (b) $\{a, e, o, i, u\}$
   - (c) $\{1, 4, 9, 16, 25, 36\}$
   - (d) $\{-2, -1, 0, 1, 2\}$

6. Let $A = \{1, 2, 3, 4, 5\}$. Which of the following sets equal A?
   - (a) $\{4, 1, 2, 3, 5\}$
   - (b) $\{2, 3, 4\}$
   - (c) $\{1, 2, 3, 4, 5, 6\}$
   - (d) $\{x \mid x$ is an integer and $x^2 \leq 25\}$
   - (e) $\{x \mid x$ is a positive integer and $x \leq 5\}$
   - (f) $\{x \mid x$ is a positive rational number and $x \leq 5\}$

7. Which of the following sets are empty?
   - (a) $\{x \mid x$ is a real number and $x^2 - 1 = 0\}$
   - (b) $\{x \mid x$ is a real number and $x^2 + 1 = 0\}$
   - (c) $\{x \mid x$ is a real number and $x^2 = -9\}$
   - (d) $\{x \mid x$ is a real number and $x = 2x + 1\}$
   - (e) $\{x \mid x$ is a real number and $x = x + 1\}$

8. List all subsets of the set $\{a, b\}$.

9. List all subsets of the set $\{$ALGOL, BASIC, FORTRAN$\}$.

10. List all subsets of the set $\varnothing$.

11. Let $A = \{1, 2, 5, 8, 11\}$. Answer each of the following as true or false.
    - (a) $\{5, 1\} \subseteq A$
    - (b) $\{8, 1\} \in A$
    - (c) $\{1, 6\} \nsubseteq A$
    - (d) $\{1, 8, 2, 11, 5\} \nsubseteq A$
    - (e) $\varnothing \subseteq A$
    - (f) $\{2\} \subseteq A$
    - (g) $A \subseteq \{11, 2, 5, 1, 8, 4\}$
    - (h) $\{3\} \notin A$

12. Let $A = \{x \mid x$ is an integer and $x^2 < 16\}$. Answer each of the following as true or false.
    - (a) $\{0, 1, 2, 3\} \subseteq A$
    - (b) $\{-3, -2, -1\} \subseteq A$
    - (c) $\varnothing \subseteq A$
    - (d) $\{x \mid x$ is an integer and $|x| < 4\} \subseteq A$
    - (e) $A \subseteq \{-3, -2, -1, 0, 1, 2, 3\}$

13. Let
    $$A = \{1\}, \quad B = \{1, a, 2, b, c\}, \quad C = \{b, c\},$$
    $$D = \{a, b\}, \quad E = \{1, a, 2, b, c, d\}$$
    In each part, replace the symbol ⬚ by the symbol $\subseteq$ or $\nsubseteq$ to give a correct statement.
    - (a) $A \ \square \ B$
    - (b) $\varnothing \ \square \ A$
    - (c) $B \ \square \ C$
    - (d) $C \ \square \ E$
    - (e) $D \ \square \ C$
    - (f) $B \ \square \ E$

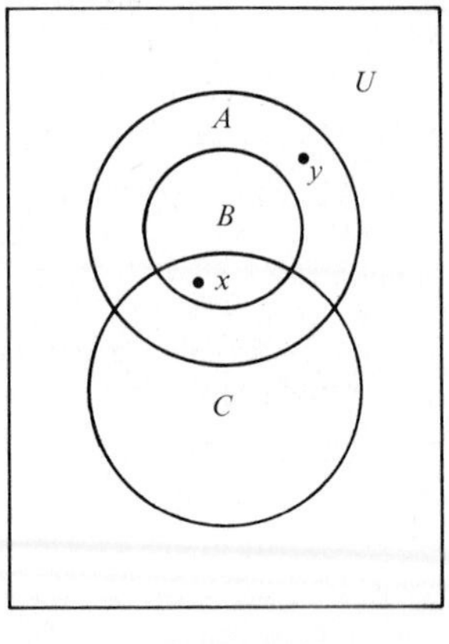

Figure 3

14. In each part, find the set with the smallest number of elements that contains the given sets as subsets.
    (a) $\{a, b, c\}, \{a, d, e, f\}, \{b, c, e, g\}$
    (b) $\{1, 2\}, \{1, 3\}, \varnothing$
    (c) $\{1, a\}, \{2, b\}$

15. Use Fig. 3 to answer each of the following as true or false.

    (a) $A \subseteq B$ (b) $B \subseteq A$
    (c) $C \subseteq B$ (d) $x \in B$
    (e) $x \in A$ (f) $y \in B$

16. If $A = \{1, 2\}$, find $P(A)$.

17. Show that $A = B$ if and only if $A \subseteq B$ and $B \subseteq A$.

1.2 Sequences

Some of the most important sets arise in connection with sequences. **A sequence** is simply a list of objects, one after another, and numbered in natural increasing order by the positive integers. The list may be finite; that is, it may stop after a certain number of items or it may, in principle at least, go on forever.

Example 1   The list

$$1, 4, 9, 16, 25, \ldots, n^2, \ldots$$

is the infinite sequence of all the positive square integers. The three dots in the expression above mean "and so on." Using the natural numbering of the positive integers, the first element in this sequence is 1, the second is 4, and so on.

In the sequence given in Example 1, all the items listed are distinct. However, this need not be true for every sequence.

Example 2   The sequence

$$1, 0, 0, 1, 0, 1, 0, 0, 1, 1, 1$$

is a finite sequence with repeated items. The digit zero, for example, occurs as the second, third, fifth, seventh, and eighth elements of the sequence.

A general sequence, that is, one where we do not specify the entries, can be written as $a_1, a_2, a_3, \ldots$, or sometimes as $(a_i)_{1 \leqslant i < \infty}$. The elements a_i need not be numbers. They can be elements from any set. If the sequence is finite, say having n terms, we may write it as $(a_i)_{1 \leqslant i \leqslant n}$.

Example 3

(a) A word such as *abacabcd* can be viewed as the (finite) sequence or list of elements a, b, a, c, a, b, c, d from the ordinary alphabet.

(b) The word *abababab* ... defines an infinite sequence of letters, *a, b, a, b, a, b,* ... .

As illustrated in Example 3, it is frequently convenient to omit the commas in a sequence when no confusion results.

The **set corresponding to a sequence** is simply the set of all distinct elements in the sequence. Note that an essential feature of a sequence is the order in which the elements are listed in the sequence. However, the order in which the elements of a set are listed in the set is of no significance at all.

Example 4

(a) The set corresponding to the sequence in Example 1 is

$$\{1, 4, 9, 16, 25, \dots\}.$$

(b) The set corresponding to the sequence in Example 3(b) is simply $\{a, b\}$.

The idea of a sequence is also important in computer science, where a sequence is sometimes called a **linear array** or **list**. We will make a slight but useful distinction between a sequence and an array, and use a slightly different notation. If we have a sequence $S = s_1, s_2, \dots$, we think of all the elements of S as completely determined. The element s_4, for example, is some fixed element of S, located in position four. Moreover, if we change any of the elements s_i, we think of the result as a new sequence, and we probably will rename it. Thus if we begin with the finite sequence

$$S = 0, 1, 2, 3, 2, 1, 1$$

and we change the 3 to 4, getting

$$0, 1, 2, 4, 2, 1, 1$$

we would think of this as a different sequence, say S'.

An array, on the other hand, may be viewed as a "sequence of positions," which we represent below as boxes.

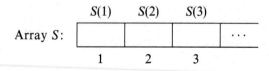

The positions form a finite or infinite list, depending on the desired size of the array. Elements from some set may be assigned to the positions of the array S. The element assigned to position n will be denoted by $S(n)$, and the sequence $S(1)$, $S(2)$,

$S(3)$, ... will be called the **sequence of values** of the array S. The point is that S is considered to be a well-defined object, even if some of the positions have not been assigned values, or if some values are changed during the discussion.

A set is called **countable** if it is the set corresponding to some sequence. Informally, this means that the members of the set can be arranged in a list, with a first, second, third, ..., element, and the set can therefore be "counted." We shall show in Section 1.6 that all finite sets are countable. However, not all infinite sets are countable. A set that is not countable is called **uncountable**.

The most accessible example of an uncountable set is the set of all real numbers between zero and one, that is, the set of all real numbers in the open interval $(0, 1)$. Depending on your definition of real numbers, it is either an axiom or a theorem that a real number in $(0, 1)$ is just an infinite decimal $.a_1 a_2 a_3 \ldots$, where a_i is an integer such that $0 \leq a_i \leq 9$. We shall now show that this set is uncountable. We will prove this result by contradiction; that is, we will show that the countability of this set implies an impossible situation (see the Appendix).

Assume that the set of all such decimals is countable. Then we can form the following list (sequence), containing all such decimals:

$$d_1 = .a_1 a_2 a_3 \ldots$$
$$d_2 = .b_1 b_2 b_3 \ldots$$
$$d_3 = .c_1 c_2 c_3 \ldots$$
$$\vdots$$

Then every infinite decimal must appear somewhere on this list. We shall establish a contradiction by constructing an infinite decimal that is not on this list. Now construct a decimal x as follows: $x = .x_1 x_2 x_3 \ldots$, where x_1 is 1 if $a_1 = 2$, otherwise $x_1 = 2$; x_2 is 1 if $b_2 = 2$, otherwise $x_2 = 2$; $x_3 = 1$ if $c_3 = 2$, otherwise $x_3 = 2$. This process can clearly be continued indefinitely. The resulting number x is certainly an infinite decimal of 1's and 2's and is certainly in $(0, 1)$, but by its construction it differs from each number in the list at some digit. Thus x is not on the list, a contradiction to our assumption. Hence no matter how the list is constructed, there is some real number in $(0, 1)$ that is not in the list.

It can be shown that the set of all rational numbers is countable.

Given a set S, we can construct the set S^* consisting of all finite sequences of elements of S. Often, the set S is not a set of numbers but some set of symbols. 'n this case S is called an **alphabet**, and the finite sequences in S^* are called **words** from S, or sometimes **strings** from S. For this case in particular, the sequences in S^* are *not* written with commas between the elements. We assume that S^* contains the **empty sequence** or **empty string**, containing no symbols, and we denote this string by Λ. This string will be useful in Chapters 5 and 6.

Example 5   Let $S = \{a, b, c, \ldots, z\}$ be the usual English alphabet. Then S^* consists of all ordinary words, such as ape, sequence, antidisestablishmentarianism, and so on, as well as "words" such as yxaloble, zigadongdong, cccaaaa, and pqrst. All finite sequences from S are in S^*, whether they have meaning or not.

If $w_1 = s_1 s_2 \cdots s_n$ and $w_2 = t_1 t_2 \cdots t_k$ are elements of S^*, for some set S, we define the **catenation** of w_1 with w_2 as the sequence $s_1 s_2 s_3 \cdots s_n t_1 t_2 \cdots t_k$. The catenation of w_1 with w_2 is written $w_1 \cdot w_2$, and is another element of S^*. Note that if w belongs to S^*, then $w \cdot \Lambda = w$ and $\Lambda \cdot w = w$. This property is convenient and is one of the main reasons for introducing the empty string Λ.

Example 6 Let $S = \{\text{John, Sam, Jane, swims, runs, well, quickly, slowly}\}$. Then S^* consists of finite sequences formed from these words. Thus S^* contains legitimate sentences, such as "Jane swims quickly" and "Sam runs well," as well as nonsense sentences such as "Well swims Jane slowly John." Here we separate the elements in each sequence by blanks. This is often done when the elements of S are "words."

EXERCISE SET 1.2

In Exercises 1–6, give the set corresponding to the sequence.

1. 2, 1, 2, 1, 2, 1, 2, 1
2. 0, 2, 4, 6, 8, 10, ...
3. 1, 3, 5, 7, 9, 11, 13, ...
4. $a\,a\,b\,b\,c\,c\,d\,d\,e\,e\cdots z\,z$
5. $a\,b\,b\,c\,c\,c\,d\,d\,d\,d$
6. 1, 2, 2, 3, 3, 3, 4, 4, 4, 4, 5, 5, 5, 5, 5, 6, ...

In Exercises 7–10, write out the first four, beginning with $n = 1$, terms of the sequence whose general term is given.

7. 2^n
8. $n!$ (see p. 46)

9. $3n^2 + 2n - 6$
10. $n/n + 1$

11. Give three different sequences that have $\{x, y, z\}$ as a corresponding set.
12. Give three different sequences that have $\{1, 2, 3, \ldots\}$ as a corresponding set.

In Exercises 13–18, write a formula for the nth term of the sequence.

13. 1, 3, 5, 7, ...
14. 0, 3, 8, 15, 24, 35, ...
15. 1, −1, 1, −1, 1, −1, ...
16. 0, 2, 0, 2, 0, 2, ...
17. 1, 4, 7, 10, 13, 16
18. $1, \frac{1}{2}, \frac{1}{4}, \frac{1}{8}, \frac{1}{16}, \ldots$

1.3 Operations on Sets

In this section we discuss several operations which will combine given sets to yield new sets. These operations, which are analogous to the familiar operations on the real numbers, will play a key role in the many applications and ideas that follow.

If A and B are sets, we define their **union** as the set consisting of all elements that belong to A *or* B and denote it by $A \cup B$. Thus

$$A \cup B = \{x \mid x \in A \text{ or } x \in B\}$$

Observe that $x \in A \cup B$ if $x \in A$ or $x \in B$ or x belongs to both A and B.

Example 1   Let

$$A = \{a, b, c, e, f\}, \qquad B = \{b, d, r, s\}$$

Find $A \cup B$.

  Solution. Since $A \cup B$ consists of all the elements that belong to either A or B, then

$$A \cup B = \{a, b, c, d, e, f, r, s\}$$

  We can illustrate the union of two sets by a Venn diagram as follows. If A and B are the sets given in Fig. 1(a), then $A \cup B$ is the set of points in the shaded region as indicated in Fig. 1(b).

  If A and B are sets, we define their **intersection** as the set consisting of all elements that belong to *both* A and B and denote it by $A \cap B$. Thus

$$A \cap B = \{x \mid x \in A \text{ and } x \in B\}$$

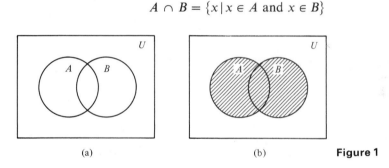

    (a)   (b)   **Figure 1**

Example 2   Let

$$A = \{a, b, c, e, f\}, \qquad B = \{b, e, f, r, s\}, \qquad C = \{a, t, u, v\}$$

Find $A \cap B$, $A \cap C$, and $B \cap C$.

  Solution. The elements b, e, and f are the only ones that belong to both A and B, so

$$A \cap B = \{b, e, f\}$$

Similarly,

$$A \cap C = \{a\}$$

There are no elements that belong to both B and C, so

$$B \cap C = \varnothing$$

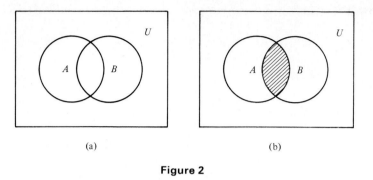

(a)   (b)

Figure 2

Two sets that have no common elements, such as B and C in Example 2, are called **disjoint** sets.

We can illustrate the intersection of two sets by a Venn diagram as follows. If A and B are the sets given in Fig. 2(a), then $A \cap B$ is the set of points in the shaded region as indicated in Fig. 2(b). Figure 3 illustrates a Venn diagram of two disjoint sets.

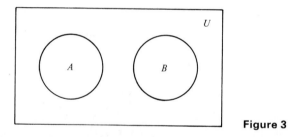

Figure 3

The operations of union and intersection can be defined for three or more sets in the obvious manner. Thus

$$A \cup B \cup C = \{x \mid x \in A \text{ or } x \in B \text{ or } x \in C\}$$

and

$$A \cap B \cap C = \{x \mid x \in A \text{ and } x \in B \text{ and } x \in C\}$$

The shaded region in Fig. 4(b) is the union of the sets A, B, and C shown in Fig. 4(a), and the shaded region in Fig. 4(c) is the intersection of the sets A, B, and C shown in Fig. 4(a).

In general, if $A_1, A_2, \ldots, A_n$ are subsets of U, then $A_1 \cup A_2 \cup \cdots \cup A_n$ will be denoted by $\bigcup_{k=1}^{n} A_k$, and $A_1 \cap A_2 \cap \cdots \cap A_n$ will be denoted by $\bigcap_{k=1}^{n} A_k$.

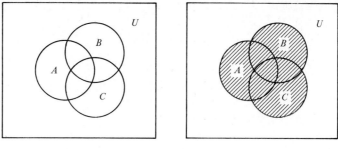

(a)   (b)

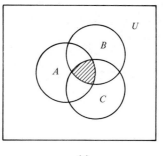

(c)

Figure 4

Example 3   Let

$$A = \{1, 2, 3, 4, 5, 7\}, \qquad B = \{1, 3, 8, 9\}, \qquad C = \{1, 3, 6, 8\}$$

Then $A \cap B \cap C$ is the set of elements that belong to A, B, and C. Thus

$$A \cap B \cap C = \{1, 3\}$$

If A and B are two sets, we define the **complement of B with respect to A** as the set of all elements that belong to A but not to B and denote it by $A - B$. Thus

$$A - B = \{x \mid x \in A \text{ and } x \notin B\}$$

Example 4   Let

$$A = \{a, b, c\} \qquad \text{and} \qquad B = \{b, c, d, e\}$$

Then

$$A - B = \{a\} \quad \text{and} \quad B - A = \{d, e\}$$

If A and B are the sets in Fig. 5(a), then $A - B$ and $B - A$ are the sets of points in the shaded regions in Fig. 5(b) and (c), respectively.

If U is a universal set containing A, then $U - A$ is called the **complement** of A and is denoted by $\bar{A}$. Thus

$$\bar{A} = \{x \mid x \notin A\}$$

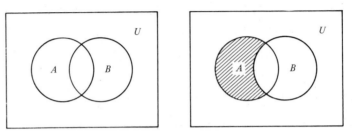

(a)   (b)

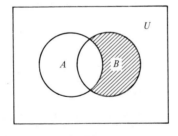

(c)

Figure 5

Example 5   Let

$$A = \{x \mid x \text{ is an integer and } x \geq 4\}$$

Then

$$\bar{A} = \{x \mid x \text{ is an integer and } x < 4\}$$

If A is the set in Fig. 6, its complement is the shaded region in that figure.

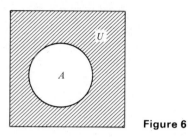

Figure 6

If A and B are two sets, we define their **symmetric difference** as the set consisting of all elements that belong to A or to B, but not to both A and B, and we denote it by $A \oplus B$. Thus

$$A \oplus B = \{x \mid (x \in A \text{ and } x \notin B) \text{ or } (x \in B \text{ and } x \notin A)\}$$

Example 6   Let

$$A = \{a, b, c, d\} \qquad \text{and} \qquad B = \{a, c, e, f, g\}$$

Then

$$A \oplus B = \{b, d, e, f, g\}$$

If A and B are as indicated in Fig. 7(a), their symmetric difference is the shaded region shown in Fig. 7(b). It is easy to see that

$$A \oplus B = (A - B) \cup (B - A)$$

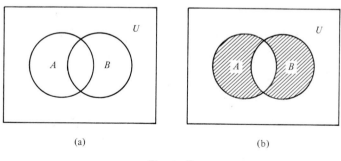

Figure 7

Algebraic Properties of Set Operations

The operations on sets that we have just defined satisfy many algebraic properties, some of which resemble the algebraic properties satisfied by the real number system.

All the principal properties listed here can be proved using the definitions given and the rules of logic. We shall prove only some of the properties and leave proofs of the remaining ones as exercises for the reader. Venn diagrams are often useful to suggest or to justify the method of proof.

Theorem 1 The operations on sets defined above satisfy the following properties:

Commutative Properties
1. $A \cup B = B \cup A$
2. $A \cap B = B \cap A$

Associative Properties
3. $A \cup (B \cup C) = (A \cup B) \cup C$
4. $A \cap (B \cap C) = (A \cap B) \cap C$

Distributive Properties
5. $A \cap (B \cup C) = (A \cap B) \cup (A \cap C)$
6. $A \cup (B \cap C) = (A \cup B) \cap (A \cup C)$

Idempotent Properties
7. $A \cup A = A$
8. $A \cap A = A$

Properties of the Complement
9. $(\overline{\overline{A}}) = A$
10. $A \cup \overline{A} = U$
11. $A \cap \overline{A} = \emptyset$
12. $\overline{\emptyset} = U$
13. $\overline{U} = \emptyset$
14. $\overline{A \cup B} = \overline{A} \cap \overline{B}$ $\left.\right\}$ DeMorgan's laws
15. $\overline{A \cap B} = \overline{A} \cup \overline{B}$

Properties of the Universal Set
16. $A \cup U = U$
17. $A \cap U = A$

Properties of the Null Set
18. $A \cup \emptyset = A$
19. $A \cap \emptyset = \emptyset$

Proof. We prove property 14 and leave the proofs of the remaining properties as an exercise for the reader.

Suppose that $x \in \overline{A \cup B}$. Then $x \notin A \cup B$, so $x \notin A$ and $x \notin B$. This means that $x \in \bar{A} \cap \bar{B}$, so

$$\overline{A \cup B} \subseteq \bar{A} \cap \bar{B}$$

Conversely, suppose that $x \in \bar{A} \cap \bar{B}$. Then $x \notin A$ and $x \notin B$, so $x \notin A \cup B$, or $x \in \overline{A \cup B}$. Hence

$$\bar{A} \cap \bar{B} \subseteq \overline{A \cup B}$$

It then follows that $\overline{A \cup B} = \bar{A} \cap \bar{B}$.

Suppose now that A and B are finite subsets of a universal set U. It is frequently useful to have a formula for $|A \cup B|$, the cardinality of the union. If A and B are disjoint, that is, if $A \cap B = \varnothing$, then each element of $A \cup B$ appears in either A or B but not both; therefore, $|A \cup B| = |A| + |B|$. If A and B overlap, as shown in Fig. 8, then $A \cap B$ belongs to both sets, and the sum $|A| + |B|$ includes the number of elements in $A \cap B$ twice. To correct for this duplication, we subtract $|A \cap B|$. Thus, we have the following theorem, sometimes called the **Addition Principle**.

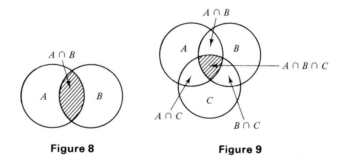

Figure 8   Figure 9

Theorem 2   If A and B are finite sets, then $|A \cup B| = |A| + |B| - |A \cap B|$.

The situation for three sets is more complicated, as we show in Fig. 9. We state the following three-set addition principle without comment.

Theorem 3   Let A, B, and C be finite sets. Then

$$|A \cup B \cup C|$$
$$= |A| + |B| + |C| - |A \cap B| - |B \cap C| - |A \cap C| + |A \cap B \cap C|$$

Example 7   A computer company must hire 25 programmers to handle systems programming tasks and 40 programmers for applications programming. Of those hired, 10 will be expected to perform tasks of each type. How many programmers must be hired?

    Solution. Let A be the number of systems programmers hired, and let B be the number of applications programmers hired. We must have $|A| = 25$ and $|B| = 40$, and $|A \cap B| = 10$. Thus the number that must be hired is $|A \cup B| = |A| + |B| - |A \cap B| = 25 + 40 - 10 = 55$.

Example 8   A survey is taken on methods of commuter travel. Each respondent is asked to check BUS, TRAIN, or AUTOMOBILE, as a major method of traveling to work. More than one answer is permitted. The results reported were as follows:

  (a)   30 people checked BUS.
  (b)   35 people checked TRAIN.
  (c) 100 people checked AUTOMOBILE.
  (d)   15 people checked BUS and TRAIN.
  (e)   15 people checked BUS and AUTOMOBILE.
  (f)   20 people checked TRAIN and AUTOMOBILE.
  (g)   5 people checked all three methods.

    How many respondents completed their surveys?

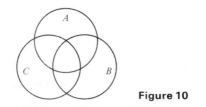

Figure 10

    Solution. Let A, B, and C be the sets of people who checked BUS, TRAIN, and AUTOMOBILE, respectively (see Fig. 10). We know that $|A| = 30, |B| = 35, |C| = 100, |A \cap B| = 15, |A \cap C| = 15, |B \cap C| = 20,$ and $|A \cap B \cap C| = 5$. The total number of people responding is then $|A \cup B \cup C| = (30 + 35 + 100) - (15 + 15 + 20) + 5 = 120$.

Characteristic Functions

A very useful concept for sets is the characteristic function. We treat the general concept of function in Section 3.1, but for now we can proceed intuitively, and think of a function on a set as a rule that assigns some "value" to each element of the set.

If A is a subset of a universal set U, the **characteristic function** f_A of A is defined as follows:

$$f_A(x) = \begin{cases} 1 & \text{if } x \in A \\ 0 & \text{if } x \notin A \end{cases}$$

We may add and multiply characteristic functions, since their values are numbers, and these operations sometimes help us prove theorems about properties of subsets.

Theorem 4   Characteristic functions of subsets satisfy the following properties.

(a) $f_{A \cap B} = f_A f_B$; that is, $f_{A \cap B}(x) = f_A(x) f_B(x)$ for all x.
(b) $f_{A \cup B} = f_A + f_B - f_A f_B$; that is, $f_{A \cup B}(x) = f_A(x) + f_B(x) - f_A(x) f_B(x)$ for all x.
(c) $f_{A \oplus B} = f_A + f_B - 2 f_A f_B$; that is, $f_{A \oplus B}(x) = f_A(x) + f_B(x) - 2 f_A(x) f_B(x)$ for all x.

Proof.
(a) $f_A(x) f_B(x)$ equals 1 if and only if both $f_A(x)$ and $f_B(x)$ are equal to 1, and this happens if and only if x is in A and in B, that is in $A \cap B$. Since $f_A f_B$ is 1 on $A \cap B$ and 0 otherwise, it must equal $f_{A \cap B}$.
(b) If $x \in A$, then $f_A(x) = 1$, so $f_A(x) + f_B(x) - f_A(x) f_B(x) = 1 + f_B(x) - f_B(x) = 1$. Similarly, when $x \in B$, $f_A(x) + f_B(x) - f_A(x) f_B(x) = 1$. If x is not in A or B, then $f_A(x)$ and $f_B(x)$ are 0, so $f_A(x) + f_B(x) - f_A(x) f_B(x) = 0$. Thus $f_A + f_B - f_A f_B$ is 1 on $A \cup B$ and 0 otherwise, so it equals $f_{A \cup B}$.
We leave (c) as an exercise.

Computer Representation of Sets and Subsets

To represent a set in a computer, the elements of the set must be arranged in a sequence. The particular sequence selected is of no importance. When we list the set $A = \{a, b, c, \ldots, r\}$, we normally assume no particular ordering of the elements in A. In the remainder of this section, we identify the set A with the sequence $a, b, c, \ldots, r$. That is, we assume that a is the first element, b is the second one, c is the third one, and so on.

When a universal set U is finite, say $U = \{x_1, x_2, \ldots, x_n\}$, and A is a subset of U, then the characteristic function f_A assigns 1 to an element x_i that belongs to A and 0 to an element x_i that does not belong to A. Thus f_A can be represented by a sequence of 0's and 1's of length n.

Example 9   Let $U = \{1, 2, 3, 4, 5, 6\}$, $A = \{1, 2\}$, $B = \{2, 4, 6\}$, and $C = \{4, 5, 6\}$. Then $f_A(x)$ has value 1 for $x = 1, 2$, and is otherwise 0. Hence f_A corresponds to the sequence 1, 1, 0, 0, 0, 0. In a similar way, the finite sequence 0, 1, 0, 1, 0, 1 represents f_B and 0, 0, 0, 1, 1, 1 represents f_C.

Any set with n elements can be arranged in a sequence of length n (as we show in Section 1.6), so each of its subsets corresponds to a sequence of zeros and ones of length n, representing the characteristic function of that subset. This fact allows us to represent a universal set in a computer as an array A of length n. Assignment of a zero or one to each location $A[n]$ of the array specifies a unique subset of U.

Example 10 Let $U = \{a, b, e, g, h, r, s, w\}$. The array A of length 8, shown in Fig. 11, represents U, since $A[k] = 1$ for $1 \le k \le 8$.

| 1 | 1 | 1 | 1 | 1 | 1 | 1 | 1 |

| 1 | 0 | 1 | 0 | 0 | 1 | 0 | 1 |

Figure 11   Figure 12

If $A = \{a, e, r, w\}$, then

$$A[k] = \begin{cases} 1 & \text{if } k = 1, 3, 6, 8 \\ 0 & \text{if } k = 2, 4, 5, 7 \end{cases}$$

Hence the array shown in Fig. 12 represents the subset A.

EXERCISE SET 1.3

In Exercises 1–4, let

$U = \{a, b, c, d, e, f, g, h, k\},$ $\qquad A = \{a, b, c, g\},$

$B = \{d, e, f, g\},$ $\qquad C = \{a, c, f\},$ $\qquad D = \{f, h, k\}$

1. Compute
- (a) $A \cup B$
- (b) $B \cup C$
- (c) $A \cap C$
- (d) $B \cap D$
- (e) $A - B$
- (f) $\bar{A}$
- (g) $A \oplus B$
- (h) $A \oplus C$

2. Compute
- (a) $A \cup D$
- (b) $B \cup D$
- (c) $C \cap D$
- (d) $A \cap D$
- (e) $B - C$
- (f) $C - B$
- (g) $\bar{B}$
- (h) $C \oplus D$

3. Compute
- (a) $A \cup B \cup C$
- (b) $A \cap B \cap C$
- (c) $A \cap (B \cup C)$
- (d) $(A \cup B) \cap C$
- (e) $\overline{A \cup B}$
- (f) $\overline{A \cap B}$

4. Compute
- (a) $A \cup \varnothing$
- (b) $A \cap U$
- (c) $B \cup B$
- (d) $C \cap \varnothing$
- (e) $\overline{C \cup D}$
- (f) $\overline{C \cap D}$

In Exercises 5–8, let

$$U = \{1, 2, 3, 4, 5, 6, 7, 8, 9\}$$

$$A = \{1, 2, 4, 6, 8\}, \qquad B = \{2, 4, 5, 9\}$$

$$C = \{x \mid x \text{ is a positive integer and } x^2 \le 16\},$$

$$D = \{7, 8\}$$

5. Compute
- (a) $A \cup B$
- (b) $A \cup C$
- (c) $A \cap C$
- (d) $C \cap D$
- (e) $A - B$
- (f) $B - A$
- (g) $\bar{A}$
- (h) $A \oplus B$

6. Compute
- (a) $A \cup D$
- (b) $B \cup C$

(c) $A \cap D$ (d) $B \cap C$
(e) $C - D$ (f) $\bar{C}$
(g) $C \oplus D$ (h) $B \oplus C$

7. Compute
   (a) $A \cup B \cup C$ (b) $B \cup C \cup D$
   (c) $A \cap B \cap C$ (d) $B \cap C \cap D$
   (e) $A \cup A$ (f) $A \cap \bar{A}$

8. Compute
   (a) $\overline{A \cup B}$ (b) $\overline{A \cap B}$
   (c) $A \cap (B \cup C)$ (d) $(A \cup B) \cap D$
   (e) $A \cup \bar{A}$ (f) $A \cap (\bar{C} \cup D)$

In Exercises 9 and 10, let

$$U = \{a, b, c, d, e, f, g, h\}$$
$$A = \{a, c, f, g\}, \quad B = \{a, e\}, \quad C = \{b, h\}$$

9. Compute
   (a) $\bar{A}$ (b) $\bar{B}$
   (c) $\overline{A \cup B}$ (d) $\overline{A \cap B}$
   (e) $\bar{U}$ (f) $A - B$

10. Compute
    (a) $\bar{A} \cap \bar{B}$ (b) $\bar{B} \cup \bar{C}$
    (c) $\overline{A \cup A}$ (d) $\overline{C \cap C}$
    (e) $A \oplus B$ (f) $B \oplus C$

11. Let

    $$U = \text{set of all real numbers}$$
    $$A = \{x \mid x \text{ is a solution of } x^2 - 1 = 0\}$$
    $$B = \{-1, 4\}$$

    Compute
    (a) $\bar{A}$ (b) $\bar{B}$
    (c) $\overline{A \cup B}$ (d) $\overline{A \cap B}$

In Exercises 12 and 13, refer to Fig. 13.

12. Answer the following as true or false.
    (a) $y \in A \cap B$ (b) $x \in B \cup C$
    (c) $w \in B \cap C$ (d) $u \notin C$

13. Answer the following as true or false.
    (a) $x \in A \cap B \cap C$ (b) $y \in A \cup B \cup C$
    (b) $z \in A \cap C$ (d) $v \in B \cap C$

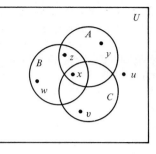

Figure 13

14. Verify Theorem 2 for the following sets.
    (a) $A = \{1, 2, 3, 4\}$, $B = \{2, 3, 5, 6, 8\}$
    (b) $A = \{a, b, c, d, e, f\}$, $B = \{a, c, f, g, h, i, r\}$
    (c) $A = \{x \mid x$ is a positive integer $< 8\}$, $B = \{x \mid x$ is an integer such that $2 \le x \le 5\}$

15. Verify Theorem 2 for the following disjoint sets.
    (a) $A = \{1, 2, 3, 4\}$, $B = \{5, 6, 7, 8, 9\}$
    (b) $A = \{a, b, c, d, e\}$, $B = \{f, g, r, s, t, u\}$
    (c) $A = \{x \mid x$ is a positive integer and $x^2 \le 16\}$, $B = \{x \mid x$ is a negative integer and $x^2 \le 25\}$

16. If A and B are disjoint sets such that $|A \cup B| = |A|$, what can we say about B?

17. Verify Theorem 3 for the following sets.
    (a) $A = \{a, b, c, d, e\}$, $B = \{d, e, f, g, h, i, k\}$, $C = \{a, c, d, e, k, r, s, t\}$
    (b) $A = \{1, 2, 3, 4, 5, 6,\}$, $B = \{2, 4, 7, 8, 9\}$, $C = \{1, 2, 4, 7, 10, 12\}$
    (c) $A = \{x \mid x$ is a positive integer $< 8\}$, $B = \{x \mid x$ is an integer such that $2 \le x \le 4\}$, $C = \{x \mid x$ is an integer such that $x^2 < 16\}$

18. In a survey of 260 college students, the following data were obtained:

    64 had taken a mathematics course.

    94 had taken a computer science course.

    58 had taken a business course.

    28 had taken both a mathematics and a business course.

    26 had taken both a mathematics and a computer science course.

    22 had taken both a computer science and a business course.

    14 had taken all three courses.

(a) How many students were surveyed who had taken none of the three courses?

(b) Of the students surveyed, how many had taken only a computer science course?

19. A survey of 500 television watchers produced the following information:

285 watch football games.

195 watch hockey games.

115 watch basketball games.

45 watch football and basketball games.

70 watch football and hockey games.

50 watch hockey and basketball games.

50 do not watch any of the three games.

(a) How many people in the survey watch all three games?

(b) How many people watch exactly one of the three games?

20. In a psychology experiment, the subjects under study were classified according to body type and sex as follows:

	Endomorph	Ectomorph	Mesomorph
Male	72	54	36
Female	62	64	38

(a) How many male subjects are there?

(b) How many subjects are ectomorphs?

(c) How many subjects are either female or endomorphs?

(d) How many subjects are not male mesomorphs?

(e) How many subjects are either male, ectomorph, or mesomorph?

21. Let

$U = \{$ALGOL, FORTRAN, BASIC, COBOL, ADA, PL1, PASCAL, LISP$\}$,

$B = \{$ALGOL, BASIC, ADA$\}$,

$C = \{$ALGOL, ADA, PASCAL, LISP$\}$,

$D = \{$FORTRAN, BASIC, ADA, PL1, PASCAL$\}$,

$E = \{$ALGOL, COBOL, ADA, PASCAL, LISP$\}$

In each of the following, represent the given set by an array of zeros and ones.

(a) $B \cup C$ (b) $C \cap D$

(c) $B \cup (D \cap E)$ (d) $\bar{B} \cup E$

(e) $\bar{C} \cap (B \cup E)$

22. Let

$$U = \{b, d, e, f, h, k, m, n\}$$

$$B = \{b\}, \qquad C = \{d, f, m, n\}, \qquad D = \{d, k, n\}$$

(a) What is $f_B(b)$?

(b) What is $f_C(e)$?

(c) Find the sequences of length 8 that correspond to f_B, f_C, and f_D.

(d) Represent $B \cup C$, $C \cup D$, and $C \cap D$ by arrays of zeros and ones.

23. Prove that $A \subseteq A \cup B$.

24. Prove that $A \cap B \subseteq A$.

25. Prove that if $C \subseteq A$ and $C \subseteq B$, then $C \subseteq A \cup B$.

26. Prove that if $A \subseteq C$ and $B \subseteq C$, then $A \cup B \subseteq C$.

27. Prove that $A - A = \varnothing$.

28. Prove that $A - B = A \cap \bar{B}$.

29. Prove that $A - (A - B) \subseteq B$.

30. Prove that $A \oplus B = B \oplus A$.

31. If $A \cup B = A \cup C$, must $B = C$? Explain.

32. If $A \cap B = A \cap C$, must $B = C$? Explain.

33. Prove that if $A \subseteq B$ and $C \subseteq D$, then $A \cup C \subseteq B \cup D$ and $A \cap C \subseteq B \cap D$.

34. Prove property (c) of Theorem 4.

35. Using characteristic functions, prove that

$$(A \oplus B) \oplus C = A \oplus (B \oplus C)$$

36. When is $A - B = B - A$? Explain.

1.4 Counting Sequences and Subsets

We begin with a simple but general result which we will use frequently in this section and elsewhere.

Theorem 1   If two independent tasks T_1 and T_2 are to be performed in sequence, and if T_1 can be performed in n_1 ways and T_2 in n_2 ways, the sequence $T_1 T_2$ can be performed in $n_1 n_2$ ways.

Proof. Each choice of a method for performing T_1 will result in a different way of performing the task sequence. There are n_1 such methods, and for each of these, we may choose n_2 methods of performing T_2. Thus, in all, there will be $n_1 n_2$ ways of performing the sequence $T_1 T_2$.

Theorem 1 is sometimes called the **multiplication principle** of combinatorics. It is an easy matter to extend the result as follows.

Theorem 2   If independent tasks $T_1, T_2, \ldots, T_k$ are to be performed in sequence, and T_1 can be performed in n_1 ways, T_2 in n_2 ways, $\ldots$, T_k in n_k ways, then the sequence $T_1 T_2 \cdots T_k$ can be performed in exactly $n_1 n_2 \cdots n_k$ ways.

We omit the proof.

Example 1   A label identifier, for a computer program, consists of one letter followed by three digits. If repetitions are allowed, how many distinct label identifiers are possible?

Solution. There are 26 possibilities for the beginning letter and there are 10 possibilities for each of the three digits. Thus, by the extended multiplication principle, there are

$$26 \times 10 \times 10 \times 10 = 26{,}000 \text{ possible label identifiers}$$

Example 2   Let S be a set with n elements. How many subsets does S have?

Solution. We know from Section 1.3 that each subset of S is determined by its characteristic function, and if S has n elements, this function may be described as an array of 0's and 1's, having length n. The first element of the array can be filled in two ways (with a 0 or a 1), and this is true for all succeeding elements as well. Thus, by the extended multiplication principle, there are $2 \cdot 2 \cdots 2 = 2^n$ ways of filling in the array, and therefore 2^n subsets of S.

Suppose now that S is a set and that $|S| = n$, where n is a positive integer. A particular arrangement of the elements of S into a sequence of length n is usually called a **permutation** of S. In Chapter 3 we use the term "permutation" in a slightly more complicated way, in order to increase its utility.

Example 3   Let $S = \{a, b, c\}$. Then the possible permutations of S are the sequences abc, acb, bac, bca, cab, and cba.

We can generalize the idea of permutation in the following way. If $0 \leq r \leq n$, we may form a sequence of length r using r distinct elements from S. When $r = n$, this sequence is just a permutation of the elements of S. When $r < n$, the sequence is called a **permutation of S taken r at a time**. This terminology is standard, and therefore we adopt it, but it is confusing. A better terminology might be "a permutation of r elements chosen from S." Many sequences of interest are permutations of some set of n objects taken r at a time.

Example 4   Let $S = \{1, 2, 3, 4\}$. Then the sequences 124, 421, 341, and 243 are some permutations of S taken three at a time. The sequences 12, 43, 31, 24, and 21 are examples of different permutations of S taken two at a time.

Example 5   Let S consist of all 52 cards in an ordinary deck of playing cards. Suppose that this deck is shuffled and a hand of five cards is dealt. A list of the cards in this hand, in the order in which they were dealt, is a permutation of S, taken five at a time. Examples would include AH, 3D, 5C, 2H, JS; 2H, 3H, 5H, QH, KD; JH, JD, JS, 4H, 4C; and 3D, 2H, AH, JS, 5C. Note that the first and last hands are the same, but they represent different permutations since they were dealt in a different order.

An important problem is to determine the number of permutations, taken r at a time, of a set S with n elements (where $0 \leq r \leq n$). This number is one when $r = 0$, since the only sequence of zero elements is the empty sequence. When $0 < r \leq n$, Theorem 3 gives the solution to the problem.

Theorem 3   Let S be a set with $|S| = n$, and let $1 \leq r \leq n$. Then the number of permutations of S taken r at a time is $n \cdot (n - 1) \cdot (n - 2) \cdots (n - r + 1)$.

Proof. Let A be an array of length r, and suppose that we specify a permutation of S, taken r at a time, in the following way. Choose any element from S and assign it to $A(1)$. Then choose any remaining element and assign it to $A(2)$. Continue in this way, always choosing unassigned elements, until all positions from $A(1)$ to $A(r)$ have been assigned. Then any permutation of S, taken r at a time, can be specified by this process.

Now $A(1)$ may be assigned in n ways, since we may choose any element of S. Having made this choice, we have $n - 1$ elements of S remaining to be assigned to $A(2)$. This means that there are $n(n - 1)$ ways to choose $A(1)$ and $A(2)$. We can then choose $A(3)$ in $n - 2$ ways, then $A(4)$ in $n - 3$ ways, and so on. Thus, by the multiplication principle, there are $n \cdot (n - 1) \cdot (n - 2) \cdots (n - r + 1)$ ways of forming the sequence.

Theorem 3 also shows that if $|S| = n$, the number of permutations of S, taken r at a time, depends only on n not on S. We denote this number by P_r^n and refer to it as the **number of permutations of n objects taken r at a time.** If $r = n$, then $P_r^n = n \cdot (n - 1) \cdots 2 \cdot 1$ by Theorem 3, as long as $n \geq 1$. This number is written $n!$ and called n **factorial**. If we agree to define $0!$ as 1, then for every n we see that the number of permutations of n objects is $n!$ If $n \geq 1$ and $1 \leq r \leq n$, we can give a more compact form for P_r^n as follows.

$$P_r^n = n \cdot (n - 1) \cdots (n - r + 1) = \frac{n \cdot (n - 1) \cdots (n - r + 1) \cdot (n - r) \cdot (n - r - 1) \cdots 2 \cdot 1}{(n - r) \cdot (n - r - 1) \cdots 2 \cdot 1}$$

$$= \frac{n!}{(n - r)!}$$

Our convention regarding $0!$ results in this formula being true also for $r = 0$ and $n = 0$. Thus

$$P_r^n = \frac{n!}{(n - r)!} \qquad \text{when } n \geq 0, \quad 0 \leq r \leq n$$

Example 6   If S is the set of Example 3, then $n = 3$ and the number of permutations of S is $3! = 6$. Thus all permutations of S are listed in Example 3, as stated there.

If S is the set of Example 5, consisting of 52 playing cards, then the number of permutations of S taken five at a time is

$$P_5^{52} = \frac{52!}{47!} = 52 \cdot 51 \cdot 50 \cdot 49 \cdot 48 = 311,875,200$$

This is the number of five-card hands that can be dealt, if we count the order in which they are dealt.

A second important problem is to determine how many r-element subsets there are in a set S with $|S| = n$. Again we assume that $0 \leq r \leq n$. The traditional combinatorial name for an r-element subset of an n-element set S is a **combination of S, taken r at a time.**

Example 7   Let $S = \{1, 2, 3, 4\}$. The following are all the distinct combinations of S, taken three at a time: $A_1 = \{1, 2, 3\}$, $A_2 = \{1, 2, 4\}$, $A_3 = \{1, 3, 4\}$, $A_4 = \{2, 3, 4\}$. Note that these are subsets not sequences. Thus $A_1 = \{2, 1, 3\} = \{2, 3, 1\} = \{1, 3, 2\} = \{3, 2, 1\} = \{3, 1, 2\}$. In other words, when it comes to combinations, unlike permutations, the order of the elements is irrelevant.

Example 8   Let S be the set of Example 5. Then a combination of S taken five at a time is just a hand of five cards regardless of how these cards were dealt.

We now want to count the number of r-element subsets of an n-element set S. This is most easily accomplished by using the information on permutations that we have developed so far.

Theorem 4 Let S be a set with $|S| = n$, and let $0 \leq r \leq n$. Then the number of combinations of the elements of S, taken r at a time, that is, the number of r-element subsets of S, is

$$\frac{n!}{r!(n-r)!}$$

Proof. Observe that each permutation of the elements of S, taken r at a time, can be produced by performing the following two tasks in sequence.

Task 1. Choose a subset A of S containing r elements of S.

Task 2. Choose a particular permutation of A.

We are trying to compute the number of ways of choosing A. Call this number C. Then task 1 can be performed in C ways, and task 2 can be performed in $r!$ ways. Thus the total number of ways of performing both tasks, which is P_r^n, is seen, by the multiplication principle, to equal $Cr!$. Hence

$$Cr! = P_r^n = \frac{n!}{(n-r)!}$$

Therefore,

$$C = \frac{n!}{r!(n-r)!}$$

as claimed.

Note again that the number of combinations of S, taken r at a time, does not depend on S, but only on n. This number is written C_r^n or sometimes $\binom{n}{r}$, and is called the **number of combinations of n objects taken r at a time**. We have

$$C_r^n = \frac{n!}{r!(n-r)!}$$

Example 9 Compute the number of distinct five card hands which can be dealt from a deck of 52 cards.

Solution. This number is

$$C_5^{52} = \frac{52!}{5!47!} = \frac{52 \cdot 51 \cdot 50 \cdot 49 \cdot 48}{1 \cdot 2 \cdot 3 \cdot 4 \cdot 5} = 2{,}598{,}960$$

Compare this number with the number computed in Example 6.

EXERCISE SET 1.4

1. A bank password consists of two letters of the English alphabet followed by two digits from 0 to 9. How many different passwords are there?

2. In a psychological experiment a subject must arrange a cube, square rectangle, triangle, and circle in a row. How many different arrangements are possible?

3. A coin is tossed four times and the result of each toss is recorded. How many different sequences of heads and tails are possible?

4. A menu is to consist of a soup dish, a main course, a dessert, and a beverage. Suppose that the chef can select from four soup dishes, five main courses, three desserts, and two beverages. How many different menus can be prepared?

5. A die is tossed four times and the numbers shown are arranged in a sequence. How many different sequences are there?

6. Compute
   (a) P_4^4 (b) P_5^6 (c) P_2^7
   (d) P_{n-1}^n (e) P_{n-2}^n (f) P_{n-1}^{n+1}

7. How many permutations are there of each of the following sets?
   (a) $\{r, s, t, u\}$ (b) $\{1, 2, 3, 4, 5\}$
   (c) $\{a, b, 1, 2, 3, c\}$

8. For each set S, find the number of permutations of S taken r at a time.
   (a) $S = \{1, 2, 3, 4, 5, 6, 7\}, r = 3$
   (b) $S = \{a, b, c, d, e, f\}, r = 2$
   (c) $S = \{x \mid x \text{ is an integer and } x^2 < 16\}, r = 4$

9. In how many ways can six men and six women be seated in a row if
   (a) any person may sit next to any other person?
   (b) men and women must occupy alternate seats?

10. How many different arrangements of the letters in the word "bought" can be formed if the vowels must be kept next to each other?

11. Find the number of different permutations of the letters in the word "group."

12. Find the number of distinguishable permutations of the letters in the word "Boolean." (*Hint:* For example, a permutation that only interchanges the

two o's is not distinguishable from the permutation that leaves the word unchanged.)

13. In how many ways can six people be seated in a circle?

14. A bookshelf is to be used to display six new books. Suppose that there are eight computer science books and five French books for display. If we are to show four computer science books and two French books, how many different displays are there if we require that the books in each subject must be kept together?

15. Compute
   (a) C_7^7 (b) C_4^7 (c) C_5^6
   (d) C_{n-1}^n (e) C_{n-2}^n (f) C_{n-1}^{n+1}

16. Show that $C_r^n = C_{n-r}^n$.

17. In how many ways can a committee of three faculty members and two students be selected from seven faculty members and eight students?

18. In how many ways can a six-card hand be dealt from a deck of 52 cards?

19. At a certain college, the housing office has decided to appoint for each floor one male and one female residential advisor. How many different pairs of advisors can be selected for a seven-story building from 12 male candidates and 15 female candidates?

20. A microcomputer manufacturer who is designing an advertising campaign is considering six magazines, three newspapers, two television stations, and four radio stations. In how many ways can six advertisements be run if
   (a) all six are to be in magazines?
   (b) two are to be in magazines, two are to be in newspapers, one is to be on television, and one is to be on radio?

21. How many different eight-card hands with five red cards and three black cards can be dealt from a deck of 52 cards?

22. (a) Find the number of subsets of a set containing four elements.
    (b) Find the number of subsets of a set containing n elements.

23. An urn contains 15 balls, eight of which are red

and seven are black. In how many ways can five balls be chosen so that

(a) all five are red?

(b) all five are black?

(c) two are red and three are black?

(d) three are red and two are black?

24. In how many ways can a committee of six people be selected from a group of 10 people if one person is to be designated as chair?

25. Show that $C_r^{n+1} = C_{r-1}^n + C_r^n$.

1.5 Algorithms and Pseudocode

Algorithms

An **algorithm** is a complete list of the steps necessary to perform a task or computation. The steps in an algorithm may be general descriptions, leaving much detail to be filled in, or they may be totally precise descriptions of every detail.

Example 1   A recipe for baking a cake can be viewed as an algorithm. It might be written as follows.

1. Add milk to cake mix.
2. Add egg to cake mix and milk.
3. Beat mixture for 2 minutes.
4. Pour mixture into pan and cook in oven for 35 minutes at 350°F.

END OF ALGORITHM

It is a good idea to add the last line so that there can be no mistake about where the algorithm ends. The algorithm above is fairly general, and assumes that the user understands how to pour milk, break an egg, set controls on an oven, and perform a host of other unspecified actions. If these steps were all included, the algorithm would be much more detailed, but long and unwieldy. One possible solution, if the added detail is necessary, is to group collections of related steps into other algorithms which we call **subroutines**, and simply refer to these subroutines at appropriate points in the main algorithm. We hasten to point out that we are using the term "subroutine" in the general sense of an algorithm whose primary purpose is to form part of a more general algorithm. We do not give the term the precise meaning that it would have in a computer programming language. Subroutines are given names, and when an algorithm wishes the steps in a subroutine to be performed, it signifies this by *calling* the subroutine. We will specify this by a statement CALL NAME, where NAME is the name of the subroutine.

Example 2   Consider the following version of Example 1, which uses subroutines to add detail. Let us title this algorithm BAKECAKE.

ALGORITHM BAKECAKE
1. CALL ADDMILK
2. CALL ADDEGG
3. CALL BEAT(2)
4. CALL COOK(OVEN,350)
END OF ALGORITHM BAKECAKE

The subroutines of this example will give the details of each step. For example, subroutine ADDEGG might consist of the following general steps.

SUBROUTINE ADDEGG
1. Remove egg from carton.
2. Break egg on edge of bowl.
3. Drop egg, without shell, into bowl.
4. RETURN
END OF SUBROUTINE ADDEGG

Of course, these steps could be broken into substeps, which themselves could be implemented as subroutines. The purpose of step 4, the "return" statement, is to signify that one should continue with the original algorithm that "called" the subroutine.

Our primary concern is with algorithms to implement mathematical computations, investigate mathematical questions, manipulate "strings" or sequences of symbols and numbers, move data from place to place in arrays, and so on. Sometimes the algorithms will be of a general nature, suitable for human use, and sometimes they will be stated in a formal, detailed way suitable for programming in a computer language. Later in this section we will describe a reasonable language for stating algorithms.

It often happens that a test is performed at some point in an algorithm, and the result of this test determines which of two sets of steps will be performed next. Such a test, and the resulting decision to begin performing a certain set of instructions, will be called a **branch**.

Example 3   Consider the following algorithm for deciding whether to study for a "discrete structures" test.

ALGORITHM FLIP
1. Toss a coin.
2. IF the result is "heads," GO TO 5.
3. Study for test.

4. GO TO 6.

5. See a show.

6. Take test next day.

END OF ALGORITHM FLIP

Note that the branching is accomplished by GO TO statements, which direct the user to the next instruction to be performed, in case it is not the next instruction in sequence. In the past, especially for algorithms written in computer programming languages such as FORTRAN, the GO TO statement was universally used to describe branches. The past few years have seen many advances in the art of algorithm and computer program design. Out of this experience has come the view that the indiscriminate use of GO TO statements, to branch from one instruction to any other instruction, leads to algorithms (and computer programs) which are difficult to understand, hard to modify, and prone to error. Also, recent techniques for actually proving that an algorithm or program does what it is supposed to do will not work in the presence of unrestricted GO TO statements.

In light of the foregoing remarks, it is a widely held view that algorithms should be **structured.** This term refers to a variety of restrictions on branching, which help to overcome difficulties posed by the GO TO statement. In a structured branch, the test condition follows an IF statement. When the test is true, the instructions following a THEN statement are performed. Otherwise, the instructions following an ELSE statement are performed.

Example 4   Consider again the algorithm FLIP, described in Example 3. The following is a structured version of FLIP.

ALGORITHM FLIP

1. Toss a coin.

2. IF (heads results) THEN
   a. Study for test.

3. ELSE
   a. See a show.

4. Take test next day.

END OF ALGORITHM FLIP

This algorithm is easy to read and is formulated without GO TO statements. In fact, it does not require numbering or lettering of the steps, but we keep these to set off and emphasize the instructions. Of course, the algorithm FLIP of Example 3 is not very different from that of Example 4. The point is that the GO TO statement has the potential for abuse, which is eliminated in the structured form.

Another commonly encountered situation that calls for a branch is the **loop**, in which a set of instructions is repeatedly followed, either for a definite number of times or until some condition is encountered.

In structured algorithms, a loop may be formulated as shown in the following example.

Example 5   The following algorithm describes the process of machining 50 bolts.

ALGORITHM BOLTS
1. COUNT ← 50
2. WHILE (COUNT > 0)
   a. Cut rod.
   b. Machine threads.
   c. Stamp head.
   d. COUNT ← COUNT − 1
END OF ALGORITHM BOLTS

In this algorithm, the variable COUNT is first assigned the value 50. The loop is handled by the WHILE statement. The condition, COUNT > 0, is checked, and as long as it is true, statements a through d are performed. When COUNT = 0 (after 50 steps), the looping stops. Later in this section we will give the details of this and other methods of looping, which are generally considered to be structured. In structured algorithms, the only deviations permitted from a normal, sequential execution of steps are those given by loops or iterations, and those resulting from the use of the IF, THEN, ELSE construction. Use of the latter construction for branching is called **selection**.

In this text we will need to describe numerous algorithms, many of which are quite technical. Description of these algorithms in ordinary English may be feasible, and in many cases, we will give such descriptions. However, it is often easier to get an overview of an algorithm if it is presented in a concise, symbolic form. Some authors use diagrammatic representations called **flowcharts** for this purpose. Figure 1 shows a flowchart for the algorithm given in Example 4. These diagrams have a certain appeal and are still widely used in the computer programming field, but many believe that they are undesirable since they are more in accord with older programming practice than with the newer structured programming ideas.

The other alternative is to express algorithms in a way that resembles a computer programming language, or to use an actual programming language such as Pascal. We choose to use a **pseudocode** language rather than an actual programming language, and the earlier examples of this section provide a hint as to the structure of this pseudocode form. There are several reasons for making this choice. First, knowledge of a programming language is not necessary for the understanding of the contents of this book. The fine details of a programming language are necessary for communication with a computer, but may serve only to obscure the description of an algorithm. Moreover, we feel that the algorithms should be expressed in such a way that an easy translation to any desired computer programming language is possible. Pseudocode is very simple to learn and easy to use, and it in no way interferes with one's learning of an actual programming language.

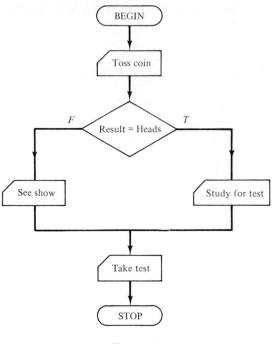

Figure 1

The second reason for using pseudocode is the fact that many professional programmers believe that developing and maintaining pseudocode versions of a program, before and after translation to an actual programming language, encourages good programming practice and aids in developing and modifying programs. We feel that the student should see a pseudocode in use for this reason. The pseudocode described below is largely taken from Rader (see Further Readings), and has seen service in a practical programming environment. We have made certain cosmetic changes in the interest of pedagogy.

One warning is in order. An algorithm written in pseudocode may, if it is finely detailed, be very reminiscent of a computer program. This is deliberate, even to the use of terms like SUBROUTINE and the statement RETURN at the end of a subroutine to signify that we should return to the steps of the main algorithm. Also, the actual programming of algorithms is facilitated by the similarity of pseudocode to a programming language. However, always remember that a pseudocode algorithm is *not* a computer program. It is meant for human beings, not machines, and we are only obliged to include sufficient detail to make the algorithm clear to human readers. The details of pseudocode are given below.

Pseudocode

In pseudocode, successive steps are usually labeled with consecutive numbers. If a step begins a selection or a loop, several succeeding steps may be considered subordinate to this step (for example, the body of a loop). Subordinate lines are indented several spaces and labeled with consecutive letters instead of numbers. If these steps had subordinates, they in turn would be indented and labeled with numbers. We use only consecutive numbers or letters as labels, and we alternate them in succeeding levels of subordination. A typical structuring of steps with subordinate steps is illustrated in the following.

1. line 1
   a. line 2
   b. line 3
      1. line 4
      2. line 5
   c. line 6
2. line 7
3. line 8
   a. line 9
      1. line 10
   b. line 11
4. line 12

Steps that have the same degree of indentation will be said to be at the **same level**. Thus the next line at the level of line 1 is line 7, while the next line at the level of line 3 is line 6, and so on.

Selection in pseudocode is expressed with the form IF–THEN–ELSE, as shown below:

1. IF (CONDITION) THEN

   | true-block |

2. ELSE

   | false-block |

The true- and false-blocks (to be executed respectively when CONDITION is true and CONDITION is false) may contain any legitimate pseudocode including selections or iterations. Sometimes we will omit statement 2, the ELSE statement, and the false-block. In this case, the true-block is executed only when CONDITION is true and then, whether CONDITION is true or false, control passes to the next statement that is at the same level as statement 1.

Example 6   Consider the following statements in pseudocode. Assume that X is a rational number.

1. IF $(X > 13,000)$ THEN
   a. $Y \leftarrow X + .02(X - 13,000)$

2. ELSE
   a. $Y \leftarrow X + .03X$

In statement 1, CONDITION is: $X > 13{,}000$. If $X > 13{,}000$, then Y is computed by the formula

$$X + .02(X - 13{,}000)$$

while if $X \leq 13{,}000$, then Y is computed by the formula

$$X + .03X$$

We will use the ordinary symbols of mathematics to express algebraic relationships and conditions in pseudocode. The symbols $+$, $-$, $\times$, and $/$ will be used for the basic arithmetic operations, and the symbols $<$, $>$, $\leq$, $\geq$, $=$, and $\neq$ (does not equal) will be used for testing conditions. The number X raised to the power Y will be denoted by X^Y, the square of a number A will be denoted by A^2, and so on. Moreover, products such as $3 \times A$ will be denoted by $3A$, and so on, when no confusion is possible.

We will use a left arrow, $\leftarrow$, rather than the equal sign, for assignments of values to variables. Thus, as in Example 6, the expression $Y \leftarrow X + .03X$ means that Y is assigned the value specified by the right-hand side. The use of "$=$" for this purpose conflicts with the use of this symbol for testing conditions. Thus "$X = X + 1$" would either be an assignment or a question about the number X. The use of "$\leftarrow$" avoids this problem.

A fundamental way to express iteration expressions in pseudocode is the WHILE form shown below:

1. WHILE (CONDITION)

   | repeat-block |

Here the CONDITION is tested, and if true the block of pseudocode following it is executed. This process is repeated until CONDITION becomes false, after which control passes to the next statement that is at the same level as statement 1.

Example 7 Consider the following algorithm in pseudocode; N is assumed to be a positive integer.

1. $X \leftarrow 0$
2. $Y \leftarrow 0$
3. WHILE $(X < N)$
   a. $X \leftarrow X + 1$
   b. $Y \leftarrow Y + X$

4. $Y = Y/2$

END OF ALGORITHM

In this algorithm, CONDITION is $X < N$. As long as CONDITION is true, that is, as long as $X < N$, statements a and b will be executed repeatedly. As soon as CONDITION is false, that is, as soon as $X = N$, statement 4 will be executed. This means that the WHILE loop is executed N times and the algorithm computes

$$\frac{1 + 2 + \cdots + N}{2}$$

which is the value of variable Y at the completion of the algorithm.

A simple modification of the WHILE form called the UNTIL form is useful, and we include it, although it could be replaced by completely equivalent statements using WHILE. This construction is

1. UNTIL (CONDITION)

    ┌─────────────────┐
    │ repeat-block   │
    └─────────────────┘

Here the loop continues to be executed *until* the condition is true; that is, it continues only as long as the condition is false. Also, CONDITION is tested *after* the repeat-block rather than *before*, so that the block must be repeated at least once.

Example 8 The algorithm given in Example 7 could also be written with an UNTIL statement as follows:

1. $X \leftarrow 0$
2. $Y \leftarrow 0$
3. UNTIL $(X \geq N)$
    a. $X \leftarrow X + 1$
    b. $Y \leftarrow Y + X$
4. $Y \leftarrow Y/2$

END OF ALGORITHM

In this algorithm, the CONDITION $X \geq N$ is tested at the completion of step 3. If it is false, the body of step 3 is repeated. This process continues until the test reveals that CONDITION is true (when $X = N$). At that time step 4 is immediately executed.

The UNTIL form of iteration is a convenience, and could be formulated with a WHILE statement. The form

1. UNTIL (CONDITION)

$$\boxed{\text{block } 1}$$

is actually equivalent to the form

1. $\boxed{\text{block } 1}$

2. WHILE (CONDITION = FALSE)

$$\boxed{\text{block } 1}$$

In each case, the instructions in block 1 are followed once, regardless of CONDITION. After this, CONDITION is checked, and if it is true, the process stops; otherwise, block 1 instructions are followed again. This procedure of checking CONDITION, then repeating instructions in block 1 if CONDITION is false, is continued until CONDITION is true. Since both forms produce the same results, they are equivalent.

The other form of iteration is the one most like a traditional do loop, and we express it as a FOR statement:

1. FOR VAR $= X$ THRU Y [BY Z]

$$\boxed{\text{repeat-block}}$$

In this form, VAR is an integer variable used to count the number of times the instructions in repeat-block have been followed. X, Y, and Z, if desired, are either integers or expressions whose computed values are integers. The variable VAR begins at X and increases Z units at a time (Z is 1 if not specified). After each increase in X, the repeat-block is executed as long as the new value of X is not greater than Y. The conditions on VAR, specified by X, Y, and Z, are checked before each repetition of the instructions in the block. The block is repeated only if those conditions are true. The brackets around "BY Z" are not part of the statement, but simply mean that this part is optional and may be omitted. Note that the repeat block is always executed at least once, since no check is made until X is changed.

Example 9 The pseudocode statement

$$\text{FOR VAR} = 2 \text{ THRU } 10 \text{ BY } 3$$

will cause the repeat-block to be executed three times, corresponding to VAR $= 2, 5,$

8. The process ends then, since the next value of VAR would be 11, which is greater than 10.

We will use lines of pseudocode by themselves to illustrate different parts of a computation. However, when the code represents a complete thought, we may choose to designate it as an algorithm, a subroutine, or a function.

A set of instructions that will primarily be used at various places by other algorithms is often designated as a **subroutine**. A subroutine is given a name for reference, a list of input variables, which it will receive from other algorithms, and output variables, which it will pass on to the algorithms that use it. A typical title of a subroutine is

SUBROUTINE NAME $(A, B, \ldots; X, Y, \ldots)$

The values of the input variables are assumed to be supplied to the subroutine when it is used. Here NAME is a name generally chosen as a memory aid for the task performed by the subroutine; A, B, and so on, are input variables; X, Y, and so on, are output variables. The semicolon is used to separate input variables from output variables.

A subroutine will end with the statement RETURN. As we remarked earlier in this section, this simply reminds us to return to the algorithm (if any) that is using the subroutine.

An algorithm uses a subroutine by including the statement

CALL NAME $(A, B, \ldots; X, Y \ldots)$

where NAME is a subroutine and the input variables A, B, and so on, have all been assigned values. This process was also illustrated in earlier examples.

Example 10   The following subroutine computes the square of a positive integer N by successive additions.

SUBROUTINE SQR $(N; X)$
1. $X \leftarrow N$
2. $Y \leftarrow 1$
3. WHILE $(Y \neq N)$
   a. $X \leftarrow X + N$
   b. $Y \leftarrow Y + 1$
4. RETURN
END OF SUBROUTINE SQR

If the result of the steps performed by a subroutine is a single number, we may call the subroutine a FUNCTION. In this case, we title such a program as follows:

FUNCTION NAME($A, B, C, \ldots$)

where NAME is the name of the function, $A, B, C, \ldots$, are input variables. We also specify the value to be returned as follows:

RETURN(Y)

where Y is the value to be returned.

The name **function** is used because such subroutines remind us of familiar functions such as sin (x), log (x), and so on. When an algorithm requires the use of a function defined elsewhere, it simply uses the function in the familiar way and does not use the phrase CALL. Thus if a function FN1(X) has been defined, the following steps of pseudocode will compute 1 plus the value of the function FN1 at $3X + 1$.

1. $Y \leftarrow 3X + 1$
2. $Y \leftarrow 1 + \text{FN1}(Y)$

Example 11   The program given in Example 10 can be written as a function as follows:

FUNCTION SQR(N)
1. $X \leftarrow N$
2. $Y \leftarrow 1$
3. WHILE $(Y \neq N)$
   a. $X \leftarrow X + N$
   b. $Y \leftarrow Y + 1$
4. RETURN (X)
END OF FUNCTION SQR

Variables such as Y in Examples 10 and 11 are called **local variables**, since they are used only by the algorithm in its computations and are not part of input or output.

We will have many occasions to use linear arrays, and we need to be able to incorporate them into algorithms written in pseudocode. An array A will have locations indicated by $A[1], A[2], A[3], \ldots$ (as we noted in Section 1.3) and we will use this notation in pseudocode statements. Later, we will introduce arrays with more "dimensions." In most actual programming languages, such arrays must be introduced by dimension statements or declarations, which indicate the maximum number of locations that may be used in the array and the nature of the data to be stored. In pseudocode we will not require such statements, and the presence of brackets after a variable will serve to indicate that it is an array.

Example 12   Suppose that $X[1]$, $X[2]$, ..., $X[N]$ contain real numbers and that we want to exhibit the maximum such number. The following instructions will do that.

1. MAX ← $X[1]$
2. FOR $I = 2$ THRU N
   a. IF (MAX < $X[I]$) THEN
   b. MAX ← $X[I]$

Example 13   Suppose that $A[1]$, $A[2]$, ..., $A[N]$ contain zeros and ones, so that A represents a subset (which we will also call A) of a universal set U with N elements (see Section 1.3). Similarly, a subset B of U is represented by another array, $B[1]$, $B[2]$, ..., $B[N]$. The following pseudocode will compute the representation of the union $C = A \cup B$, and store it in locations $C[1]$, $C[2]$, ..., $C[N]$ of an array C.

1. FOR $I = 1$ THRU N
   a. IF (($A[I] = 1$) OR ($B[I] = 1$)) THEN
      1. $C[I] \leftarrow 1$
   b. ELSE
      1. $C[I] \leftarrow 0$

Finally, we do include a GO TO statement to direct attention to some other point in the algorithm. The usage would be GO TO LABEL, where LABEL is a name assigned to some line of the algorithm. If that line had the number 1, for example, then the line would have to begin

LABEL: 1 ...

We avoid the GO TO statement when possible, but there are times when the GO TO statement is extremely useful, for example, to exit a loop prematurely if certain conditions are detected.

EXERCISE SET 1.5

In Exercises 1–8, write out the steps (in pseudocode) needed to perform the task described.

1. In a western country, the tax structure is as follows. An income of $30,000 or more results in $6000 tax, an income of $20,000 to $30,000 pays $2500 tax, and an income of less than $20,000 pays a 10% tax. Write a function (in pseudocode) TAX that accepts a variable INCOME and outputs the tax appropriate to that income.
2. The table below shows brokerage commissions for firm X based on both price per share and number of shares purchased.

Commission Schedule (per share)

	Less than $150/share	$150/share or more
Less than 100 shares	$3.25	$2.75
100 shares or more	$2.75	$2.50

Write the pseudocode for a subroutine COMM with input variables NUMBER and PRICE (giving number of shares purchased and price per share), and output variable FEE giving the total commission for the transaction (not the per share commission).

3. Let X_1, X_2, ..., X_N be a given set of numbers. Write out the steps (in pseudocode) needed to compute the sum and the average of the numbers.

4. Write an algorithm to compute the sum of cubes of all numbers from 1 to N (that is, $1^3 + 2^3 + 3^3 + \cdots + N^3$).

5. Suppose that the array X consists of the real numbers $X[1]$, $X[2]$, $X[3]$; and the array Y consists of the real numbers $Y[1]$, $Y[2]$, $Y[3]$. Write an algorithm to compute

$$X[1]Y[1] + X[2]Y[2] + X[3]Y[3]$$

6. Let the array $A[1]$, $A[2]$, ..., $A[N]$ contain the coefficients a_1, a_2, ..., a_N of a polynomial $\sum_{i=1}^{N} a_i x^i$. Write the pseudocode for a subroutine that has the array A and variables N and X as inputs, and as output has the value of the polynomial at X.

7. Let $A[1]$, $A[2]$, $A[3]$ be the coefficients of a quadratic equation $ax^2 + bx + c = 0$ (that is, $A[1]$ contains a, $A[2]$ contains b, and $A[3]$ contains c). Write the pseudocode for an algorithm that computes the roots $R1$ and $R2$ of the equation, if they are real and distinct. If the roots are real and equal, the value should be assigned to $R1$ and a message printed. If the roots are not real, an appropriate message should be printed and computation halted. You may use functions $SQ(X)$ (which returns the square root of any nonnegative number X) and PRINT(TEXT), which outputs the message in TEXT.

8. Let $[a_1, a_2), [a_2, a_3), ..., [a_{N-1}, a_N]$ be N adjacent intervals on the real line. If $A[1], ..., A[N]$ contain the numbers $a_1, ..., a_n$, respectively, and X is a real number, write an algorithm that computes a variable INTERVAL as follows: If X is not between a_1 and a_N, INTERVAL = 0, but if X is in the ith interval, then INTERVAL = i. Thus INTERVAL specifies which interval (if any) contains the number X.

In Exercises 9–12, let A and B be arrays of length N that contain zeros and ones, and suppose they represent subsets (which we also call A and B) of some universal set U with N elements. Write algorithms that specify an array C representing the set indicated.

9. $C = A \oplus B$

10. $C = A \cap \bar{B}$

11. $C = \bar{A} \cap \bar{B}$

12. $C = A \cap (A \oplus \bar{B})$

In Exercises 13–20, write pseudocode programs to compute the quantity specified. Here N is a positive integer.

13. The sum of the first N nonnegative even integers.

14. The sum of the first N nonnegative odd integers.

15. The product of the first N positive even integers.

16. The product of the first N positive odd integers.

17. The sum of the squares of the first 77 positive integers.

18. The sum of the cubes of the first 23 positive integers.

19. The sum of the first 10 terms of the series

$$\sum_{n=1}^{\infty} \frac{1}{3n+1}$$

20. The smallest number of terms of the series

$$\sum_{n=1}^{\infty} \frac{1}{n+1}$$

whose sum exceeds 5.

In Exercises 21–25, describe what is accomplished by the pseudocode. Unspecified inputs or variables X and Y will represent rational numbers, while N and M will represent integers.

21. SUBROUTINE MAX($X,Y;Z$)
   1. $Z \leftarrow X$
   2. IF ($X < Y$) THEN
      a. $Z \leftarrow Y$
   3. RETURN
   END OF SUBROUTINE MAX

22. 1. $X \leftarrow 0$
   2. $I \leftarrow 1$
   3. WHILE ($X < 10$)
      a. $X \leftarrow X + 1/I$
      b. $I \leftarrow I + 1$

23.   FUNCTION $F(X)$
1.   IF $(X < 0)$ THEN
   a. $R \leftarrow -X$
2.   ELSE
   a. $R \leftarrow X$
3.   RETURN (R)
END OF FUNCTION F

24.   FUNCTION $F(X)$
1.   IF $(X < 1)$ THEN
   a. $R \leftarrow X^2 + 1$
2.   ELSE
   a.   IF $(X < 3)$ THEN
      1. $R \leftarrow 2X + 6$
   b.   ELSE
      1. $R \leftarrow X + 7$
3.   RETURN (R)
END OF FUNCTION F

25.   IF $(M < N)$ THEN
   a. $R \leftarrow 0$
2.   ELSE
   a. $K \leftarrow N$
   b.   WHILE $(K < M)$
      1. $K \leftarrow K + N$
   c.   IF $(K = M)$ THEN
      1. $R \leftarrow 1$
   d.   ELSE
      1. $R \leftarrow 0$

In Exercises 26–30, give the value of all variables at the time when the given set of instructions terminates. N will always represent a positive integer.

26.   1. $I \leftarrow 1$
2. $X \leftarrow 0$
3.   WHILE $(I \leq N)$
   a. $X \leftarrow X + 1$
   b. $I \leftarrow I + 1$

27.   1. $I \leftarrow 1$
2. $X \leftarrow 0$
3.   WHILE $(I \leq N)$
   a. $X \leftarrow X + I$
   b. $I \leftarrow I + 1$

28.   1. $A \leftarrow 1$
2. $B \leftarrow 1$
3.   UNTIL $(B > 100)$
   a. $B \leftarrow 2A - 2$
   b. $A \leftarrow A + 3$

29.   1.   FOR $I = 1$ THRU 50 BY 2
   a. $X \leftarrow 0$
   b. $X \leftarrow X + I$
2.   IF $(X < 50)$ THEN
   a. $X \leftarrow 25$
3.   ELSE
   a. $X \leftarrow 0$

30.   1. $X \leftarrow 1$
2. $Y \leftarrow 100$
3.   WHILE $(X < Y)$
   a. $X \leftarrow X + 2$
   b. $Y \leftarrow \frac{1}{2} Y$

1.6 Induction and Recursion

Mathematical Induction

Let k be a fixed integer (positive, negative, or zero). Suppose that for each integer $n \geq k$ we have a corresponding proposition $P(n)$ and that we wish to show $P(n)$ true for all $n \geq k$. Suppose that:

(a) $P(k)$ is true.
(b) For all $n \geq k$, whenever $P(n)$ is true, it follows that $P(n + 1)$ is true.

Then the **principle of mathematical induction** states that $P(n)$ is true for all $n \geq k$.

Thus, to prove the validity of a proposition $P(n)$ for all $n \geq k$, we must first prove that $P(k)$, the proposition corresponding to the first integer k, is true. We next have to show that if the result is assumed to be true for some integer $n \geq k$, it is also true for $n + 1$. The first step is called the **basis step** of the induction; the second is called the **induction step**. In general, the basis step is easy; work is required in showing that if $P(n)$ is assumed to be true, it then follows that $P(n + 1)$ is true.

Example 1   Show, by mathematical induction, that for all $n \geq 1$,

$$1 + 2 + \cdots + n = \frac{n(n + 1)}{2}$$

Solution. Let $P(n)$ be the statement

$$1 + 2 + \cdots + n = \frac{n(n + 1)}{2}$$

In this example, $k = 1$.

Basis Step. We must first show that $P(1)$ is true. We see that $P(1)$ is

$$1 = \frac{1(1 + 1)}{2}$$

which is clearly true.

Induction Step. Assume now that we know $P(n)$ to be true for some $n \geq 1$. That is, we assume that for some fixed $n \geq 1$,

$$1 + 2 + \cdots + n = \frac{n(n + 1)}{2} \tag{1}$$

We now wish to prove the validity of the statement $P(n + 1)$. We have

$$1 + 2 + \cdots + n + (n + 1) = [1 + 2 + \cdots + n] + (n + 1)$$

$$= \frac{n(n + 1)}{2} + (n + 1) \qquad \text{[the expression in brackets is replaced by}$$

$$= \frac{n(n + 1) + 2(n + 1)}{2} \qquad \frac{n(n + 1)}{2} \text{ using equation (1)]}$$

$$= \frac{(n + 1)(n + 2)}{2} \qquad \text{(factoring)}$$

which proves the validity of $P(n + 1)$. By the principle of mathematical induction, it follows that $P(n)$ is true for all $n \geq 1$.

The following example shows one way in which induction can be useful in computer programming.

Example 2   Consider the following function in pseudocode.

FUNCTION SQ(A)
1. $C \leftarrow 0$
2. $D \leftarrow 0$
3. WHILE $(D \neq A)$
   a. $C \leftarrow C + A$
   b. $D \leftarrow D + 1$
4. RETURN (C)
END OF FUNCTION SQ

The name "SQ" of the function suggests that it computes the square of a positive integer A. A few trials with particular values of A will convince you that the program does carry out this task. However, suppose we now want to prove that this function always computes the square of the positive integer A, no matter how large A might be. We shall give a proof by induction. Let C_n and D_n be the values of the variables C and D, respectively, after passing through the WHILE loop n times. Let $P(n)$ be the statement

$$C_n = A \times D_n$$

We shall prove by induction that $P(n)$ is valid for all $n \geq 0$. Here $k = 0$.

Basis Step. $P(0)$ is the statement $C_0 = A \times D_0$, which is true since $C_0 = D_0 = 0$ (the values of C and D are zero after zero passes through the WHILE loop).

Induction Step. Suppose that $P(n)$ is true for some $n \geq 0$. That is, we assume that after n passes through the WHILE loop,

$$C_n = A \times D_n \tag{2}$$

After one more pass through the loop, C is increased by A, and D is increased by 1, so

$$C_{n+1} = C_n + A$$
$$D_{n+1} = D_n + 1$$

Then

$$C_{n+1} = C_n + A = (A \times D_n) + A \qquad \text{[using equation (2)]}$$
$$= A \times (D_n + 1)$$
$$= A \times D_{n+1}$$

so $P(n + 1)$ is true. By the principle of mathematical induction, it follows that as long as looping occurs

$$C_n = A \times D_n$$

When the loop terminates, $D = A$, so

$$C = A \times A = A^2$$

which is the value returned by the function SQ.

Example 2 illustrates the use of a so-called **loop invariant**, a relationship between variables which persists through all iterations of the loop. This technique for proving that loops (and programs) do what they claim to do is an important part of the theory of algorithm verification. In Example 2 it is clear that the looping stops, but in more complex cases, this may also be proved by induction.

Example 3   Consider the following subroutine.

SUBROUTINE EXP($N,M;R$)
1. $R \leftarrow 1$
2. WHILE ($M > 0$)
   a. $R \leftarrow R \times N$
   b. $M \leftarrow M - 1$
3. RETURN
END OF SUBROUTINE EXP

We claim that if N and M are nonnegative integers, then EXP returns N^M. To prove this, let R_n and M_n be respective values of R and M after $n(\geq 0)$ passes through the WHILE loop.

We claim that $P(n)$: $R_n \times N^{M_n} = N^M$ is true for all $n \geq 0$, and we prove this by mathematical induction. Here $k = 0$.

Basis Step. $R_0 = 1$, $M_0 = M$, since these are the values of the variables before looping begins; thus $P(0)$ is the statement $1 \times N^M = N^M$, which is true.

Induction Step. Suppose that $P(n)$ is true for some $n \geq 0$. That is, we assume that after n passes through the loop.

$$R_n \times N^{M_n} = N^M$$

After one additional pass through the loop,

$$R_{n+1} = R_n \times N \qquad \text{and} \qquad M_{n+1} = M_n - 1$$

Thus

$$R_{n+1} \times N^{M_{n+1}} = (R_n \times N) \times N^{M_n - 1} = R_n \times N^{M_n} = N^M$$

so $P(n + 1)$ is true. Hence, by the principle of mathematical induction, $P(n)$ is true for all $n \geq 0$. When looping ends, $M_n = 0$ and so $R_n = N^M$. Thus the subroutine always returns the value N^M.

Example 4   Let $A_1, A_2, A_3, \ldots, A_n$ be any n subsets of a set U. We show by induction that

$$\overline{\left(\bigcup_{k=1}^{n} A_k \right)} = \bigcap_{k=1}^{n} \bar{A}_k$$

[extended De Morgan's law (see Section 1.3)].

Let $P(n)$ be the statement that the given equality holds for any n subsets of U. We prove by induction that $P(n)$ is always true, for $n \geq 1$.

Basis Step. If $n = 1$, the result holds trivially.

Induction Step. Suppose that $P(n)$ is true for some $n \geq 1$ and let $A_1, \ldots, A_n, A_{n+1}$ be any $n + 1$ subsets of U. Then let $B = A_1 \cup A_2 \cup \cdots \cup A_n$. We have

$$\overline{\left(\bigcup_{k=1}^{n+1} A_k \right)} = \overline{B \cup A_{n+1}}$$

$$= \bar{B} \cap \bar{A}_{n+1} \qquad \text{(by Section 1.3)}$$

$$= \left(\bigcap_{k=1}^{n} \bar{A}_k \right) \cap \bar{A}_{n+1} \qquad \text{(by induction assumption)}$$

$$= \bigcap_{k=1}^{n+1} \bar{A}_k$$

Thus $P(n + 1)$ is true, so by induction, $P(n)$ is true for all $n \geq 1$.

Example 5   We show by induction that any finite, nonempty set is countable (that is, it can be arranged in a list).

Basis Step. If $S = \{b\}$ is a set with one element, then b forms a sequence all by itself whose set is S.

Induction Step. Suppose that the result is true for all sets S with $|S| = n \geq 1$ (see Section 1.1). Now let S be a set with $|S| = n + 1$ and choose an element b in S. Since $S - \{b\}$ has n members, the induction hypothesis tells us that there is a sequence x_1, $x_2, \ldots, x_n$ having corresponding set $S - \{b\}$. The sequence $x_1, x_2, \ldots, x_n, b$ then has corresponding set S, and we are done.

Warning. In proving results by induction, you should not start by assuming that $P(n + 1)$ is true and attempting to manipulate this result until you arrive at a true statement. This common mistake is always an incorrect use of the principle of mathematical induction.

Recursion

It often happens that a whole set of objects is defined at once, for example the elements in a sequence, the members of a set, or the operations that an algorithm or program performs on its inputs. Sometimes the definition, for certain items, makes reference to earlier or simpler versions of itself. If so, we say that the definition (or algorithm, or program) is **recursive**.

Example 6   Consider the following definition of the factorial function:

$$1! = 1$$

$$n! = n(n - 1)! \qquad \text{for } n > 1$$

Notice that in the second part of the definition, factorial is used as part of the definition of $n!$. Thus this definition is recursive. Note that it is not circular because by the time one wants to define $n!$ (if we look at the definition in increasing order of n), we have already defined $(n - 1)!$, so we can use its value.

     This definition leads naturally to an algorithm for computing $n!$ as follows:

```
FUNCTION FAC(N)
1. IF (N = 1) THEN
   a. A ← 1
2. ELSE
   a. A ← N × FAC(N − 1)
3. RETURN (A)
END OF FUNCTION FAC
```

The algorithm, or program, given in Example 6 is also called **recursive** since it uses a call to itself in its definition. Some programming languages allow recursive subroutines and functions, and some do not, but recursion as a programming technique (possibly followed by a recursion elimination technique) is useful in any case.

Often, a recursive definition of an object will lead naturally to a recursive algorithm. Sometimes (as with factorial) it would be better to do the computation in another way, but often, as with tree searching, which we discuss later, the recursive algorithm is preferred, because it is easier to construct and understand.

We now consider another slightly different example of recursion.

Example 7   Suppose that we define a sequence (F_n), $1 \leq n < \infty$, of integers as follows:

$$F_1 = 1$$
$$F_2 = 1$$
$$F_n = F_{n-1} + F_{n-2} \qquad \text{if } n > 2$$

Since, in the third line, the definition refers to earlier versions of itself, it is recursive. It is easy to compute the first few terms in this sequence: 1, 1, 2, 3, 5, 8, ..., called the **Fibonacci sequence**. A naturally recursive algorithm for computing these numbers is:

```
FUNCTION FIB(N)
1. IF (N = 1) THEN
   a. A ← 1
2. ELSE
   a. IF (N = 2) THEN
      1. A ← 1
   b. ELSE
      1. A ← FIB(N − 1) + FIB(N − 2)
3. RETURN (A)
END OF FUNCTION FIB
```

This algorithm is deceptively simple in appearance. If the reader will choose an integer (say 3 or 4) and trace through the sequence of computations and calls, the complexity will be apparent. Again, we are using this example to illustrate recursion. Other methods of computing the Fibonacci numbers are more efficient.

The main connection between recursion and induction is that objects that are defined recursively are often defined by means of a natural sequence, so induction is frequently the best (possibly the only) way to prove results about recursively defined objects.

Example 8   Consider our recursive definition of $n!$ given in Example 6. Suppose we wish to prove that for all $n \geq 1$,

$$n! \geq 2^{n-1} \tag{3}$$

We proceed by mathematical induction. Let $P(n)$ be the statement given by equation (3). Here $k = 1$.

Basis Step. $P(1)$ is the statement

$$1! \geq 2^0 = 1$$

Since $1!$ was defined to be 1, this statement is true.

Induction Step. Suppose that $P(n)$ is true for some $n \geq 1$. That is, for this value of n,

$$n! \geq 2^{n-1}$$

Then by the recursive definition,

$$
\begin{aligned}
(n + 1)! &= (n + 1) \times n! \\
&\geq (n + 1) \times 2^{n-1} \\
&\geq 2 \times 2^{n-1} \qquad \text{(since } n \geq 1 \text{, we have} \\
&= 2^n \qquad\qquad\quad n + 1 \geq 2)
\end{aligned}
$$

and $P(n + 1)$ is true. By the principle of mathematical induction, it follows that $P(n)$ is true for all $n \geq 1$.

Regular expressions and sets

In the formal languages that we discuss in Chapter 5, and in the finite state machines of Chapter 7, the concept of regular expressions plays an important role. Let I be a set. A **regular expression over I** is a string constructed from the elements of I and the symbols $(,), \vee, *, \lambda$, according to the following recursive definition.

RE1.   The symbol λ is a regular expression.
RE2.   If $x \in I$, the symbol x is a regular expression.
RE3.   If α and β are regular expressions, then the expression $\alpha\beta$ is regular.
RE4.   If α and β are regular expressions, then the expression $(\alpha \vee \beta)$ is regular.
RE5.   If α is a regular expression, then the expression $(\alpha)^*$ is regular.

By convention, if the regular expression α consists of a single symbol x, where $x \in I$, or if α begins and ends with parentheses, then we will write $(\alpha)^*$ simply as α^*.

When no confusion results, we will refer to a regular expression over I simply as a **regular expression** (omitting reference to I).

Example 9   Let $I = \{0,1\}$. Show that the expressions,

(a) 0* (0 ∨ 1)*

(b) 00* (0 ∨ 1)*1

(c) (01)* (01 ∨ 1*}

are all regular expressions over I.

Solution. (a) By RE2, 0 and 1 are regular expressions. Thus (0 ∨ 1) is regular by RE4, and so 0* and (0 ∨ 1)* are regular by RE5 (and the convention stated above). Finally, we see that 0* (0 ∨ 1)* is regular by RE3.

(b) We know that 0, 1, and 0* (0 ∨ 1)* are all regular. Thus by RE3, used twice, 00* (0 ∨ 1)* 1 must be regular.

(c) By RE3, 01 is a regular expression. Since 1* is regular, (01 ∨ 1*) is regular by RE4, and (01)* is regular by RE5. Then the regularity of (01)* (01 ∨ 1*) follows from RE3.

Associated with each regular expression over I, there is a corresponding subset of I^*, where we recall that I^* is the set of all strings, or sequences of finite length, of elements of I. Such sets are called **regular subsets** of I^* or just **regular sets**, if no reference to I is needed. To compute the regular set corresponding to a regular expression, we use the following correspondence rules.

1. The expression λ corresponds to the set $\{\Lambda\}$, where Λ is the empty string in I^*.
2. If $x \in I$, the regular expression x corresponds to the set $\{x\}$.
3. If α and β are regular expressions corresponding to subsets A and B of I^*, then $\alpha\beta$ corresponds to $A \cdot B = \{s \cdot t \mid s \in A \text{ and } t \in B\}$. Thus $A \cdot B$ is the set of all catenations of strings in A with strings in B.
4. If the regular expressions α and β correspond to subsets A and B of I^*, then $\alpha \vee \beta$ corresponds to $A \cup B$.
5. If the regular expression α corresponds to the subset A of I^*, then $(\alpha)^*$ corresponds to the set A^*. Note that A is a set of strings from I. Elements from A^* are finite sequences of such strings, and thus may themselves be interpreted as strings from I. Note also that we always have $\Lambda \in A^*$.

Example 10   Let $I = \{a, b, c\}$. Then the regular expression a^* corresponds to the set of all finite sequences of a's, such as aaa, $aaaaaaa$, and so on. The regular expression $a(b \vee c)$ corresponds to the set $\{ab, ac\} \subset I^*$. Finally, the regular expression $ab(bc)^*$ corresponds to the set of all strings which begin with ab, and then repeat the symbols bc n times, where $n \geq 0$. This set includes the strings ab, $abbcbc$, $abbcbcbcbc$, and so on.

Example 11   Let $I = \{0, 1\}$. Find the regular sets corresponding to the three regular expressions of Example 9.

Solution. (a) The set corresponding to $0*(0 \vee 1)*$ consists of all sequences of 0's and 1's. Thus, this set is I^*.

(b) The expression $00* (0 \vee 1)* 1$ corresponds to the set of all sequences of 0's and 1's which begin with at least one 0 and end with at least one 1.

(c) The expression $(01)* (01 \vee 1*)$ corresponds to the set of all sequences of 0's and 1's which either repeat the string 01 a total of $n \geq 1$ times, or begin with a total of $n \geq 0$ repetitions of 01 and end with some number $k \geq 0$ of 1's. This set includes, for example, the strings 1111, 01, 010101, 0101010111111, and 011.

EXERCISE SET 1.6

In Exercises 1–17, prove that the statement is true by using mathematical induction.

1. $2 + 4 + 6 + \cdots + 2n = n(n + 1)$

2. $1^2 + 3^2 + 5^2 + \cdots + (2n - 1)^2$
$$= \frac{n(2n + 1)(2n - 1)}{3}$$

3. $2 + 5 + 8 + \cdots + (3n - 1) = \dfrac{n(3n + 1)}{2}$

4. $4 + 8 + 12 + \cdots + 4n = 2n(n + 1)$

5. $1 + 2 + 2^2 + 2^3 + \cdots + 2^n = 2^{n+1} - 1$

6. $5 + 10 + 15 + \cdots + 5n = \dfrac{5n(n + 1)}{2}$

7. $1^2 + 2^2 + 3^2 + \cdots + n^2 = \dfrac{n(n + 1)(2n + 1)}{6}$

8. $1 \cdot 2 + 2 \cdot 3 + 3 \cdot 4 + \cdots + n(n + 1)$
$$= \frac{n(n + 1)(n + 2)}{3}$$

9. $1^3 + 2^3 + 3^3 + \cdots + n^3 = \dfrac{n^2(n + 1)^2}{4}$

10. $1 + 5 + 9 + \cdots + (4n - 3) = n(2n - 1)$

11. $2 + 2^2 + 2^3 + \cdots + 2^n = 2^{n+1} - 2$

12. $a + ar + ar^2 + \cdots + ar^{n-1} = \dfrac{a(1 - r^n)}{1 - r}$ $(r \neq 1)$

13. $1 + 2^n < 3^n$ $(n \geq 2)$

14. $n < 2^n$ $(n > 1)$

15. $1 + 2 + 3 + \cdots + n < \frac{1}{8}(2n + 1)^2$

16. $\dfrac{1}{1 \cdot 2} + \dfrac{1}{2 \cdot 3} + \dfrac{1}{3 \cdot 4} + \cdots + \dfrac{1}{n(n + 1)} = \dfrac{n}{n + 1}$

17. $1 + a + a^2 + \cdots + a^{n-1} = \dfrac{a^n - 1}{a - 1}$

18. Prove by mathematical induction that if a set A has k elements, then $P(A)$ has 2^k elements.

19. Show by mathematical induction that if $A_1, \ldots, A_n$ are any n subsets of a set U, then
$$\overline{\left(\bigcap_{k=1}^{n} A_k\right)} = \bigcup_{k=1}^{n} \bar{A}_k$$

20. Show by mathematical induction that if $A_1, \ldots, A_n$, and B are any subsets of a set U, then
$$\left(\bigcup_{k=1}^{n} A_k\right) \cap B = \bigcup_{k=1}^{n} (A_k \cap B)$$

21. Show by mathematical induction that if $A_1, A_2, \ldots, A_n$, and B are any subsets of a set U, then
$$\left(\bigcap_{k=1}^{n} A_k\right) \cup B = \bigcap_{k=1}^{n} (A_k \cup B)$$

In Exercises 22–26, show that the given algorithm, correctly used, produces the output stated, by showing (using mathematical induction) that the relationship indicated is a loop invariant and checking its value when looping ceases. All variables represent non-negative integers.

22. SUBROUTINE COMP$(X, Y; Z)$
   1. $Z \leftarrow X$
   2. $W \leftarrow Y$

3.   WHILE $(W > 0)$
   a. $Z \leftarrow Z + Y$
   b. $W \leftarrow W - 1$
4.   RETURN
END OF SUBROUTINE COMP

COMPUTES: $Z = X + Y^2$
LOOP INVARIANT: $(Y \times W) + Z = X + Y^2$

23.   SUBROUTINE DIFF$(X, Y; Z)$
1. $Z \leftarrow X$
2. $W \leftarrow Y$
3.   WHILE $(W > 0)$
   a. $Z \leftarrow Z - 1$
   b. $W \leftarrow W - 1$
4.   RETURN
END OF SUBROUTINE DIFF

COMPUTES: $Z = X - Y$
LOOP INVARIANT: $X - Z + W = Y$

24.   SUBROUTINE POWER$(X, Y; Z)$
1. $Z \leftarrow 0$
2. $W \leftarrow Y$
3.   WHILE $(W > 0)$
   a. $Z \leftarrow Z + X$
   b. $W \leftarrow W - 1$
4. $W \leftarrow Y - 1$
5. $U \leftarrow Z$
6.   WHILE $(W > 0)$
   a. $Z \leftarrow Z + U$
   b. $W \leftarrow W - 1$
7.   RETURN
END OF SUBROUTINE POWER

COMPUTES: $Z = X \times Y^2$
LOOP INVARIANT (*First Loop*):
$Z + (X \times W) = X \times Y$
LOOP INVARIANT (*Second Loop*):
$Z + (X \times Y \times W) = X \times Y^2$
(*Hint:* Recall the value of Z $(= U)$ after loop 1, and use that in loop 2.)

25.   SUBROUTINE CUB$(X; Z)$
1. $A \leftarrow 1$
2. $B \leftarrow 0$
3. $C \leftarrow X$

4. $Z \leftarrow 0$
5.   WHILE $(C > 0)$
   a. $Z \leftarrow Z + A + B$
   b. $B \leftarrow B + 2A + 1$
   c. $A \leftarrow A + 3$
   d. $C \leftarrow C - 1$
6.   RETURN
END OF SUBROUTINE CUB

COMPUTES: $Z = X^3$
LOOP INVARIANTS: (show that all remain true as the loop proceeds)
  (i) $A = 3(X - C) + 1$
  (ii) $B = 3(X - C)^2$
  (iii) $Z = (X - C)^3$

26.   SUBROUTINE EXP2$(N, M; R)$
1. $R \leftarrow 1$
2. $K \leftarrow 2M$
3.   WHILE $(K > 0)$
   a. $R \leftarrow R \times N$
   b. $K \leftarrow K - 1$
4.   RETURN
END OF SUBROUTINE EXP 2

COMPUTES: $R = N^{2M}$
LOOP INVARIANT: $R \times N^K = N^{2M}$

27.   We define T-numbers recursively as follows:
(1)   0 is a T-number.
(2)   If X is a T-number, $X + 3$ is a T-number.

Write out a description of the set of T-numbers.

28.   Define an S-number by:
(1)   8 is an S-number.
(2)   If X is an S-number and Y is a multiple of X, then Y is an S-number.
(3)   If X is an S-number and X is a multiple of Y, then Y is an S-number.

Describe the set of S-numbers.

29.   Let $F(N)$ be a function defined for all nonnegative integers by the following recursive definition.

$$F(0) = 0$$

$$F(1) = 1$$

$$F(N + 2) = 2F(N) + F(N + 1), N \geq 0$$

Write out the first six values of F.

30. Let $G(N)$ be a function defined for all nonnegative integers by the following recursive definition.

$$G(0) = 1$$

$$G(1) = 2$$

$$G(N + 2) = G(N)^2 \times G(N + 1), N \geq 0$$

Write out the first five values of G.

31. Let $I = \{+, \times, a, b\}$. Show that the following expressions are regular over I.

(a) $a + b\,(ab)^* \,(a \times b \vee a)$

(b) $a + b \times (a^* \vee b)$

(c) $((a^*b \vee +)^* \vee \times ab^*)$

32. Let $I = \{0, 1\}$. Describe the regular subsets of I^*

corresponding to the following regular expressions.

(a) $0^*\, 1\, 0^*\, 1\, 0^*$

(b) $(0^*\, 1\, 0^*\, 1\, 0^*)^*$

(c) $(0 \vee 1)^*\, 111$

(d) $(00 \vee 111)^*\, 1$

33. Let $I = \{a, b, c\}$. In each part below is listed a string in I^*, and a regular expression over I. In each case, state whether or not the string on the left belongs to the regular set corresponding to the expression on the right.

(a) ac a^*b^*c

(b) $abcc$ $(abc \vee c)^*$

(c) $aaabc$ $((a \vee b) \vee c)^*$

(d) ac $(a^*b \vee c)$

(e) $abab$ $(ab)^*c$

1.7   Division in the Integers

We shall now discuss some needed results on division and factoring in the integers. If n and m are integers and n is not zero, we can plot the integer multiples of n on the real line, and locate m. Figure 1 shows a typical example when $n > 0$, and Fig. 2 shows an example when $n < 0$, and m is different from that in Fig. 1.

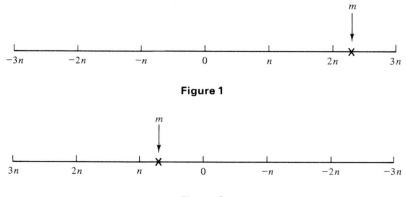

Figure 1

Figure 2

In either case, the distance between consecutive points is $|n|$, the absolute value of n. If m is a multiple of n, say $m = qn$, then we can write $m = qn + r$, where $r = 0$. If m is not a multiple of n, then let qn be the first multiple of n to the left of m, and let r be the distance (positive) between m and qn. Then $m = qn + r$, and $0 < r < |n|$ (see Fig. 3). Thus, in all cases, $m = qn + r, 0 \leq r < |n|$.

Figure 3

Example 1

(a)   If $n = 4$ and $m = -6$, then

$$-6 = (-2) \cdot 4 + 2$$

so $r = 2$.

(b)   If $n = -3$ and $m = -11$, then

$$-11 = 4(-3) + 1$$

so $r = 1$. We could also write

$$-11 = 3(-3) + (-2)$$

but this would not be the decomposition described in Theorem 1 below, since r would be -2, and according to Theorem 1, r must always be nonnegative. To avoid confusion when n is negative, just remember the geometric picture.

We state these observations as a theorem.

Theorem 1   If $n \neq 0$ and m are integers, we can write $m = qn + r$ for some integer q, with $0 \leq r < |n|$, in one and only one way.

If r is zero in Theorem 1, so that m is a multiple of n, we write $n \mid m$, which is read "n divides m." If m is not a multiple of n, we write $n \nmid m$, which is read "n does not divide m." We now prove some simple properties of divisibility.

Theorem 2   Let a, b, and c be integers.

(a)   If $a \mid b$ and $a \mid c$, then $a \mid (b \pm c)$ (here and later, we use $\pm$ if the result is equally true for $+$ or $-$).

(b)   If $a \mid b$ or $a \mid c$, then $a \mid bc$.

(c)   If $a \mid b$ and $b \mid c$, then $a \mid c$.

Proof.

(a)   If $a \mid b$ and $a \mid c$, we have $b = k_1 a$ and $c = k_2 a$ for some integers k_1 and k_2. Then

$$b \pm c = (k_1 \pm k_2)a$$

so $a \mid (b \pm c)$.

(b) as in (a), we have $b = k_1 a$ or $c = k_2 a$. Then $bc = k_1 ac$ or $bc = k_2 ab$, so $a \mid bc$.

(c) If $a \mid b$ and $b \mid c$, we have $b = k_1 a$ and $c = k_2 b$, so

$$c = k_2 b = k_2(k_1 a) = (k_2 k_1)a$$

and hence $a \mid c$.

Let us now restrict our attention to the set of positive integers Z^+. A number $p > 1$ in Z^+ is called **prime** if the only positive integers that divide p are p and 1.

Example 2   The numbers 2, 3, 5, 7, 11, and 13 are prime, while 4, 10, 16, and 21 are not prime. It is easy to write an algorithm to determine if a positive integer $n > 1$ is a prime number. We could divide by every integer from 2 to $n - 1$, and if none of these is a divisor of n, then n is a prime. To make this process more efficient, we note that if $mk = n$, then either m or k is less than or equal to $\sqrt{n}$. This means that if n is not prime, it has a divisor k satisfying the inequality $1 < k \le \sqrt{n}$, so we need only test for divisors in this range. Also, if n has any even number as a divisor, it must have 2 as a divisor. Thus, after checking for division by 2, we may skip all even integers. Pseudocode for the resulting algorithm is shown below. We assume the existence of functions SQR and INT, where SQR(n) returns the greatest integer not exceeding $\sqrt{n}$, and INT(X), X a rational number, returns the greatest integer not exceeding X. For instance, SQR(10) = 3, SQR(25) = 5, INT(7.124) = 7, and INT(8) = 8. We also assume a print function PRINT.

```
SUBROUTINE PRIME(N)
1. IF (N/2 = INT(N/2)) THEN
   a. PRINT ('NOT PRIME')
   b. RETURN
2. ELSE
   a. FOR D = 3 THRU SQR(N) BY 2
      1. IF (N/D = INT(N/D)) THEN
         a. PRINT ('NOT PRIME')
         b. RETURN
   b. PRINT ('PRIME')
   c. RETURN
END OF SUBROUTINE PRIME
```

Theorem 3   Every positive integer $n > 1$ can be uniquely written as

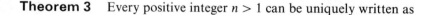

$$n = p_1^{k_1} p_2^{k_2} \cdots p_s^{k_s}$$

where $p_1 < p_2 < \cdots < p_s$ are the distinct primes that divide n, and the k's are positive integers giving the number of times each prime occurs as a factor of n.

We omit the proof of Theorem 3, but we give several illustrations.

Example 3

$$9 = 3 \cdot 3 = 3^2$$
$$24 = 12 \cdot 2 = 2 \cdot 2 \cdot 2 \cdot 3$$
$$= 2^3 \cdot 3$$
$$30 = 2 \cdot 3 \cdot 5$$

Greatest Common Divisor

If a, b, and k are in Z^+, and $k \mid a$, $k \mid b$, we say that k is a **common divisor** of a and b. If d is the largest such k, d is called the **greatest common divisor**, or GCD of a and b, and we write

$$d = \text{GCD } (a, b)$$

This number has some interesting properties. It can be written as a combination of a and b, and it is not only larger than all other common divisors, it is a multiple of each of them.

Theorem 4   If $d = \text{GCD } (a, b)$, then

    (a) $d = sa + tb$ for some integers s and t (not necessarily positive).

    (b) If c is any other common divisor of a and b, then $c \mid d$.

Proof. Let x be the smallest positive integer that can be written as $x = sa + tb$ for some integers s and t, and let c be a common divisor of a and b. Since $c \mid a$ and $c \mid b$, we see by Theorem 2 that $c \mid x$, so $c \le x$. If we can prove that x is a common divisor of a and b, it will then be the greatest common divisor of a and b and both parts of the theorem will have been proved. By Theorem 1, $a = qx + r$ with $0 \le r < x$. Solving for r, we see that $r = a - qx = a - q(sa + tb) = a - qsa - qtb = (1 - qs)a + (-qt)b$. If r is not zero, then since $r < x$ and r is a multiple of a, plus a multiple of b, we will have a contradiction to the fact that x is the smallest positive number that is a sum of multiples of a and b. Thus $r = 0$, so $x \mid a$. In the same way, we can show that $x \mid b$, and this completes the proof.

Suppose now that a, b, and d are in Z^+ and that d is a common divisor of a and b which is a multiple of every other common divisor of a and b. Then d is the greatest common divisor of a and b.

This result and Theorem 4(b) have shown the following result:

Let a, b, and d be in Z^+. The integer d is the greatest common divisor of a and b if and only if:

(a) $d \mid a$ and $d \mid b$.

(b) whenever $c \mid a$ and $c \mid b$, then $c \mid d$.

Example 4

(a) The common divisors of 12 and 30 are 2, 3, and 6, so that

$$\text{GCD } (12, 30) = 6$$

and

$$6 = 1 \cdot 30 + (-2) \cdot 12$$

(b) It is clear that

$$\text{GCD } (17, 95) = 1$$

since 17 is a prime, and the reader may verify that

$$1 = 28 \cdot 17 + (-5) \cdot 95$$

One remaining question is that of how to compute the GCD in general. Repeated application of Theorem 1 provides the key.

We now present an algorithm, called the **Euclidean algorithm**, for finding GCD (a, b). Suppose that $a > b > 0$ (otherwise, interchange a and b), and write

$$a = k_1 b + r_1, \qquad \text{where } 0 \leq r_1 < b \tag{1}$$

We now continue using Theorem 1 as follows:

$$
\begin{aligned}
b &= k_2 r_1 + r_2 & 0 &\leq r_2 < r_1 \\
r_1 &= k_3 r_2 + r_3 & 0 &\leq r_3 < r_2 \\
r_2 &= k_4 r_3 + r_4 & 0 &\leq r_4 < r_3 \\
&\;\;\vdots & &\;\;\vdots \\
r_{n-2} &= k_n r_{n-1} + r_n & 0 &\leq r_n < r_{n-1} \\
r_{n-1} &= k_{n+1} r_n + r_{n+1} & 0 &\leq r_{n+1} < r_n
\end{aligned}
\tag{2}
$$

Since

$$a > b > r_1 > r_2 > r_3 > r_4, \ldots$$

the remainders will eventually become zero, so at some point we obtain

$$r_{n+1} = 0$$

We now show that $r_n = \text{GCD}\,(a, b)$. If c is a common divisor of a and b, then from equation (1) it follows that $c \mid r_1$, so c is a common divisor of b and r_1. Conversely, if c is a common divisor of b and r_1, then from equation (1) it follows that $c \mid a$, so c is a common divisor of a and b. Thus the common divisors of a and b are the same as the common divisors of b and r_1, and therefore

$$\text{GCD}\,(a, b) = \text{GCD}\,(b, r_1)$$

Repeating this argument with b and r_1, we see that

$$\text{GCD}\,(b, r_1) = \text{GCD}\,(r_1, r_2)$$

Continuing, we have

$$\text{GCD}\,(a, b) = \text{GCD}\,(b, r_1) = \text{GCD}\,(r_1, r_2) = \cdots = \text{GCD}\,(r_{n-1}, r_n)$$

Since $r_{n-1} = k_{n+1} r_n$, we see that $\text{GCD}\,(r_{n-1}, r_n) = r_n$. Hence $r_n = \text{GCD}\,(a, b)$.

Example 5

(a) Let $a = 190$ and $b = 34$. Then

$$190 = 5 \cdot 34 + 20$$
$$34 = 1 \cdot 20 + 14$$
$$20 = 1 \cdot 14 + 6$$
$$14 = 2 \cdot 6 + 2$$
$$6 = 3 \cdot 2 + 0$$

so

$$\text{GCD}\,(190, 34) = 2$$

(b) Let $a = 108$ and $b = 60$. Then

$$108 = 1 \cdot 60 + 48$$
$$60 = 1 \cdot 48 + 12$$
$$48 = 4 \cdot 12$$

so

$$\text{GCD } (108, 60) = 12$$

In Theorem 4(a) we observed that if $d = \text{GCD } (a, b)$, we can find integers s and t such that

$$d = sa + tb$$

The integers s and t can be found as follows. Solve the next-to-last equation in (2) for r_n:

$$r_n = r_{n-2} - k_n r_{n-1} \tag{3}$$

Now solve the second-to-last equation in (2),

$$r_{n-3} = k_{n-1} r_{n-2} + r_{n-1}$$

for r_{n-1}:

$$r_{n-1} = r_{n-3} - k_{n-1} r_{n-2}$$

and substitute this expression for r_{n-1} in (3):

$$r_n = r_{n-2} - k_n[r_{n-3} - k_{n-1} r_{n-2}]$$

Continue up the equations in (2) and (1) replacing r_i by an expression involving r_{i-2} and r_{i-1}, finally arriving at an expression involving only a and b.

Example 6

(a) Let $a = 190$ and $b = 34$ as in Example 5(a). Then

$$\text{GCD } (190, 34) = 2 = 14 - 2(6)$$
$$= 14 - 2[20 - 1(14)] \qquad 6 = 20 - 1 \cdot 14$$
$$= 3(14) - 2 \cdot 20$$

$$= 3[34 - 1(20)] - 2(20) \qquad 14 = 34 - 1 \cdot 20$$
$$= 3(34) - 5(20)$$
$$= 3(34) - 5(190 - 5 \cdot 34) \qquad 20 = 190 - 5 \cdot 34$$
$$= 28(34) - 5(190)$$

Hence $s = 28$ and $t = -5$.

    (b) Let $a = 108$ and $b = 60$ as in Example 5(b). Then

$$GCD\ (108,\ 60) = 12 = 60 - 1(48)$$
$$= 60 - 1[108 - 1(60)] \qquad 48 = 108 - 1 \cdot 60$$
$$= 2(60) - 108$$

Hence $s = 2$ and $t = -1$.

    The following theorem is used in Example 7 below.

Theorem 5   If a and b are in Z^+, then GCD $(a, b) =$ GCD $(b, b \pm a)$.

    Proof. If c divides a and b, it divides $b \pm a$ by Theorem 2. Since $a = b - (b - a) = -b + (b + a)$, we see, by Theorem 2, that a common divisor of b and $b \pm a$ also divides a and b. Since a and b have the same common divisors as b and $b \pm a$, they must have the same GCD.

Example 7   Consider the following program for finding the greatest common divisor of two positive integers.

```
FUNCTION GCD(X, Y)
1. WHILE(X ≠ Y)
   A. If (X > Y) THEN
      1. X ← X − Y
   B. ELSE
      1. Y ← Y − X
2. RETURN(X)
END OF FUNCTION GCD
```

    We now prove that this algorithm does compute the greatest common divisor of X and Y, and we do so by induction. Let X_n and Y_n be the values of X and Y, respectively, after $n \geq 0$ passes through the WHILE loop. Let $P(n)$ be the statement

$$GCD\ (X_n,\ Y_n) = GCD\ (X,\ Y)$$

We shall prove, by induction, that $P(n)$ is valid for all $n \geq 0$.

Basis Step. Clearly, $X_0 = X$ and $Y_0 = Y$ (since the WHILE loop has been traversed zero times). Thus GCD $(X_0, Y_0) = $ GCD (X, Y) and $P(0)$ is true.

Induction Step. Suppose that $P(n)$ is true for some value of $n \geq 0$. That is, we assume that after n passes through the WHILE loop,

$$\text{GCD } (X_n, Y_n) = \text{GCD } (X, Y)$$

After one more pass through the loop, one of two cases is possible. Either

$$X_{n+1} = X_n \quad \text{and} \quad Y_{n+1} = Y_n - X_n$$

or

$$X_{n+1} = X_n - Y_n \quad \text{and} \quad Y_{n+1} = Y_n$$

In either case, Theorem 5 shows that

$$\text{GCD } (X_{n+1}, Y_{n+1}) = \text{GCD } (X_n, Y_n)$$
$$= \text{GCD } (X, Y)$$

by our assumption that $P(n)$ is true. Hence $P(n + 1)$ is true. By the principle of mathematical induction, it follows that $P(n)$ is valid for all $n \geq 0$.

When the function exits, $X = Y$. Thus, if it exits after n passes through the WHILE loop, $X_n = Y_n$, so GCD $(X_n, Y_n) = X_n$. By the induction proof above,

$$\text{GCD } (X, Y) = X_n$$

and this is the value returned by the function.

Least Common Multiple

If a, b, and k are in Z^+, and $a \mid k$, $b \mid k$, we say that k is a **common multiple** of a and b. The smallest such k, call it c, is called the **least common multiple**, or LCM, of a and b, and we write

$$c = \text{LCM } (a, b)$$

The following result shows that we can obtain the least common multiple from the greatest common divisor, so we do not need a separate procedure for finding the least common multiple.

Theorem 6   If a and b are two positive integers, then

$$\text{GCD } (a, b) \cdot \text{LCM } (a, b) = ab$$

Proof. Let $p_1, p_2, \ldots, p_k$ be all the prime numbers that are factors of either a or b. Then we can write

$$a = p_1^{a_1} p_2^{a_2} \cdots p_k^{a_k}$$

and

$$b = p_1^{b_1} p_2^{b_2} \cdots p_k^{b_k}$$

where some of the a_i and b_i may be zero.

It then follows that

$$\text{GCD } (a, b) = p_1^{\min\{a_1, \, b_1\}} p_2^{\min\{a_2, \, b_2\}} \cdots p_k^{\min\{a_k, \, b_k\}}$$

and

$$\text{LCM } (a, b) = p_1^{\max\{a_1, \, b_1\}} p_2^{\max\{a_2, \, b_2\}} \cdots p_k^{\max\{a_k, \, b_k\}}$$

Hence

$$\text{GCD } (a, b) \cdot \text{LCM } (a, b) = p_1^{a_1 + b_1} p_2^{a_2 + b_2} \cdots p_k^{a_k + b_k}$$
$$= (p_1^{a_1} p_2^{a_2} \cdots p_k^{a_k}) \cdot (p_1^{b_1} p_2^{b_2} \cdots p_k^{b_k})$$
$$= ab$$

Example 8   Let

$$a = 540 \qquad \text{and} \qquad b = 504$$

Factoring a and b into primes, we obtain

$$a = 540 = 2^2 \cdot 3^3 \cdot 5$$
$$b = 504 = 2^3 \cdot 3^2 \cdot 7$$

Thus all the prime numbers that are factors of either a or b are

$$p_1 = 2, \qquad p_2 = 3, \qquad p_3 = 5, \qquad p_4 = 7$$

Then

$$a = 540 = 2^2 \cdot 3^3 \cdot 5^1 \cdot 7^0$$
$$b = 504 = 2^3 \cdot 3^2 \cdot 5^0 \cdot 7^1$$

We then have

$$\text{GCD } (540, 504) = 2^{\min\{2,\ 3\}} \cdot 3^{\min\{3,\ 2\}} \cdot 5^{\min\{1,\ 0\}} \cdot 7^{\min\{0,\ 1\}}$$

$$= 2^2 \cdot 3^2 \cdot 5^0 \cdot 7^0$$

$$= 2^2 \cdot 3^2$$

$$= 36$$

Also,

$$\text{LCM } (540, 504) = 2^{\max\{2,\ 3\}} \cdot 3^{\max\{3,\ 2\}} \cdot 5^{\max\{1,\ 0\}} \cdot 7^{\max\{0,\ 1\}}$$

$$= 2^3 \cdot 3^3 \cdot 5^1 \cdot 7^1$$

$$= 7560$$

Then

$$\text{GCD } (540, 504) \cdot \text{LCM } (540, 504)$$

$$= 36 \cdot 7560$$

$$= 272{,}160 = 504 \cdot 540$$

As a verification, we can also compute GCD $(540, 504)$ by the Euclidean algorithm, obtaining the same result (verify).

Exercise Set 1.7

In Exercises 1–6, for the given integers n and m, write m as $qn + r$, with $0 \leq r < |n|$.

1. $m = 20, n = 3$ 2. $m = 64, n = 37$
3. $m = 3, n = 22$ 4. $m = 48, n = 12$
5. $m = 8, n = -3$ 6. $m = -12, n = -140$

7. Which of the following numbers are prime?
 (a) 22   (b) 29
 (c) 47   (d) 81
 (e) 527   (f) 247
 (g) 1180   (h) 1201

8. Write each integer as a product of powers of primes (as indicated by Theorem 3).
 (a) 60   (b) 350
 (c) 858   (d) 1666
 (e) 1125   (f) 210

In Exercises 9–14, find the greatest common divisor d of the integers a and b, and write d as $sa + tb$.

9. 32, 27   10. 45, 33
11. 40, 88   12. 60, 100
13. 34, 58   14. 77, 128

In Exercises 15–18, find the least common multiple of the integers.

15. 72, 108   16. 150, 70

17. 175, 245   18. 32, 27

19. Let a and b be integers. Prove that if p is a prime and $p \mid ab$ then $p \mid a$ or $p \mid b$. (*Hint*: If $p \nmid a$, then $1 = \text{GCD } (a, p)$; use Theorem 4 to write $1 = sa + tp$.)

20. Show that if GCD $(a, c) = 1$ and $c \mid ab$, then $c \mid b$.

21. Show that if GCD $(a, c) = 1$, $a \mid m$, and $c \mid m$, then $ac \mid m$. (*Hint*: Use Exercise 20.)

22. Show that if $d = $ GCD (a, b), $a \mid b$, and $c \mid b$, then $ac \mid bd$.

23. Show that GCD $(ca, cb) = c$ GCD (a, b).

24. Use induction to show that if p is a prime and $p \mid a^n$ ($n \geq 1$), then $p \mid a$.

25. Prove that if GCD $(a, b) = 1$, then GCD $(a^n, b^n) = 1$ for all $n \geq 1$. (*Hint*: Use Exercise 24.)

26. Show that LCM $(a, ab) = ab$.

27. Show that if GCD $(a, b) = 1$, then LCM$(a, b) = ab$.

28. Let $c = $ LCM (a, b). Show that if $a \mid k$ and $b \mid k$, then $c \mid k$.

1.8 Matrices

A **matrix** is a rectangular array of numbers arranged in m horizontal **rows** and n vertical **columns**:

$$A = \begin{bmatrix} a_{11} & a_{12} & \cdots & a_{1n} \\ a_{21} & a_{22} & \cdots & a_{2n} \\ \vdots & \vdots & & \vdots \\ a_{m1} & a_{m2} & \cdots & a_{mn} \end{bmatrix} \tag{1}$$

The *i***th row** of **A** is

$$[a_{i1} \quad a_{i2} \quad \cdots \quad a_{in}] \qquad (1 \leq i \leq m)$$

the *j***th column** of **A** is

$$\begin{bmatrix} a_{1j} \\ a_{2j} \\ \vdots \\ a_{mj} \end{bmatrix} \qquad (1 \leq j \leq n)$$

We shall say that **A** is *m* **by** *n*, written as $m \times n$. If $m = n$, we say that **A** is a **square matrix** of order n and that the numbers $a_{11}, a_{22}, \ldots, a_{nn}$ form the **main diagonal** of **A**. We refer to the number a_{ij}, which is in the ith row and jth column of **A**, as the *i,j***th element** of **A** or as the *(i, j)* **entry** of **A**, and we often write (1) as

$$A = [a_{ij}]$$

Example 1   Let

$$A = \begin{bmatrix} 2 & 3 & 5 \\ 0 & -1 & 2 \end{bmatrix}, \qquad B = \begin{bmatrix} 2 & 3 \\ 4 & 6 \end{bmatrix}, \qquad C = \begin{bmatrix} -1 \\ 2 \\ 0 \end{bmatrix},$$

$$D = \begin{bmatrix} 1 & 0 & -1 \\ -1 & 2 & 3 \\ 2 & 4 & 5 \end{bmatrix}, \qquad E = \begin{bmatrix} 1 & -1 & 3 & 4 \end{bmatrix}$$

Then $\mathbf{A}$ is 2×3 with $a_{12} = 3$ and $a_{23} = 2$, $\mathbf{B}$ is 2×2 with $b_{21} = 4$, $\mathbf{C}$ is 3×1, $\mathbf{D}$ is 3×3, and $\mathbf{E}$ is 1×4.

A square matrix $\mathbf{A} = [a_{ij}]$ for which every entry off the main diagonal is zero, that is, $a_{ij} = 0$ for $i \neq j$, is called a **diagonal matrix**.

Example 2 The following are diagonal matrices:

$$\mathbf{F} = \begin{bmatrix} 4 & 0 \\ 0 & 3 \end{bmatrix} \quad \text{and} \quad \mathbf{G} = \begin{bmatrix} 2 & 0 & 0 \\ 0 & -3 & 0 \\ 0 & 0 & 5 \end{bmatrix}$$

Matrices are used in many applications in computer science and we shall see them in our study of relations and graphs. At this point we present the following simple application showing that matrices can be used to display data in a tabular form.

Example 3 The following matrix gives the airline distances between the cities indicated (in statute miles).

	London	Madrid	New York	Tokyo
London	0	785	3469	5959
Madrid	785	0	3593	6706
New York	3469	3593	0	6757
Tokyo	5959	6707	6757	0

Two $m \times n$ matrices $\mathbf{A} = [a_{ij}]$ and $\mathbf{B} = [b_{ij}]$ are said to be **equal** if $a_{ij} = b_{ij}$, $1 \leq i \leq m$, $1 \leq j \leq n$, that is, if corresponding elements agree.

Example 4 If

$$\mathbf{A} = \begin{bmatrix} 2 & -3 & -1 \\ 0 & 5 & 2 \\ 4 & -4 & 6 \end{bmatrix} \quad \text{and} \quad \mathbf{B} = \begin{bmatrix} 2 & x & -1 \\ y & 5 & 2 \\ 4 & -4 & z \end{bmatrix}$$

Then $\mathbf{A} = \mathbf{B}$ if and only if $x = -3$, $y = 0$, and $z = 6$.

If $\mathbf{A} = [a_{ij}]$ and $\mathbf{B} = [b_{ij}]$ are $m \times n$ matrices, then the **sum** of $\mathbf{A}$ and $\mathbf{B}$ is the matrix $\mathbf{C} = [c_{ij}]$ defined by

$$c_{ij} = a_{ij} + b_{ij} \qquad (1 \le i \le m,\ 1 \le j \le n)$$

That is, $\mathbf{C}$ is obtained by adding corresponding elements of $\mathbf{A}$ and $\mathbf{B}$.

Example 5   Let

$$\mathbf{A} = \begin{bmatrix} 3 & 4 & -1 \\ 5 & 0 & -2 \end{bmatrix} \quad \text{and} \quad \mathbf{B} = \begin{bmatrix} 4 & 5 & 3 \\ 0 & -3 & 2 \end{bmatrix}$$

Then

$$\mathbf{A} + \mathbf{B} = \begin{bmatrix} 3+4 & 4+5 & -1+3 \\ 5+0 & 0+(-3) & -2+2 \end{bmatrix} = \begin{bmatrix} 7 & 9 & 2 \\ 5 & -3 & 0 \end{bmatrix}$$

Observe that the sum of the matrices $\mathbf{A}$ and $\mathbf{B}$ is defined only when $\mathbf{A}$ and $\mathbf{B}$ have the same number of rows and the same number of columns. We agree to write $\mathbf{A} + \mathbf{B}$ only when the sum is defined.

A matrix all of whose entries are zero is called a **zero matrix** and is denoted by **0**.

Example 6   The following are all zero matrices:

$$\begin{bmatrix} 0 & 0 \\ 0 & 0 \end{bmatrix} \qquad \begin{bmatrix} 0 & 0 & 0 \\ 0 & 0 & 0 \end{bmatrix} \qquad \begin{bmatrix} 0 & 0 & 0 \\ 0 & 0 & 0 \\ 0 & 0 & 0 \end{bmatrix}$$

The following theorem gives some basic properties of matrix addition; the proofs are omitted.

Theorem 1

(a) $\mathbf{A} + \mathbf{B} = \mathbf{B} + \mathbf{A}$

(b) $(\mathbf{A} + \mathbf{B}) + \mathbf{C} = \mathbf{A} + (\mathbf{B} + \mathbf{C})$

(c) $\mathbf{A} + \mathbf{0} = \mathbf{0} + \mathbf{A} = \mathbf{A}$

If $\mathbf{A} = [a_{ij}]$ is an $m \times p$ matrix and $\mathbf{B} = [b_{ij}]$ is a $p \times n$ matrix, then the **product** of $\mathbf{A}$ and $\mathbf{B}$, denoted $\mathbf{AB}$, is the $m \times n$ matrix $\mathbf{C} = [c_{ij}]$ defined by

$$c_{ij} = a_{i1}b_{1j} + a_{i2}b_{2j} + \cdots + a_{ip}b_{pj} \qquad (1 \le i \le m,\ 1 \le j \le n) \tag{1}$$

Equation (1) shows that the i,jth element in the product matrix is computed by adding all products obtained by multiplying each entry in the ith row of **A** by the corresponding entry in the jth column of **B**, as shown in Fig. 1.

$$\begin{bmatrix} a_{11} & a_{12} & \cdots & a_{1p} \\ a_{21} & a_{22} & \cdots & a_{2p} \\ \vdots & \vdots & & \vdots \\ \boxed{a_{i1} \quad a_{i2} \quad \cdots \quad a_{ip}} \\ \vdots & \vdots & & \vdots \\ a_{m1} & a_{m2} & \cdots & a_{mp} \end{bmatrix} \begin{bmatrix} b_{11} & b_{12} & \cdots & \boxed{b_{1j}} & \cdots & b_{1n} \\ b_{21} & b_{22} & \cdots & \boxed{b_{2j}} & \cdots & b_{2n} \\ \vdots & \vdots & & \vdots & & \vdots \\ b_{p1} & b_{p2} & \cdots & \boxed{b_{pj}} & \cdots & b_{pn} \end{bmatrix} = \begin{bmatrix} c_{11} & c_{12} & \cdots & c_{1n} \\ c_{21} & c_{22} & \cdots & c_{2n} \\ \vdots & \vdots & \boxed{c_{ij}} & \vdots \\ c_{m1} & c_{m2} & \cdots & c_{mn} \end{bmatrix}$$

Figure 1

Example 7   Let

$$\mathbf{A} = \begin{bmatrix} 2 & 3 & -4 \\ 1 & 2 & 3 \end{bmatrix} \quad \text{and} \quad \mathbf{B} = \begin{bmatrix} 3 & 1 \\ -2 & 2 \\ 5 & -3 \end{bmatrix}$$

Then

$$\mathbf{AB} = \begin{bmatrix} (2)(3) + (3)(-2) + (-4)(5) & (2)(1) + (3)(2) + (-4)(-3) \\ (1)(3) + (2)(-2) + (3)(5) & (1)(1) + (2)(2) + (3)(-3) \end{bmatrix}$$

$$= \begin{bmatrix} -20 & 20 \\ 14 & -4 \end{bmatrix}$$

An **array of dimension two** is a modification of the idea of a matrix, in the same way that a linear array is a modification of the idea of a sequence (see Section 1.3). By an **$n \times m$ array A** we will mean an $n \times m$ matrix **A** of positions. We may assign numbers to these positions later, and make future changes in these assignments, and we will still refer to the array as **A**. This is a model for two-dimensional storage of information in a computer. The number assigned to row i and column j of an array **A** will be denoted $A[i, j]$.

The following algorithm tests two matrices $\mathbf{A}(n \times m)$ and $\mathbf{B}(p \times q)$ for multiplication compatability and returns the product if possible.

SUBROUTINE MATMUL $(A, B, N, M, P, Q; C)$
1. IF $(M = P)$ THEN
    a. FOR $I = 1$ THRU N
        1. FOR $J = 1$ THRU Q
            a. $C[I, J] \leftarrow 0$
            b. FOR $K = 1$ THRU M
                1. $C[I, J] \leftarrow C[I, J] + (A[I, K] \times B[K, J])$

2. ELSE

    a. CALL PRINT ('INCOMPATIBLE')

3. RETURN

END OF SUBROUTINE MATMUL

As we have seen earlier (see p. 65), the properties of matrix addition resemble familiar properties for the addition of real numbers. However, most properties of matrix multiplication do not resemble those of real number multiplication. First, observe that if A is an $m \times p$ matrix and B is a $p \times n$ matrix, then AB can be computed and is an $m \times n$ matrix. As for BA, we have any of the following four possibilities:

1. BA may not be defined (we may have $n \neq m$).
2. BA may be defined ($m = n$) and then BA is $p \times p$, while AB is $m \times m$, $m \neq p$. Thus AB and BA are not equal.
3. AB and BA may both be of the same sizes, but unequal as matrices.
4. $AB = BA$

We now agree to write AB only when the product is defined.

Example 8   Let

$$A = \begin{bmatrix} 2 & 1 \\ 3 & -2 \end{bmatrix} \quad \text{and} \quad B = \begin{bmatrix} 1 & -1 \\ 2 & -3 \end{bmatrix}$$

Then

$$AB = \begin{bmatrix} 4 & -5 \\ -1 & 3 \end{bmatrix} \quad \text{and} \quad BA = \begin{bmatrix} -1 & 3 \\ -5 & 8 \end{bmatrix}$$

The basic properties of matrix multiplication are given by the following theorem.

Theorem 2

(a) $A(BC) = (AB)C$

(b) $A(B + C) = AB + AC$

(c) $(A + B)C = AC + BC$

The $n \times n$ diagonal matrix

$$\mathbf{I}_n = \begin{bmatrix} 1 & 0 & \cdot & \cdot & \cdot & 0 \\ 0 & 1 & \cdot & \cdot & \cdot & 0 \\ \cdot & \cdot & & & & \cdot \\ \cdot & \cdot & & & & \cdot \\ \cdot & \cdot & & & & \cdot \\ 0 & 0 & \cdot & \cdot & \cdot & 1 \end{bmatrix}$$

all of whose diagonal elements are 1, is called the **identity matrix** of order n. If $\mathbf{A}$ is an $m \times n$ matrix, it is easy to verify (Exercise 11) that

$$\mathbf{I}_m \mathbf{A} = \mathbf{A}\mathbf{I}_n = \mathbf{A}$$

If $\mathbf{A}$ is an $n \times n$ matrix and p is a positive integer, we define

$$\mathbf{A}^p = \underbrace{\mathbf{A} \cdot \mathbf{A} \cdots \mathbf{A}}_{p \text{ factors}}$$

and

$$\mathbf{A}^0 = \mathbf{I}_n$$

If p and q are nonnegative integers, we can prove the following laws of exponents:

$$\mathbf{A}^p \mathbf{A}^q = \mathbf{A}^{p+q}$$

and

$$(\mathbf{A}^p)^q = \mathbf{A}^{pq}$$

Observe that the rule

$$(\mathbf{AB})^p = \mathbf{A}^p \mathbf{B}^p$$

does not hold for square matrices. However, if $\mathbf{AB} = \mathbf{BA}$, then $(\mathbf{AB})^p = \mathbf{A}^p\mathbf{B}^p$ (Exercise 23).

Let r be a real number and $\mathbf{A} = [a_{ij}]$ an $m \times n$ matrix. The **scalar multiple** of $\mathbf{A}$ by r, $r\mathbf{A}$, is the $m \times n$ matrix $\mathbf{B} = [b_{ij}]$, where

$$b_{ij} = ra_{ij} \qquad (1 \le i \le m, \, 1 \le j \le n)$$

That is, $\mathbf{B}$ is obtained from $\mathbf{A}$ by multiplying each entry of $\mathbf{A}$ by r.

Example 9   If $r = -2$ and

$$A = \begin{bmatrix} 2 & -3 & 4 \\ 5 & 1 & -6 \end{bmatrix}$$

then

$$rA = \begin{bmatrix} -4 & 6 & -8 \\ -10 & -2 & 12 \end{bmatrix}$$

The following theorem summarizes the basic properties of scalar multiplication.

Theorem 3   If r and s are real numbers and A and B are matrices, then

(a) $r(sA) = (rs)A$
(b) $(r + s)A = rA + sA$
(c) $r(A + B) = rA + rB$
(d) $A(rB) = r(AB) = (rA)B$

If $A = [a_{ij}]$ is an $m \times n$ matrix, then the $n \times m$ matrix $A^T = [a_{ij}^T]$, where

$$a_{ij}^T = a_{ji} \qquad (1 \le i \le m, i \le j \le n)$$

is called the **transpose** of A. Thus the transpose of A is obtained by interchanging the rows and columns of A.

Example 10   Let

$$A = \begin{bmatrix} 2 & -3 & 5 \\ 6 & 1 & 3 \end{bmatrix} \quad \text{and} \quad B = \begin{bmatrix} 3 & 4 & 5 \\ 2 & -1 & 0 \\ 1 & 6 & -2 \end{bmatrix}$$

Then

$$A^T = \begin{bmatrix} 2 & 6 \\ -3 & 1 \\ 5 & 3 \end{bmatrix} \quad \text{and} \quad B^T = \begin{bmatrix} 3 & 2 & 1 \\ 4 & -1 & 6 \\ 5 & 0 & -2 \end{bmatrix}$$

The following theorem summarizes the basic properties of transpose.

Theorem 4   If r is a real number and $\mathbf{A}$ and $\mathbf{B}$ are matrices, then

(a) $(\mathbf{A}^T)^T = \mathbf{A}$

(b) $(\mathbf{A} + \mathbf{B})^T = \mathbf{A}^T + \mathbf{B}^T$

(c) $(\mathbf{AB})^T = \mathbf{B}^T\mathbf{A}^T$

(d) $(r\mathbf{A})^T = r\mathbf{A}^T$

A matrix $\mathbf{A} = [a_{ij}]$ is called **symmetric** if

$$\mathbf{A}^T = \mathbf{A}$$

Thus, if A is symmetric, it must be a square matrix. It is easy to show (Exercise 24) that $\mathbf{A}$ is symmetric if and only if

$$a_{ij} = a_{ji}$$

That is, $\mathbf{A}$ is symmetric if and only if the entries of $\mathbf{A}$ are symmetric with respect to the main diagonal of $\mathbf{A}$.

Example 11   If

$$\mathbf{A} = \begin{bmatrix} 1 & 2 & -3 \\ 2 & 4 & 5 \\ -3 & 5 & 6 \end{bmatrix} \quad \text{and} \quad \mathbf{B} = \begin{bmatrix} 1 & 2 & -3 \\ 2 & 4 & 0 \\ 3 & 2 & 1 \end{bmatrix}$$

Then $\mathbf{A}$ is symmetric and $\mathbf{B}$ is not symmetric.

Boolean Matrix Operations

A **Boolean matrix** is an $m \times n$ matrix whose entries are either zero or one. We shall soon define three operations on Boolean matrices that have useful applications in Chapter 2. We first define the Boolean operations $\vee$ and $\wedge$ for the numbers 0 and 1 by the following tables:

$\vee$	0	1
0	0	1
1	1	1

$\wedge$	0	1
0	0	0
1	0	1

One way to interpret these operations is to think of 0 as representing all false statements and 1 as representing all true statements. Then if p and q are statements, $p \vee q$ can be viewed as the compound statement "p or q," and $p \wedge q$ can be viewed as the compound statement "p and q." For example, if p is the statement "2 is a positive number" and q is the statement "3 is an even number," then $p \lor q$ is the statement "2 is a positive number or 3 is an even number." The tables above can be interpreted as giving the truth or falsity of a compound statement based on the truth or falsity of the component statements. Since p is true (1) and q is false (0), the table shows that the compound statement $p \lor q$ is true $(1 \lor 0 = 1)$. Similar interpretations hold for the other entries of the tables.

It is easy to verify that these operations satisfy some of the familiar properties of ordinary addition and multiplication. Thus, if a, b, and $c \in \{0, 1\}$, then

$$a \lor b = b \lor a; \qquad a \land b = b \land a$$

$$a \lor (b \lor c) = (a \lor b) \lor c; \qquad a \land (b \land c) = (a \land b) \land c$$

$$a \land (b \lor c) = (a \land b) \lor (a \land c); \qquad a \lor (b \land c) = (a \lor b) \land (a \lor c)$$

For simplicity, we shall write $a \lor (b \lor c)$ simply as $a \lor b \lor c$, and this idea can be generalized to $a_1 \lor a_2 \lor \cdots \lor a_n$. A similar generalization applies to $\land$.

Let $\mathbf{A} = [a_{ij}]$ and $\mathbf{B} = [b_{ij}]$ be $m \times n$ Boolean matrices. We define $\mathbf{A} \lor \mathbf{B} = [c_{ij}]$, the **join** of $\mathbf{A}$ and $\mathbf{B}$, by

$$c_{ij} = a_{ij} \lor b_{ij}$$

That is, we use the Boolean operation $\lor$ on the corresponding entries of the matrices $\mathbf{A}$ and $\mathbf{B}$. We define $\mathbf{A} \land \mathbf{B} = [d_{ij}]$, the **meet** of $\mathbf{A}$ and $\mathbf{B}$, by

$$d_{ij} = a_{ij} \land b_{ij}$$

That is, we use the Boolean operation $\land$ on the corresponding entries of $\mathbf{A}$ and $\mathbf{B}$.

Finally, if $\mathbf{A}$ is an $m \times p$ and $\mathbf{B}$ is a $p \times n$ Boolean matrix, we define $\mathbf{A} \odot \mathbf{B} = [e_{ij}]$, the **Boolean product** of $\mathbf{A}$ and $\mathbf{B}$, by

$$e_{ij} = (a_{i1} \land b_{1j}) \lor (a_{i2} \land b_{2j}) \lor \cdots \lor (a_{ip} \land b_{pj})$$

That is, the Boolean product of $\mathbf{A}$ and $\mathbf{B}$ is obtained in the same manner as the ordinary product of $\mathbf{A}$ and $\mathbf{B}$ except that we use Boolean operations $\lor$ and $\land$ instead of ordinary addition and multiplication, respectively. As before, we only write operations for Boolean matrices when they are defined.

Example 12   Let

$$\mathbf{A} = \begin{bmatrix} 1 & 0 & 1 \\ 0 & 1 & 1 \\ 1 & 1 & 0 \\ 0 & 0 & 0 \end{bmatrix}, \qquad \mathbf{B} = \begin{bmatrix} 1 & 1 & 0 \\ 1 & 0 & 1 \\ 0 & 0 & 1 \\ 1 & 1 & 0 \end{bmatrix}, \qquad \mathbf{C} = \begin{bmatrix} 1 & 0 & 0 \\ 0 & 1 & 1 \\ 1 & 0 & 1 \end{bmatrix}$$

Compute (a) $\mathbf{A} \vee \mathbf{B}$; (b) $\mathbf{A} \wedge \mathbf{B}$; (c) $\mathbf{B} \odot \mathbf{C}$.

Solution.

(a) $\mathbf{A} \vee \mathbf{B} = \begin{bmatrix} 1\vee 1 & 0\vee 1 & 1\vee 0 \\ 0\vee 1 & 1\vee 0 & 1\vee 1 \\ 1\vee 0 & 1\vee 0 & 0\vee 1 \\ 0\vee 1 & 0\vee 1 & 0\vee 0 \end{bmatrix} = \begin{bmatrix} 1 & 1 & 1 \\ 1 & 1 & 1 \\ 1 & 1 & 1 \\ 1 & 1 & 0 \end{bmatrix}$

(b) $\mathbf{A} \wedge \mathbf{B} = \begin{bmatrix} 1\wedge 1 & 0\wedge 1 & 1\wedge 0 \\ 0\wedge 1 & 1\wedge 0 & 1\wedge 1 \\ 1\wedge 0 & 1\wedge 0 & 0\wedge 1 \\ 0\wedge 1 & 0\wedge 1 & 0\wedge 0 \end{bmatrix} = \begin{bmatrix} 1 & 0 & 0 \\ 0 & 0 & 1 \\ 0 & 0 & 0 \\ 0 & 0 & 0 \end{bmatrix}$

(c) $\mathbf{B} \odot \mathbf{C} =$

$$\begin{bmatrix} (1\wedge 1)\vee(1\wedge 0)\vee(0\wedge 1) & (1\wedge 0)\vee(1\wedge 1)\vee(0\wedge 0) & (1\wedge 0)\vee(1\wedge 1)\vee(0\wedge 1) \\ (1\wedge 1)\vee(0\wedge 0)\vee(1\wedge 1) & (1\wedge 0)\vee(0\wedge 1)\vee(1\wedge 0) & (1\wedge 0)\vee(0\wedge 1)\vee(1\wedge 1) \\ (0\wedge 1)\vee(0\wedge 0)\vee(1\wedge 1) & (0\wedge 0)\vee(0\wedge 1)\vee(1\wedge 0) & (0\wedge 0)\vee(0\wedge 1)\vee(1\wedge 1) \\ (1\wedge 1)\vee(1\wedge 0)\vee(0\wedge 1) & (1\wedge 0)\vee(1\wedge 1)\vee(0\wedge 0) & (1\wedge 0)\vee(1\wedge 1)\vee(0\wedge 1) \end{bmatrix}$$

$$= \begin{bmatrix} 1\vee 0\vee 0 & 0\vee 1\vee 0 & 0\vee 1\vee 0 \\ 1\vee 0\vee 1 & 0\vee 0\vee 0 & 0\vee 0\vee 1 \\ 0\vee 0\vee 1 & 0\vee 0\vee 0 & 0\vee 0\vee 1 \\ 1\vee 0\vee 0 & 0\vee 1\vee 0 & 0\vee 1\vee 0 \end{bmatrix} = \begin{bmatrix} 1 & 1 & 1 \\ 1 & 0 & 1 \\ 1 & 0 & 1 \\ 1 & 1 & 1 \end{bmatrix}$$

The following theorem, whose proof is left as an exercise, summarizes the basic properties of the Boolean matrix operations defined above.

Theorem 4   If $\mathbf{A}$, $\mathbf{B}$, and $\mathbf{C}$ are Boolean matrices, then

1. (a) $\mathbf{A} \vee \mathbf{B} = \mathbf{B} \vee \mathbf{A}$;
    (b) $\mathbf{A} \wedge \mathbf{B} = \mathbf{B} \wedge \mathbf{A}$
2. (a) $(\mathbf{A} \vee \mathbf{B}) \vee \mathbf{C} = \mathbf{A} \vee (\mathbf{B} \vee \mathbf{C})$;
    (b) $(\mathbf{A} \wedge \mathbf{B}) \wedge \mathbf{C} = \mathbf{A} \wedge (\mathbf{B} \wedge \mathbf{C})$

3. (a) $\mathbf{A} \wedge (\mathbf{B} \vee \mathbf{C}) = (\mathbf{A} \wedge \mathbf{B}) \vee (\mathbf{A} \wedge \mathbf{C})$
    (b) $\mathbf{A} \vee (\mathbf{B} \wedge \mathbf{C}) = (\mathbf{A} \vee \mathbf{B}) \wedge (\mathbf{A} \vee \mathbf{C})$
4. $\mathbf{A} \odot (\mathbf{B} \odot \mathbf{C}) = (\mathbf{A} \odot \mathbf{B}) \odot \mathbf{C}$

EXERCISE SET 1.8

1. Let

$$A = \begin{bmatrix} 3 & -2 & 5 \\ 4 & 1 & 2 \end{bmatrix}, \quad B = \begin{bmatrix} 3 \\ -2 \\ 4 \end{bmatrix},$$

$$C = \begin{bmatrix} 2 & 3 & 4 \\ 5 & 6 & -1 \\ 2 & 0 & 8 \end{bmatrix}$$

   (a) What is a_{12}, a_{22}, a_{23}?
   (b) What is b_{11}, b_{31}?
   (c) What is c_{13}, c_{21}, c_{33}?
   (d) List the elements on the main diagonal of **C**.

2. Which of the following are diagonal matrices?

$$A = \begin{bmatrix} 2 & 3 \\ 0 & 0 \end{bmatrix}, \quad B = \begin{bmatrix} 3 & 0 & 0 \\ 0 & -2 & 0 \\ 0 & 0 & 5 \end{bmatrix}$$

$$C = \begin{bmatrix} 0 & 0 & 0 \\ 0 & 0 & 0 \\ 0 & 0 & 0 \end{bmatrix}, D = \begin{bmatrix} 2 & 6 & -2 \\ 0 & -1 & 0 \\ 0 & 0 & 3 \end{bmatrix},$$

$$E = \begin{bmatrix} 4 & 0 & 0 \\ 0 & 4 & 0 \\ 0 & 0 & 4 \end{bmatrix}$$

3. If $\begin{bmatrix} a+b & c+d \\ c-d & a-b \end{bmatrix} = \begin{bmatrix} 4 & 6 \\ 10 & 2 \end{bmatrix}$,

   find $a, b, c,$ and d.

4. If $\begin{bmatrix} a+2b & 2a-b \\ 2c+d & c-2d \end{bmatrix} = \begin{bmatrix} 4 & -2 \\ 4 & -3 \end{bmatrix}$,

   find $a, b, c,$ and d.

In Exercises 5–10, let

$$A = \begin{bmatrix} 2 & 1 & 3 \\ 4 & 1 & -2 \end{bmatrix}, \quad B = \begin{bmatrix} 0 & 1 \\ 1 & 2 \\ 2 & 3 \end{bmatrix},$$

$$C = \begin{bmatrix} 1 & -2 & 3 \\ 4 & 2 & 5 \\ 3 & 1 & 2 \end{bmatrix}, \quad D = \begin{bmatrix} -3 & 2 \\ 4 & 1 \end{bmatrix},$$

$$E = \begin{bmatrix} 3 & 2 & -1 \\ 5 & 4 & -3 \\ 0 & 1 & 2 \end{bmatrix}, \quad F = \begin{bmatrix} -2 & 3 \\ 4 & 5 \end{bmatrix}$$

5. If possible, compute.
   (a) **C** + **E**
   (b) **AB** and **BA**
   (c) 3**D** − 2**F**
   (d) **CB** + **F**
   (e) **AB** + **DF**
   (f) 4(2**B**) and 8**B**

6. If possible, compute.
   (a) **A**(**BD**) and (**AB**)**D**
   (b) **A**(**C** + **E**) and **AC** + **AE**
   (c) 4**A** + 3**A** and 7**A**
   (d) **FD** + **AB**
   (e) **EF** + 2**A**
   (f) (−3)(2**E**) and (−6)**E**

7. If possible, compute.
   (a) 3**F** − 2**B**
   (b) **EB** + **FA**
   (c) **A**(**B** + **D**) and **AB** + **AD**
   (d) (**F** + **D**)**A**
   (e) 2**F** + 6**F** and 8**F**
   (f) **AC** + **DE**

8. If possible, compute.
   (a) A^T and $(A^T)^T$
   (b) $(C + E)^T$ and $C^T + E^T$
   (c) $(AB)^T$ and $B^T A^T$
   (d) $(3D + 2F)^T$
   (e) $(3BC + F)^T$
   (f) $(B^T C) + A$

9. If possible, compute.
   (a) $(2C - 3E)^T B$
   (b) $A^T(D + F)$
   (c) $(3E)A^T$
   (d) $(BC)^T$ and $C^T B^T$
   (e) $(B^T + A)C$
   (f) $(D^T + E)F$

10. Compute D^3.

11. Let $\mathbf{A}$ be an $m \times n$ matrix. Show that
$\mathbf{I}_m \mathbf{A} = \mathbf{A}$ and $\mathbf{A}\mathbf{I}_n = \mathbf{A}$.

12. Let

$$\mathbf{A} = \begin{bmatrix} 2 & 1 \\ 3 & -2 \end{bmatrix} \quad \text{and} \quad \mathbf{B} = \begin{bmatrix} -1 & 2 \\ 3 & 4 \end{bmatrix}$$

Show that $\mathbf{AB} \neq \mathbf{BA}$.

13. Let

$$\mathbf{A} = \begin{bmatrix} 3 & 0 & 0 \\ 0 & -2 & 0 \\ 0 & 0 & 4 \end{bmatrix}$$

(a) Compute $\mathbf{A}^3$.

(b) What is $\mathbf{A}^k$?

14. Show that $\mathbf{A}\mathbf{0} = \mathbf{0}$ for any matrix $\mathbf{A}$.

15. Show that $\mathbf{I}_n^T = \mathbf{I}_n$.

16. (a) Prove that if $\mathbf{A}$ has a row of zeros, then $\mathbf{AB}$ has a corresponding row of zeros.

(b) Prove that if $\mathbf{B}$ has a column of zeros, then $\mathbf{AB}$ has a corresponding column of zeros.

17. Prove that the sum and product of diagonal matrices is diagonal.

18. Show that the jth column of the matrix product $\mathbf{AB}$ is equal to the matrix product $\mathbf{A}\mathbf{B}_j$, where $\mathbf{B}_j$ is the jth column of $\mathbf{B}$.

19. If $\mathbf{0}$ is the 2×2 zero matrix, find two 2×2 matrices $\mathbf{A}$ and $\mathbf{B}$, $\mathbf{A} \neq \mathbf{0}$, $\mathbf{B} \neq \mathbf{0}$, such that $\mathbf{AB} = \mathbf{0}$.

20. If $\mathbf{A} = \begin{bmatrix} 0 & 1 \\ 1 & 0 \end{bmatrix}$, show that $\mathbf{A}^2 = \mathbf{I}_2$.

21. Show that $(-1)\mathbf{A} = -\mathbf{A}$.

22. Let $\mathbf{A}$ be an $m \times n$ matrix. Show that if $r\mathbf{A} = \mathbf{0}$, then $r = 0$ or $\mathbf{A} = \mathbf{0}$.

23. Let $\mathbf{A}$ and $\mathbf{B}$ be square matrices. Use mathematical induction to show that if $\mathbf{AB} = \mathbf{BA}$, then $(\mathbf{AB})^n = \mathbf{A}^n\mathbf{B}^n$, for all integers $n \geq 1$.

24 Show that $\mathbf{A} = [a_{ij}]$ is symmetric if and only if $a_{ij} = a_{ji}$.

25. Let $\mathbf{A}$ and $\mathbf{B}$ be symmetric matrices.

(a) Show that $\mathbf{A} + \mathbf{B}$ is symmetric.

(b) Show that $\mathbf{AB}$ is symmetric if and only if $\mathbf{AB} = \mathbf{BA}$.

26. Let $\mathbf{A}$ be an $n \times n$ matrix.

(a) Show that $\mathbf{AA}^T$ and $\mathbf{A}^T\mathbf{A}$ are symmetric.

(b) Show that $\mathbf{A} + \mathbf{A}^T$ is symmetric.

27. Let a, b, and $c \in \{0, 1\}$.

(a) Show that $a \vee b = b \vee a$ and $a \wedge b = b \wedge a$.

(b) Show that $a \vee (b \vee c) = (a \vee b) \vee c$ and $(a \wedge b) \wedge c = a \wedge (b \wedge c)$.

In Exercises 28–30, compute $\mathbf{A} \vee \mathbf{B}$, $\mathbf{A} \wedge \mathbf{B}$, and $\mathbf{A} \odot \mathbf{B}$ for the given matrices $\mathbf{A}$ and $\mathbf{B}$.

28. (a) $\mathbf{A} = \begin{bmatrix} 1 & 0 \\ 0 & 1 \end{bmatrix}$, $\mathbf{B} = \begin{bmatrix} 1 & 1 \\ 0 & 1 \end{bmatrix}$

(b) $\mathbf{A} = \begin{bmatrix} 1 & 0 & 0 \\ 0 & 1 & 1 \\ 1 & 0 & 0 \end{bmatrix}$, $\mathbf{B} = \begin{bmatrix} 1 & 1 & 1 \\ 0 & 0 & 1 \\ 1 & 0 & 1 \end{bmatrix}$

29. (a) $\mathbf{A} = \begin{bmatrix} 1 & 1 \\ 0 & 1 \end{bmatrix}$, $\mathbf{B} = \begin{bmatrix} 0 & 0 \\ 1 & 1 \end{bmatrix}$

(b) $\mathbf{A} = \begin{bmatrix} 0 & 0 & 1 \\ 1 & 1 & 0 \\ 1 & 0 & 0 \end{bmatrix}$, $\mathbf{B} = \begin{bmatrix} 0 & 1 & 1 \\ 1 & 1 & 0 \\ 1 & 0 & 1 \end{bmatrix}$

30. (a) $\mathbf{A} = \begin{bmatrix} 1 & 1 \\ 1 & 1 \end{bmatrix}$, $\mathbf{B} = \begin{bmatrix} 0 & 0 \\ 1 & 0 \end{bmatrix}$

(b) $\mathbf{A} = \begin{bmatrix} 1 & 0 & 0 \\ 0 & 0 & 1 \\ 1 & 0 & 1 \end{bmatrix}$, $\mathbf{B} = \begin{bmatrix} 1 & 1 & 1 \\ 1 & 1 & 1 \\ 1 & 0 & 0 \end{bmatrix}$

31. (a) Show that $\mathbf{A} \vee \mathbf{A} = \mathbf{A}$.

(b) Show that $\mathbf{A} \wedge \mathbf{A} = \mathbf{A}$.

32. (a) Show that $\mathbf{A} \vee \mathbf{B} = \mathbf{B} \vee \mathbf{A}$.

(b) Show that $\mathbf{A} \wedge \mathbf{B} = \mathbf{B} \wedge \mathbf{A}$.

33. (a) Show that $\mathbf{A} \vee (\mathbf{B} \vee \mathbf{C}) = (\mathbf{A} \vee \mathbf{B}) \vee \mathbf{C}$.

(b) Show that $\mathbf{A} \wedge (\mathbf{B} \wedge \mathbf{C}) = (\mathbf{A} \wedge \mathbf{B}) \wedge \mathbf{C}$.

(c) Show that $\mathbf{A} \odot (\mathbf{B} \odot \mathbf{C}) = (\mathbf{A} \odot \mathbf{B}) \odot \mathbf{C}$.

In Exercises 34–37, prove the indicated result by mathematical induction.

34. $(\mathbf{A}_1 + \mathbf{A}_2 + \cdots + \mathbf{A}_n)^T = \mathbf{A}_1^T + \mathbf{A}_2^T + \cdots + \mathbf{A}_n^T$

35. $(\mathbf{A}_1\mathbf{A}_2 \cdots \mathbf{A}_n)^T = \mathbf{A}_n^T\mathbf{A}_{n-1}^T \cdots \mathbf{A}_2^T\mathbf{A}_1^T$

36. $\mathbf{A}^2 \times \mathbf{A}^n = \mathbf{A}^{2+n}$

37. $(\mathbf{A}^2)^n = \mathbf{A}^{2n}$

38. Prove Theorem 4.

KEY IDEAS FOR REVIEW

☐ Set: a well-defined collection of objects.

☐ $\varnothing$ (empty or null set): the set with no elements.

☐ Equal Sets: sets with the same elements.

☐ $A \subseteq B$ (A is a subset of B): Every element of A is also an element of B.

☐ $|A|$ (cardinality of A): the number of elements in A.

☐ $P(A)$ (power set of A): the set of all subsets of A.

☐ Sequence: list of objects in a definite order.

☐ Set corresponding to a sequence: see page 8.

☐ Countable set: a set that corresponds to a sequence.

☐ $(0, 1)$ is an uncountable set.

☐ $A \cup B$ (union of A and B): $\{x \mid x \in A \text{ or } x \in B\}$.

☐ $A \cap B$ (intersection of A and B): $\{x \mid x \in A \text{ and } x \in B\}$.

☐ Disjoint sets: two sets with no elements in common.

☐ $A - B$ (complement of B with respect to A): $\{x \mid x \in A \text{ and } x \notin B\}$.

☐ $\bar{A}$ (complement of A): $\{x \mid x \notin A\}$.

☐ Algebraic properties of set operations: see page 16.

☐ Theorem (The Addition Principle): If A and B are finite sets, then

$$|A \cup B| = |A| + |B| - |A \cap B|$$

☐ Theorem (The Three-Set Addition Principle): If A, B, and C are finite sets, then

$$|A \cup B \cup C| = |A| + |B| + |C| - |A \cap B| - |B \cap C| - |A \cap C|$$
$$+ |A \cap B \cap C|$$

☐ Characteristic function of a set A:

$$f_A(x) = \begin{cases} 1 & \text{if } x \in A \\ 0 & \text{if } x \notin A \end{cases}$$

☐ Theorem (The Multiplication Principle): If two independent tasks T_1 and T_2 are to be performed in sequence, and if T_1 can be performed in n_1 ways and T_2 in n_2 ways, then the sequence $T_1 T_2$ can be performed in $n_1 n_2$ ways.

☐ Permutation: an arrangement of the n elements of a set S into a sequence of length n.

☐ Permutation of an n-element set S, taken r at a time ($0 \le r \le n$): a sequence of length r formed from distinct elements of S.

☐ Combination of an n-element set S, taken r at a time ($0 \le r \le n$): a subset of S having exactly r elements.

☐   Theorem: If S is set with $|S| = n$ and $0 \le r \le n$, then the number of permutations of S taken r at a time is

$$P_r^n = n(n-1)(n-2) \cdots (n-r+1) = \frac{n!}{(n-r)!}$$

☐   Theorem: If S is a set with $|S| = n$ and $0 \le r \le n$, then the number of combinations of the elements of S, taken r at a time, is

$$C_r^n = \frac{n!}{r!\,(n-r)!}$$

☐   Algorithm: a complete and precise description of the steps necessary to perform a task or computation.

☐   Principle of Mathematical Induction: Let k be a fixed integer. Suppose that for each integer $n \ge k$ we have a proposition $P(n)$. Suppose that

    (a) $P(k)$ is true.
    (b)   For all $n \ge k$; whenever $P(n)$ is true, it follows that $P(n+1)$ is true.

Then the principle of mathematical induction states that $P(n)$ is true for all $n \ge k$.

☐   Recursion: An algorithm is recursive if it makes reference to earlier or simpler versions of itself.

☐   Theorem: If $n \ne 0$ and m are integers, we can write $m = qn + r$, $0 \le r < |n|$ in one and only one way.

☐   GCD (a, b): $d =$ GCD (a, b) if $d\,|\,a$, $d\,|\,b$, and d is the largest common divisor of a and b.

    (a) $d = sa + tb$ for some integers s and t.
    (b)   If $c\,|\,a$ and $c\,|\,b$, then $c\,|\,d$.

☐   Euclidean algorithm: method used to find GCD (a, b); see page 56.
☐   LCM (a, b): $c =$ LCM (a, b) if $a\,|\,c$, $b\,|\,c$, and c is the smallest common multiple of a and b.
☐   GCD $(a, b) \cdot$ LCM $(a, b) = ab$.
☐   Matrix: a rectangular array of numbers.
☐   Entries of a matrix: the numbers in a matrix.
☐   Size of a matrix: $\mathbf{A}$ is $m \times n$ if it has m rows and n columns.
☐   Diagonal matrix: a square matrix with zero entries off the main diagonal.
☐   Equal matrices: matrices of the same size whose corresponding entries are equal.
☐ $\mathbf{A} + \mathbf{B}$: the matrix obtained by adding corresponding entries in $\mathbf{A}$ and $\mathbf{B}$.
☐   Zero matrix: a matrix all of whose entries are zero.

☐ **AB**: see page 65.

☐ **I**$_n$ (identity matrix): a square matrix with ones on the main diagonal and zeros elsewhere.

☐ r**A**: the matrix obtained by multiplying each entry of **A** by r.

☐ **A**^{T}: The matrix obtained from **A** by interchanging the rows and columns of **A**.

☐ Symmetric matrix: $\mathbf{A}^T = \mathbf{A}$.

☐ Array of dimension two: see page 66.

☐ Boolean matrix: a matrix whose entries are either zero or one.

☐ **A** ∨ **B**: see page 71.

☐ **A** ∧ **B**: see page 71.

☐ **A** ⊙ **B**: see page 71.

☐ Properties of Boolean matrix operations; see page 72.

FURTHER READING

ANTON, HOWARD and BERNARD KOLMAN, *Applied Finite Mathematics*, 3rd ed., Academic Press, Orlando, 1982.

FRALEIGH, JOHN B., *A First Course in Abstract Algebra*, 3rd ed., Addison-Wesley, Reading, Mass., 1981.

KNUTH, DONALD E., *The Art of Computer Programming*, v. 1, 2nd ed., Addison-Wesley, Reading, Mass., 1973.

KOLMAN, BERNARD, *Introductory Linear Algebra with Applications*, 3rd ed., Macmillan, New York, 1984.

LIPSCHUTZ, SEYMOUR, *Theory and Problems of Set Theory and Related Topics*, Schaum, New York, 1964.

RADER, ROBERT J., *Advanced Software Design Techniques*, Petrocelli, New York, 1968.

STOLL, ROBERT R., *Set Theory and Logic*, Freeman, New York, 1963.

WAND, MITCHELL, *Induction, Recursion, and Programming*, North Holland, New York, 1980.

Relations and Digraphs

Prerequisites: Chapter 1.

Relationships between people, numbers, sets, and many other entities can be formalized in the idea of a binary relation. In this chapter we develop the concept of binary relation, and we give several geometric and algebraic methods of representing such objects. We also discuss a variety of different properties that a binary relation may possess, and we introduce important examples such as equivalence relations. Finally, we introduce several useful types of algebraic manipulations which may be performed on binary relations. We discuss these manipulations from both a theoretical and computational point of view.

2.1 Product Sets and Partitions

Product Sets

An **ordered pair** (a, b) is a listing of the objects a and b in a prescribed order, with a appearing first and b appearing second. Thus an ordered pair is merely a sequence of length 2. From our earlier discussion of sequences (see Section 1.2), it follows that the ordered pairs (a_1, b_1) and (a_2, b_2) are equal if and only if $a_1 = a_2$ and $b_1 = b_2$.

If A and B are two nonempty sets, we define the **product set** or **Cartesian product** $A \times B$ as the set of all ordered pairs (a, b) with $a \in A$ and $b \in B$. Thus

$$A \times B = \{(a, b) \,|\, a \in A \quad \text{and} \quad b \in B\}$$

Example 1 Let

$$A = \{1, 2, 3\} \qquad \text{and} \qquad B = \{r, s\}$$

B	r	s	
A			
1	(1, r)	(1, s)	
2	(2, r)	(2, s)	
3	(3, r)	(3, s)	**Figure 1**

Then

$$A \times B = \{(1, r), (1, s), (2, r), (2, s), (3, r), (3, s)\}$$

Observe that the elements of $A \times B$ can be arranged in a convenient tabular array as shown in Fig. 1.

Example 2   If A and B are as in Example 1, then

$$B \times A = \{(r, 1), (s, 1), (r, 2), (s, 2), (r, 3), (s, 3)\}$$

From Examples 1 and 2, we see that $A \times B$ need not equal $B \times A$.

Theorem 1   For any two finite, nonempty sets A and B, $|A \times B| = |A||B|$. (See Section 1.1 for a definition of $|A|$.)

Proof. We prove this by mathematical induction. Let $P(n)$ be the following proposition: If A and B are finite sets and $|B| = n$, then $|A \times B| = |A| \cdot |B|$.

Basis Step. We prove $P(1)$. Let $|A| = m$ and $|B| = 1$. Then $A = \{a_1, \ldots, a_m\}$ and $B = \{b_1\}$, so

$$A \times B = \{(a_1, b_1), \ldots, (a_m, b_1)\} \qquad \text{and} \qquad |A \times B| = m = m \cdot 1 = |A| \cdot |B|$$

Induction Step. Suppose that $P(n)$ is true for some $n \geq 1$, and let A and B be finite sets with $|A| = m$ and $|B| = n + 1$. Say that $A = \{a_1, \ldots, a_m\}$ and $B = \{b_1, \ldots, b_n, b_{n+1}\}$. Let $C = \{b_1, \ldots, b_n\}$. Then if $(a, b) \in A \times B$, either $b \in C$ or $b = b_{n+1}$. The number of pairs (a, b) with $b \in C$ is $|A \times C|$, which equals mn. Also, there are m pairs $(a_1, b_{n+1}), (a_2, b_{n+1}), \ldots, (a_m, b_{n+1})$ with second element equal to b_{n+1}. Thus the total number of pairs in $A \times B$ is $mn + m$, that is, $|A \times B| = mn + m = m(n + 1) = |A| \cdot |B|$, so $P(n + 1)$ is true.

By mathematical induction, $P(n)$ is true for all $n \geq 1$, and the theorem is proved. If either A or B is empty, $A \times B$ is empty, therefore Theorem 1 holds also in these cases.

Example 3   If $A = B = \mathbb{R}$, the set of all real numbers, then $\mathbb{R} \times \mathbb{R}$, also denoted by $\mathbb{R}^2$, is the set of all points in the plane. The ordered pair (a, b) is the point (a, b) in the plane.

Example 4   A marketing research firm classifies a person according to the following two criteria:

Sex: m = male; f = female

Highest level of education completed: e = elementary school; h = high school; c = college; g = graduate school

Let $S = \{m, f\}$ and $L = \{e, h, c, g\}$. Then the product set $S \times L$ contains all the categories into which the population is classified. Thus the classification (f, g) represents a female who has completed graduate school. There are eight categories in this classification scheme.

We now define the Cartesian product of three or more nonempty sets by generalizing the earlier definition of the Cartesian product of two sets. That is the **Cartesian product** $A_1 \times A_2 \times \cdots \times A_m$ of the nonempty sets $A_1, A_2, \ldots, A_m$ is the set of all ordered m-tuples $(a_1, a_2, \ldots, a_m)$, where $a_i \in A_i$, $i = 1, 2, \ldots, m$. Thus

$$A_1 \times A_2 \times \cdots \times A_m = \{(a_1, a_2, \ldots, a_m) \mid a_i \in A_i, \ i = 1, 2, \ldots, m\}$$

Example 5   A software firm provides the following three characteristics for each program that it sells:

Language: f = FORTRAN, p = PASCAL, l = LISP

Memory: 48 = 48,000, 64 = 64,000, 128 = 128,000

Operating system: u = UNIX, c = CP/M

Let $L = \{f, p, l\}$, $M = \{48, 64, 128\}$, $O = \{u, c\}$. Then the Cartesian product $L \times M \times O$ contains all the categories that describe a program. There are $3 \cdot 3 \cdot 2 = 18$ categories in this classification scheme.

Proceeding in a manner similar to that used to prove Theorem 1, we can show that if A_1 has n_1 elements, A_2 has n_2 elements, $\ldots$, and A_m has n_m elements, then $A_1 \times A_2 \times \cdots \times A_m$ has $n_1 \cdot n_2 \cdots n_m$ elements.

Partitions

A **partition** or **quotient set** of a nonempty set S is a collection of nonempty subsets $A_1, A_2, \ldots$, of S such that:

1. $A_i \cap A_j = \varnothing$ if $i \neq j$.
2. Each element $s \in S$ belongs to one of the subsets A_i.

The subsets A_i are called the **cells** or **blocks** of the partition. Fig. 2 shows diagrammatically a partition of a set S into seven blocks.

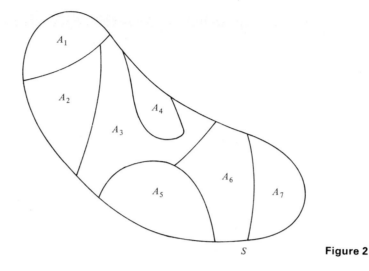

S

Figure 2

Example 6   Let

$$S = \{a, b, c, d, e, f, g, h\}$$

Consider the following subsets of S:

$$A_1 = \{a, b, c, d\}, \quad A_2 = \{a, c, e, f, g, h\}, \quad A_3 = \{a, c, e, g\},$$
$$A_4 = \{b, d\}, \quad A_5 = \{f, h\}$$

Then $\{A_1, A_2\}$ is not a partition since $A_1 \cap A_2 \neq \emptyset$. Also, $\{A_1, A_5\}$ is not a partition since $e \notin A_1$, and $e \notin A_5$. The collection $\{A_3, A_4, A_5\}$ is a partition of S.

Example 7   Consider the set S of all employees of General Motors. If we form subsets of S by grouping in one subset all employees who make exactly the same salary, we obtain a partition of S.

Example 8   Let

$$S = \text{set of all integers}$$
$$A_1 = \text{set of all even integers}$$
$$A_2 = \text{set of all odd integers}$$

Then $\{A_1, A_2\}$ is a partition of S.

Since the members of a partition of a set S are subsets of S, we see that the partition is a subset of $P(S)$. That is, partitions can be considered as particular kinds of subsets of $P(S)$.

EXERCISE SET 2.1

1. In each part, find x or y so that the ordered pairs are equal.
   (a) $(x, 3) = (4, 3)$
   (b) $(a, 3y) = (a, 9)$
   (c) $(3x + 1, 2) = (7, 2)$
   (d) $(\text{ALGOL}, \text{PASCAL}) = (y, \text{PASCAL})$

2. In each part find x and y so that the ordered pairs are equal.
   (a) $(4x, 6) = (16, 6)$
   (b) $(2x - 3, 3y - 1) = (5, 5)$
   (c) $(x^2, 25) = (49, 25)$
   (d) $(x, y) = (x^2, y^2)$

3. Let $A = \{a, b\}$ and $B = \{4, 5, 6\}$.
   (a) List the elements in $A \times B$.
   (b) List the elements in $B \times A$.
   (c) List the elements in $A \times A$.
   (d) List the elements in $B \times B$.

4. Let $A = \{\text{Fine}, \text{Smith}\}$ and $B = \{\text{president}, \text{vice-president}, \text{secretary}, \text{treasurer}\}$.
   (a) List the elements in $A \times B$.
   (b) List the elements in $B \times A$.
   (c) List the elements in $A \times A$.

5. A genetics experiment classifies fruit flies according to the following two criteria.

   Sex: $m = $ male, $f = $ female
   Wing span: $s = $ short-winged, $m = $ medium-winged, $l = $ long-winged

   (a) How many categories are there in this classification scheme?
   (b) List all the categories.

6. A car manufacturer makes three different types of frames and two types of engines:

   Frames: $s = $ sedan, $c = $ coupe, $w = $ station wagon
   Engines: $g = $ gas, $d = $ diesel

   List all possible models.

7. How many two-letter abbreviations can be formed from the 26-letter English alphabet if a letter can be repeated?

8. A medical experiment classifies each subject according to two criteria:

   Smoking pattern: $s = $ smoker, $n = $ nonsmoker
   Weight: $u = $ underweight, $a = $ average weight, $o = $ overweight

   List all possible classifications in this scheme.

9. If $A = \{a, b, c\}$, $B = \{1, 2\}$, and $C = \{r, s\}$, list all elements in $A \times B \times C$.

10. If A has three elements and B has $n \geq 1$ elements, show by mathematical induction that $A \times B$ has $3n$ elements.

11. If $A = \{x \mid x$ is real and $-2 \leq x \leq 3\}$ and $Y = \{y \mid y$ is real and $1 \leq y \leq 5\}$, sketch the set $A \times B$ in the Cartesian plane.

12. Show that if A_1 has n_1 elements, A_2 has n_2 elements, and A_3 has n_3 elements, then $A_1 \times A_2 \times A_3$ has $n_1 \cdot n_2 \cdot n_3$ elements.

13. If $A \subseteq C$ and $B \subseteq D$, show that $A \times B \subseteq C \times D$.

14. Let $S = \{1, 2, 3, 4, 5, 6, 7, 8, 9, 10\}$ and let

$$A_1 = \{1, 2, 3, 4\} \qquad A_2 = \{5, 6, 7\}$$
$$A_3 = \{4, 5, 7, 9\} \qquad A_4 = \{4, 8, 10\}$$
$$A_5 = \{8, 9, 10\} \qquad A_6 = \{1, 2, 3, 6, 8, 10\}$$

Which of the following are partitions of S?
   (a) $\{A_1, A_2, A_5\}$ (b) $\{A_1, A_3, A_5\}$
   (c) $\{A_3, A_6\}$ (d) $\{A_4, A_3, A_2\}$

15. If $A_1 = $ the set of all positive integers and $A_2 = $ the set of all negative integers, is $\{A_1, A_2\}$ a partition of Z?

16. List all partitions of the set $S = \{1, 2, 3\}$.

17. List all partitions of the set $S = \{a, b, c, d\}$.

2.2 Relations and Digraphs

The notion of a relation between two sets of objects is quite common and intuitively clear (a formal definition will be given below). If A is the set of all living human males and B is the set of all living human females, then the relation F (father) can be defined from A to B. Thus, if $x \in A$ and $y \in B$, then x is related to y by the relation F if x is the father of y, and we write $x \ F \ y$. We could also consider the relations S and H from A to B by letting $x \ S \ y$ mean that x is a son of y and $x \ H \ y$ means that x is the husband of y.

If A is the set of all real numbers, there are many commonly used relations from A to A. An example is the relation "less than," which is usually denoted by $<$, so that x is related to y if $x < y$, and the other order relations $>$, $\leq$, $\geq$, $=$, and $\neq$. We see that a relation is often described in English and is denoted by a familiar name or symbol. The problem with this approach is that we will need to discuss *any possible* relation from one abstract set to another. Most of these relations have no simple verbal description and no familiar name or symbol to remind us of their nature or properties. Furthermore, it is usually awkward, and sometimes nearly impossible, to give any precise proofs of the properties that a relation satisfies if we must deal with a verbal description of it.

To get around this problem, observe that the only thing that really matters about a relation is that we know precisely which elements in A are related to which elements in B. Thus suppose that $A = \{1, 2, 3, 4\}$ and R is a relation from A to A. If we know that $1 \ R \ 2$, $1 \ R \ 3$, $1 \ R \ 4$, $2 \ R \ 3$, $2 \ R \ 4$, $3 \ R \ 4$, then we know everything we need to know about R. Actually, R is the familiar relation $<$, "less than," but we need not know this. It would be enough to be given the foregoing list of related pairs. Thus we may say that R is completely known if we know all R-related pairs. We could then write $R = \{(1, 2), (1, 3), (1, 4), (2, 3), (2, 4), (3, 4)\}$, since R is essentially equal to or completely specified by this list of ordered pairs. Each ordered pair specifies that its first element is related to its second element, and all possible relation pairs are assumed to be given, at least in principle. This method of specifying a relation does not require any special symbol or description, and so is suitable for any relation between any two sets. Note that from this point of view, a relation from A to B is simply a subset of $A \times B$ (giving the related pairs), and conversely, any subset of $A \times B$ can be considered a relation, even if it is an unfamiliar relation for which we have no name or alternative description. We choose this approach for defining relations.

Let A and B be nonempty sets. A **relation** R **from** A **to** B is a subset of $A \times B$. If $R \subseteq A \times B$ and $(a, b) \in R$, we say that a **is related to** b **by** R, and we also write $a \ R \ b$. If a is not related to b by R, we write $a \ \not{R} \ b$. Frequently, A and B are equal. In this case, we often say that $R \subseteq A \times A$ is a **relation on** A, instead of a relation from A to A.

If $R \subseteq A \times B$ is a relation from A to B, there are two important sets associated with R. The **domain** of R, denoted by Dom (R), is the set of elements in A which are related to some element in B. In other words, Dom (R), a subset of A, is the set of all first elements in the pairs that make up R. Similarly, we define the **range** of R, denoted by Ran (R), to be the set of elements in B which are second elements of pairs in R, that is, all elements in B which are related to some element in A.

Relations are extremely important in mathematics and its applications. It is not an exaggeration to say that 90 percent of what will be discussed in the remainder of this text will concern some type of object which may be considered a relation. We now give a number of examples.

Example 1   Let

$$A = \{1, 2, 3\} \quad \text{and} \quad B = \{r, s\}$$

Then

$$R = \{(1, r)(2, s)(3, r)\}$$

is a relation from A to B.

Example 2   Let A and B be sets of real numbers. We define the following relation R (equals) from A to B:

$$a \, R \, b \quad \text{if and only if} \quad a = b$$

In Examples 1 and 2, the domain is A and the range is B.

Example 3   Let

$$A = \{1, 2, 3, 4, 5\} = B$$

Define the following relation R (less than) on A:

$$a \, R \, b \quad \text{if and only if} \quad a < b$$

Then

$$R = \{(1, 2), (1, 3), (1, 4), (1, 5), (2, 3), (2, 4), (2, 5), (3, 4), (3, 5), (4, 5)\}$$
$$\text{Dom } (R) = \{1, 2, 3, 4\}, \quad \text{Ran } (R) = \{2, 3, 4, 5\}$$

Example 4   Let $A = B = Z^+$, the set of all positive integers. Define the following relation R on Z^+

$$a \, R \, b \quad \text{if and only if} \quad a \text{ divides } b$$

Then 4 R 12, but 5 $\not{R}$ 7. Here Dom $(R) =$ Ran $(R) = Z^+$, since every number divides and is divisible by itself.

Example 5   Let A be the set of all people in the world. We define the following relation R on A:

$a\ R\ b$ if and only if   there is a sequence $a_0, a_1, \ldots, a_n$ of people such that $a_0 = a$, $a_n = b$ and a_{i-1} knows a_i, $i = 1, 2, \ldots, n$ (n will depend on a and b)

Again Dom (R) and Ran (R) are probably A, unless there is someone on earth who knows nobody or is known by nobody.

Example 6   Let $A = B = \mathbb{R}$, the set of all real numbers. We define the following relation R on A:

$$x\ R\ y\quad \text{if and only if}\quad x \text{ and } y \text{ satisfy the equation } \frac{x^2}{4} + \frac{y^2}{9} = 1$$

The set R consists of all points on the ellipse shown in Fig. 1.

$$\text{Dom } (R) = [-2, 2]\qquad \text{and}\qquad \text{Ran } (R) = [-3, 3]$$

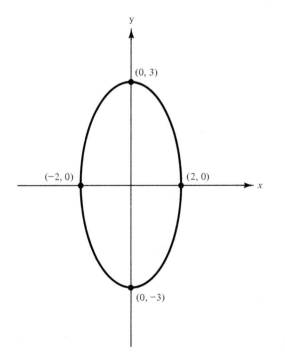

Figure 1

Example 7   Let A be the set of all possible inputs to a given computer program, and let B be the set of all possible outputs from the same program. Define the following relation R from A to B:

$a\ R\ b$ if and only if b is the output produced by the program when input a is used

Example 8   Let

$$A = B = \text{the set of all lines in the plane}$$

Define the following relation R on A:

$l_1\ R\ l_2$ if and only if l_1 is parallel to l_2

where l_1 and l_2 are lines in the plane.

Example 9   An airline services the five cities c_1, c_2, c_3, c_4, and c_5. The following table gives the cost (in dollars) of going from c_i to c_j. Thus the cost of going from c_1 to c_3 is \$100, while the cost of going from c_4 to c_2 is \$200.

From	c_1	c_2	To c_3	c_4	c_5
c_1		140	100	150	200
c_2	190		200	160	220
c_3	110	180		190	250
c_4	190	200	120		150
c_5	200	100	200	150	

We now define the following relation R on the set of cities $A = \{c_1, c_2, c_3, c_4, c_5\}$:

$c_i\ R\ c_j$ if and only if   the cost of going from c_i to c_j is less than or equal to \$180

Find R.

Solution. The relation R is the subset of $A \times A$ consisting of all cities (c_i, c_j), where the cost of going from c_i to c_j is less than or equal to \$180. Hence

$$R = \{(c_1, c_2), (c_1, c_3), (c_1, c_4), (c_2, c_4), (c_3, c_1), (c_3, c_2),$$

$$(c_4, c_3), (c_4, c_5), (c_5, c_2), (c_5, c_4)\}$$

The Matrix of a Relation

If A and B are finite sets, containing m and n elements, respectively, and R is a relation from A to B, we can represent R by the $m \times n$ matrix $\mathbf{M}_R = [m_{ij}]$, which is defined as follows:

$$m_{ij} = \begin{cases} 1 & \text{if } (a_i, b_j) \in R \\ 0 & \text{if } (a_i, b_j) \notin R \end{cases}$$

The matrix $\mathbf{M}_R$ is called the **matrix of** R.

Example 10   Let R be the relation defined in Example 1. Then the matrix of R is

$$\mathbf{M}_R = \begin{matrix} 1 \\ 2 \\ 3 \end{matrix} \begin{matrix} r & s \\ \begin{bmatrix} 1 & 0 \\ 0 & 1 \\ 1 & 0 \end{bmatrix} \end{matrix}$$

Conversely, given sets A and B with $|A| = m$ and $|B| = n$, an $m \times n$ matrix whose entries are zeros and ones determines a relation, as is illustrated in the following example.

Example 11   Consider the matrix

$$\mathbf{M} = \begin{bmatrix} 1 & 0 & 0 & 1 \\ 0 & 1 & 1 & 0 \\ 1 & 0 & 1 & 0 \end{bmatrix}$$

Since $\mathbf{M}$ is 3×4, we let

$$A = \{a_1, a_2, a_3\} \quad \text{and} \quad B = \{b_1, b_2, b_3, b_4\}$$

Then $(a_i, b_j) \in R$ if and only if $m_{ij} = 1$. Thus

$$R = \{(a_1, b_1), (a_1, b_4), (a_2, b_2), (a_2, b_3), (a_3, b_1), (a_3, b_3)\}$$

Digraphs

If A is a finite set and R is a relation on A, we can represent R pictorially as follows. Draw a small circle for each element of A and label the circle with the corresponding element of A. These circles are called **vertices**. Draw a directed line, called an

edge, from vertex a_i to vertex a_j if and only if $a_i \, R \, a_j$. The resulting pictorial representation of R is called a **directed graph** or **digraph** of R.

Thus if R is a relation on A, the edges in the digraph of R correspond exactly to the pairs in R, and the vertices correspond exactly to the elements of the set A. Sometimes, when we want to emphasize the geometric nature of some property of R, we may refer to the pairs of R themselves as edges, and the elements of A as vertices.

Example 12   Let

$$A = \{1, 2, 3, 4\}$$

$$R = \{(1, 1), (1, 2), (2, 1), (2, 2), (2, 3), (2, 4), (3, 4), (4, 1)\}$$

Then the digraph of R is shown in Fig. 2.

A collection of vertices with edges between some of the vertices determines a relation in a natural manner.

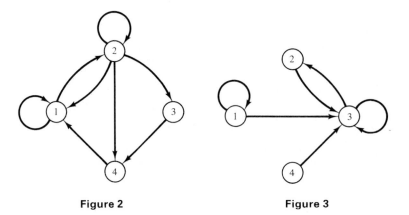

Figure 2   **Figure 3**

Example 13   Find the relation determined by Fig. 3.

Solution. Since $a_i \, R \, a_j$ if and only if there is an edge from a_i to a_j, we have

$$R = \{(1, 1), (1, 3), (2, 3), (3, 2), (3, 3), (4, 3)\}$$

In this book, digraphs are nothing but geometrical representations of relations, and any statement made about a digraph is actually a statement about the corresponding relation. This is especially important for theorems and their proofs. In some cases, it is easier or clearer to state a result in graphical terms, but a proof will always refer to the underlying relation. The reader should be aware that some authors allow more general objects as digraphs, for example by permitting several edges between the same vertices. We do not need, and will not again refer to, these generalizations, except to say that they could be covered by the idea of a function defined on a digraph (see Section 3.1).

An important concept for relations is inspired by the visual form of digraphs. If R is a relation on a set A, and $a \in A$, then the **in-degree** of a (relative to the relation R) is the number of $b \in A$ such that $(b, a) \in R$. The **out-degree** of a is the number of $b \in A$ such that $(a, b) \in R$.

What this means, in terms of the digraph of R, is that the in-degree of a vertex is the number of edges terminating at the vertex. The out-degree of a vertex is the number of edges leaving the vertex.

Example 14   Consider the digraph of Fig. 2. Vertex 1 has in-degree 3 and out-degree 2. Also consider the digraph shown in Fig. 3. Vertex 3 has in-degree 4 and out-degree 2, while vertex 4 has in-degree 0 and out-degree 1.

If R is a relation on a set A and B is a subset of A, the **restriction of R to B** is $R \cap (B \times B)$.

Example 15   Let

$$A = \{a, b, c, d, e, f\} \qquad \text{and} \qquad R = \{(a, a), (a, c), (b, c), (a, e), (b, e), (c, e)\}$$

Let $B = \{a, b, c\}$. Then

$$B \times B = \{(a, a), (a, b), (a, c), (b, a), (b, b), (b, c), (c, a), (c, b), (c, c)\}$$

and the restriction of R to B is $\{(a, a), (a, c), (b, c)\}$.

EXERCISE SET 2.2

1. Consider the relation defined in Example 4. Which of the following ordered pairs belong to R?

   (a) $(2, 3)$ (b) $(-6, 24)$ (c) $(0, 8)$

   (d) $(8, 0)$ (e) $(1, 3)$ (f) $(6, 18)$

2. Consider the relation defined in Example 6. Which of the following ordered pairs belong to R?

   (a) $(2, 0)$ (b) $(0, 2)$ (c) $(0, 3)$

   (d) $(1, 3/(2\sqrt{3}))$ (e) $(0, 0)$ (f) $(-2, 0)$

   In Exercises 3–13, find the domain, range, matrix, and, when $A = B$, the digraph of the relation.

3. $A = \{a, b, c, d\}$, $B = \{1, 2, 3\}$,
   $R = \{(a, 1), (a, 2), (b, 1), (c, 2), (d, 1)\}$

4. $A = \{4, 5, 6, 7\}$, $B = \{r, s, t, u, v\}$,
   $R = \{(4, r), (4, u), (5, r), (5, v), (7, u)\}$

5. $A = \{$IBM, Univac, Commodore, Atari, Xerox$\}$,
   $B = \{$370, 1110, 4000, 8000, 400, 800, 820$\}$,
   $R = \{$(IBM, 370), (Univac, 1110), (Commodore, 4000), (Atari, 800)$\}$

6. $A = \{1, 2, 3, 4\}$, $B = \{1, 4, 6, 8, 9\}$; $a\,R\,b$ if and only if $b = a^2$.

7. $A = \{1, 2, 3, 4, 8\} = B$; $a\,R\,b$ if and only if $a = b$.

8. $A = \{1, 2, 3, 4, 8\}$, $B = \{1, 4, 6, 9\}$; $a\,R\,b$ if and only if $a\,|\,b$.

9. $A = \{1, 2, 3, 4, 6\} = B$; $a\,R\,b$ if and only if a is a multiple of b.

10. $A = \{1, 2, 3, 4, 5\} = B$; $a\,R\,b$ if and only if $a \leq b$.

11. $A = \{1, 3, 5, 7, 9\}$, $B = \{2, 4, 6, 8\}$; $a\,R\,b$ if and only if $b < a$.

12. $A = \{1, 2, 3, 4, 8\} = B$; $a\,R\,b$ if and only if $a + b \leq 9$.

13. $A = \{1, 2, 3, 4, 5, 8\} = B$; $a\,R\,b$ if and only if $b = a + 1$.

14. Let $A = B = \mathbb{R}$. Consider the following relation R on A: $a\,R\,b$ if and only if $2a + 3b = 6$. Find Dom (R) and Ran (R).

15. Let $A = B = \mathbb{R}$. Consider the following relation R on A: $a\,R\,b$ if and only if $a^2 + b^2 = 25$. Find Dom (R) and Ran (R).

16. Let $A = B = \mathbb{R}$. Find the relation specified by the shaded region in Fig. 4.

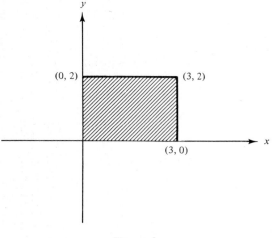

Figure 4

17. A manufacturer of automobiles has five factories $\{a_1, a_2, a_3, a_4, a_5\}$ and six distribution centers $\{b_1, b_2, b_3, b_4, b_5, b_6\}$. The following table gives the distance (in miles) from a_i to b_j.

	b_1	b_2	b_3	b_4	b_5	b_6
a_1	1200	1100	400	600	1800	700
a_2	800	700	1200	450	400	500
a_3	1000	600	1000	650	600	600
a_4	250	400	500	350	900	600
a_5	800	280	300	400	1300	2400

We define the following relation: $a_i\,R\,b_j$ if and

only if the distance from a_i to b_j is at least 800 miles. List the elements in R.

In Exercises 18 and 19, let $A = \{1, 2, 3, 4\}$. Find the relation R on A determined by the matrix.

18.
$$\mathbf{M}_R = \begin{bmatrix} 1 & 1 & 0 & 1 \\ 0 & 1 & 1 & 0 \\ 0 & 0 & 1 & 1 \\ 1 & 0 & 0 & 0 \end{bmatrix}$$

19.
$$\mathbf{M}_R = \begin{bmatrix} 1 & 0 & 1 & 0 \\ 0 & 0 & 1 & 0 \\ 1 & 0 & 0 & 0 \\ 1 & 1 & 0 & 1 \end{bmatrix}$$

20. Find the relation determined by Fig. 5.

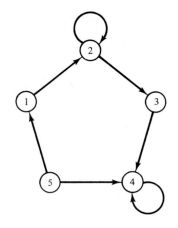

Figure 5

21. Find the relation determined by Fig. 6.

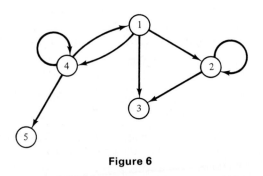

Figure 6

22. If A has n elements and B has m elements, how many different relations are there from A to B?

23. Consider the digraph of Fig. 5. Compute the in-degree and out-degree for each vertex.

24. Consider the digraph of Fig. 6. Compute the in-degree and out-degree for each vertex.

In Exercises 25 and 26, let

$A = \{1, 2, 3, 4, 5, 6, 7\}$ and
$R = \{(1, 2), (1, 4), (2, 3), (2, 5), (3, 6), (4, 7)\}$

Compute the restriction of R to B for the given subset B of A.

25. $B = \{1, 2, 4, 5\}$ **26.** $B = \{2, 3, 4, 6\}$

2.3 Paths in Relations and Digraphs

Suppose that R is a relation on a set A. A **path of length** n from a to b is a finite sequence $\pi = a, x_1, x_2, \ldots, x_{n-1}, b$, beginning with a and ending with b, such that

$$a \; R \; x_1, \quad x_1 \; R \; x_2, \quad \ldots, \quad x_{n-1} \; R \; b$$

Note that a path of length n involves $n + 1$ elements of A, although they are not necessarily distinct.

A path is most easily visualized with the aid of the digraph of the relation. It appears as a geometric "path" or succession of edges in such a digraph, where the indicated directions of the edges are followed, and in fact a path derives its name from this representation. Thus the length of a path is the number of edges in the path, where the vertices need not all be distinct.

Example 1 Consider the digraph in Fig. 1. Then $\pi_1 = 1, 2, 5, 4, 3$ is a path of length 4 from vertex 1 to vertex 3, $\pi_2 = 1, 2, 5, 1$ is a path of length 3 from vertex 1 to itself, and $\pi_3 = 2, 2$ is a path of length 1 from vertex 2 to itself.

A path that begins and ends at the same vertex is called a **cycle**. In Example 1, π_2 and π_3 are cycles of length 3 and 1, respectively. It is clear that the paths of length 1 can be identified with the ordered pairs (x, y) that belong to R. Paths in a relation R can be used to define new relations which are quite useful. If n is a fixed positive integer, we define a relation R^n on A as follows: $x \; R^n \; y$ means that there is a path of length n from x to y in R. We may also define a relation R^∞ on A, by letting

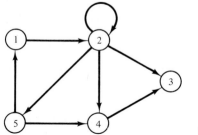

Figure 1

$x \, R^{\infty} \, y$ mean that there is some path in R from x to y. The length of such a path will depend, in general, on x and y. R^{∞} is sometimes called the **connectivity relation** for R.

Example 2   Let A be the set of all living human beings, and let R be the relation of mutual acquaintance. That is, $a \, R \, b$ means that a and b know one another. Then $a \, R^2 \, b$ means that a and b have an acquaintance in common. In general, $a \, R^n \, b$ if a knows someone x_1, who knows $x_2, \ldots$, who knows x_{n-1}, who knows b. Finally, $a \, R^{\infty} \, b$ means that some chain of acquaintances exists which begins at a and ends at b. It is interesting (and unknown) whether every two Americans, say, are related by R^{∞}.

Example 3   Let A be a set of U.S. cities, and let $x \, R \, y$ if there is a direct flight from x to y on at least one airline. Then x and y are related by R^n if one can book a flight from x to y having exactly $n - 1$ intermediate stops, and $x \, R^{\infty} \, y$ if one can get from x to y by plane.

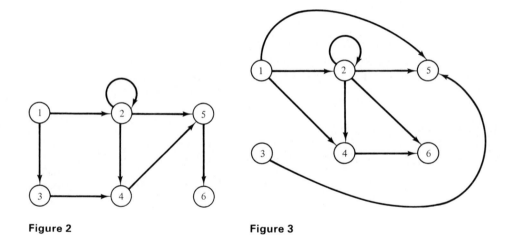

Figure 2   Figure 3

Example 4   Let $A = \{1, 2, 3, 4, 5, 6\}$. Let R be the relation whose digraph is shown in Fig. 2. Figure 3 shows the digraph of the relation R^2 on A. A line connects two vertices in Fig. 3 if and only if they are R^2-related, that is, if and only if there is a path of length two connecting those vertices in Fig. 2. Thus

$$1 \, R^2 \, 2 \quad \text{since} \quad 1 \, R \, 2 \quad \text{and} \quad 2 \, R \, 2$$

$$1 \, R^2 \, 4 \quad \text{since} \quad 1 \, R \, 2 \quad \text{and} \quad 2 \, R \, 4$$

$$1 \, R^2 \, 5 \quad \text{since} \quad 1 \, R \, 2 \quad \text{and} \quad 2 \, R \, 5$$

$$2 \, R^2 \, 2 \quad \text{since} \quad 2 \, R \, 2 \quad \text{and} \quad 2 \, R \, 2$$

$$2\ R^2\ 4 \quad \text{since} \quad 2\ R\ 2 \quad \text{and} \quad 2\ R\ 4$$

$$2\ R^2\ 5 \quad \text{since} \quad 2\ R\ 2 \quad \text{and} \quad 2\ R\ 5$$

$$2\ R^2\ 6 \quad \text{since} \quad 2\ R\ 5 \quad \text{and} \quad 5\ R\ 6$$

$$3\ R^2\ 5 \quad \text{since} \quad 3\ R\ 4 \quad \text{and} \quad 4\ R\ 5$$

$$4\ R^2\ 6 \quad \text{since} \quad 4\ R\ 5 \quad \text{and} \quad 5\ R\ 6$$

In a similar way, we can construct the digraph of R^n for any n.

Example 5   Let

$$A = \{a, b, c, d, e\}$$

$$R = \{(a, a), (a, b), (b, c), (c, e), (c, d), (d, e)\}$$

Compute (a) R^2; (b) R^∞.

Solution

(a) The digraph of R is shown in Fig. 4.

$$a\ R^2\ a \quad \text{since} \quad a\ R\ a \quad \text{and} \quad a\ R\ a$$

$$a\ R^2\ b \quad \text{since} \quad a\ R\ a \quad \text{and} \quad a\ R\ b$$

$$a\ R^2\ c \quad \text{since} \quad a\ R\ b \quad \text{and} \quad b\ R\ c$$

$$b\ R^2\ e \quad \text{since} \quad b\ R\ c \quad \text{and} \quad c\ R\ e$$

$$b\ R^2\ d \quad \text{since} \quad b\ R\ c \quad \text{and} \quad c\ R\ d$$

$$c\ R^2\ e \quad \text{since} \quad c\ R\ d \quad \text{and} \quad d\ R\ e$$

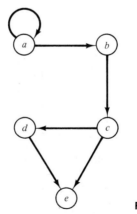

Figure 4

Hence

$$R^2 = \{(a, a), (a, b), (a, c), (b, e), (b, d), (c, e)\}$$

(b) To compute R^∞, we need all ordered pairs of vertices for which there is a path of any length from the first vertex to the second one. From Fig. 4 we see that

$$R^\infty = \{(a, a), (a, b), (a, c), (a, d), (a, e), (b, c), (b, d), (b, e), (c, d), (c, e), (d, e)\}$$

For example, $(a, d) \in R^\infty$, since there is a path of length 3 from a to d: a, b, c, d. Similarly, $(a, e) \in R^\infty$ since there is a path of length 3 from a to e: a, b, c, e as well as a path of length 4 from a to e: a, b, c, d, e.

Let R be a relation on a finite set $A = \{a_1, \ldots, a_n\}$, and let $\mathbf{M}_R = [m_{ij}]$ be the $n \times n$ matrix of R. Then $m_{ij} = 1$ if and only if $a_i \, R \, a_j$. We wish to show how the matrix $\mathbf{M}_{R^2}$, of R^2, is obtained from $\mathbf{M}_R$. We claim that $\mathbf{M}_{R^2} = \mathbf{M}_R \odot \mathbf{M}_R$ (see Section 1.8).

To prove this, let $\mathbf{N} = \mathbf{M}_R \odot \mathbf{M}_R$, and write $\mathbf{N} = [n_{ij}]$. We can have $n_{ij} = 1$ if and only if, for some k, m_{ik} and m_{kj} are both 1. This happens if and only if $a_i \, R \, a_k$ and $a_k \, R \, a_j$ for some a_k in A, or in other words, if and only if $a_i \, R^2 \, a_j$. Thus $\mathbf{N} = \mathbf{M}_{R^2}$. For brevity, we will usually denote $\mathbf{M}_R \odot \mathbf{M}_R$ simply as $(\mathbf{M}_R)^2_\odot$ (the symbol $\odot$ reminds us that this is not the usual matrix product).

Example 6 Let A and R be as in Example 5. Then

$$\mathbf{M}_R = \begin{bmatrix} 1 & 1 & 0 & 0 & 0 \\ 0 & 0 & 1 & 0 & 0 \\ 0 & 0 & 0 & 1 & 1 \\ 0 & 0 & 0 & 0 & 1 \\ 0 & 0 & 0 & 0 & 0 \end{bmatrix}$$

From the discussion above,

$$\mathbf{M}_{R^2} = \mathbf{M}_R \odot \mathbf{M}_R = \begin{bmatrix} 1 & 1 & 0 & 0 & 0 \\ 0 & 0 & 1 & 0 & 0 \\ 0 & 0 & 0 & 1 & 1 \\ 0 & 0 & 0 & 0 & 1 \\ 0 & 0 & 0 & 0 & 0 \end{bmatrix} \odot \begin{bmatrix} 1 & 1 & 0 & 0 & 0 \\ 0 & 0 & 1 & 0 & 0 \\ 0 & 0 & 0 & 1 & 1 \\ 0 & 0 & 0 & 0 & 1 \\ 0 & 0 & 0 & 0 & 0 \end{bmatrix}$$

$$= \begin{bmatrix} 1 & 1 & 1 & 0 & 0 \\ 0 & 0 & 0 & 1 & 1 \\ 0 & 0 & 0 & 0 & 1 \\ 0 & 0 & 0 & 0 & 0 \\ 0 & 0 & 0 & 0 & 0 \end{bmatrix}$$

Computing $\mathbf{M}_{R^2}$ directly from R^2, we obtain the same result.

We can see from Examples 5 and 6 how much easier it is to compute R^2 by computing $\mathbf{M}_{R^2} = \mathbf{M}_R \odot \mathbf{M}_R$ instead of searching the digraph of R for all vertices that can be joined by a path of length 2. Similarly, we can show that $\mathbf{M}_{R^3} = \mathbf{M}_R \odot (\mathbf{M}_R \odot \mathbf{M}_R) = (\mathbf{M}_R)^3_\odot$. In fact, we now show by induction that these two results can be generalized.

Theorem 1   For $n \geq 1$, and R a relation on a finite set A, we have

$$\mathbf{M}_{R^n} = \mathbf{M}_R \odot \mathbf{M}_R \odot \cdots \odot \mathbf{M}_R \quad (n \text{ factors})$$

Proof. Let $P(n)$ be the assertion that the formula above holds for an integer $n \geq 1$.

Basis Step. $P(1)$ is true since the statement in that case is $\mathbf{M}_R = \mathbf{M}_R$.

Induction Step. Suppose that $P(n)$ is true, and consider the matrix $\mathbf{M}_{R^{n+1}}$. Let $\mathbf{M}_{R^{n+1}} = [x_{ij}]$, $\mathbf{M}_{R^n} = [y_{ij}]$, $\mathbf{M}_R = [m_{ij}]$. If $x_{ij} = 1$, we must have a path of length $n + 1$ from a_i to a_j. If we let a_k be the vertex which this path reaches just before the last vertex a_j, then there is a path of length n from a_i to a_k and a path of length 1 from a_k to a_j. Thus $y_{ik} = 1$ and $m_{kj} = 1$, so $\mathbf{M}_{R^n} \odot \mathbf{M}_R$ has a 1 in position i, j. We can see, similarly, that if $\mathbf{M}_{R^n} \odot \mathbf{M}_R$ has a 1 in position i, j, then $x_{ij} = 1$. This means that $\mathbf{M}_{R^{n+1}} = \mathbf{M}_{R^n} \odot \mathbf{M}_R$.

By induction, $\mathbf{M}_{R^n} = \mathbf{M}_R \odot \cdots \odot \mathbf{M}_R$ (n factors), so by substitution,

$$\mathbf{M}_{R^{n+1}} = (\mathbf{M}_R \odot \cdots \odot \mathbf{M}_R) \odot \mathbf{M}_R$$
$$= \mathbf{M}_R \odot \cdots \odot \mathbf{M}_R \odot \mathbf{M}_R \quad (n + 1 \text{ factors})$$

and $P(n + 1)$ is true. This proves the theorem. As before, we write $\mathbf{M}_R \odot \cdots \odot \mathbf{M}_R$ (n times) as $(\mathbf{M}_R)^n_\odot$.

Now that we know how to compute the matrix of the relation R^n from the matrix of R, we would like to see how to compute the matrix of R^∞. We proceed as follows. Suppose that R is a relation on a finite set A, and $x \in A$, $y \in A$. We know that $x\ R^\infty\ y$ means that x and y are connected by a path in R of length n for some n. In general, n will depend on x and y, but clearly $x\ R^\infty\ y$ if and only if $x\ R\ y$ or $x\ R^2\ y$ or $x\ R^3\ y$ or $\ldots$. If R and S are relations on A, the relation $R \cup S$ is defined by $x\ (R \cup S)\ y$ if and only if $x\ R\ y$ or $x\ S\ y$. (The relation $R \cup S$ will be discussed in more detail in Section 2.6.) Thus our statement above tells us that $R^\infty = R \cup R^2 \cup R^3 \cup \cdots$. The reader may verify that $\mathbf{M}_{R \cup S} = \mathbf{M}_R \vee \mathbf{M}_S$, and we will show this in Section 2.6. Thus

$$\mathbf{M}_{R^\infty} = \mathbf{M}_R \vee \mathbf{M}_{R^2} \vee \mathbf{M}_{R^3} \vee \cdots$$
$$= \mathbf{M}_R \vee (\mathbf{M}_R)^2_\odot \vee (\mathbf{M}_R)^3_\odot \vee \cdots$$

The **reachability** relation R^* of a relation R on a set A that has n elements is defined as follows: $x\ R^*\ y$ means that $x = y$ or $x\ R^\infty\ y$. The idea is that y is reachable from x if either y is x or there is some path from x to y. It is easily seen that $\mathbf{M}_{R^*} = \mathbf{M}_{R^\infty} \vee \mathbf{I}_n$, where $\mathbf{I}_n$ is the $n \times n$ identity matrix. Thus our discussion above shows that

$$\mathbf{M}_{R*} = \mathbf{I}_n \vee \mathbf{M}_R \vee (\mathbf{M}_R)^2_\odot \vee (\mathbf{M}_R)^3_\odot \vee \cdots$$

Let $\pi_1 = a, x_1, x_2, \ldots, x_{n-1}, b$ be a path in a relation R of length n from a to b, and let $\pi_2 = b, y_1, y_2, \ldots, y_{m-1}, c$ be a path in R of length m from b to c. Then the **composition of π_1 and π_2** is the path $a, x_1, x_2, \ldots, b, y_1, y_2, \ldots, y_{m-1}, c$ of length $n + m$, which is denoted by $\pi_1 \circ \pi_2$. This is a path from a to c.

Example 7   Consider the relation whose digraph is given in Fig. 5, and the paths

$$\pi_1 = 1, 2, 3 \qquad \text{and} \qquad \pi_2 = 3, 5, 6, 2, 4$$

Then the composition of π_1 and π_2 is the path

$$\pi_1 \circ \pi_2 = 1, 2, 3, 5, 6, 2, 4$$

from 1 to 4 of length 6.

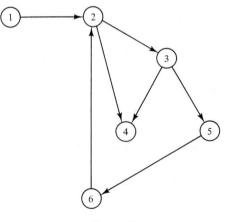

Figure 5

If $\pi = a, x_1, x_2, \ldots, x_{n-1}, b$ is a path in a relation R of length n from a to b, then the **inverse path** denoted by π^{-1} is defined by

$$\pi^{-1} = b, x_{n-1}, \ldots, x_2, x_1, a$$

Note that the inverse of a path in R may not be in R.

Example 8   Consider the relation R whose digraph is given in Fig. 5 and the paths π_1 and π_2 of Example 7. Then

$$\pi_1^{-1} = 3, 2, 1 \qquad \text{and} \qquad \pi_2^{-1} = 4, 2, 6, 5, 3$$

EXERCISE SET 2.3

For Exercises 1–11, let R be the relation whose digraph is given in Fig. 6.

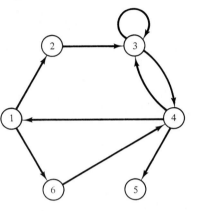

Figure 6

1. List all paths of length 1.
2. List all paths of length 2 starting from vertex 2.
3. List all paths of length 3 starting from vertex 3.
4. Find a cycle starting at vertex 2.
5. Find a cycle starting at vertex 6.
6. List all paths of length 2.
7. List all paths of length 3.
8. Draw the digraph of R^2.
9. Find $\mathbf{M}_{R^2}$.
10. Find R^∞.
11. Find $\mathbf{M}_{R^\infty}$.

For Exercises 12–22, let R be the relation whose digraph is given in Fig. 7.

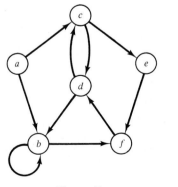

Figure 7

12. List all paths of length 1.
13. List all paths of length 2 starting from vertex c.
14. List all paths of length 3 starting from vertex a.
15. Find a cycle starting at vertex c.
16. Find a cycle starting at vertex d.
17. Find all paths of length 2.
18. Find all paths of length 3.
19. Draw the digraph of R^2.
20. Find $\mathbf{M}_{R^2}$.
21. Find R^∞.
22. Find $\mathbf{M}_{R^\infty}$.
23. Let R and S be relations on a set A. Show that

$$\mathbf{M}_{R \cup S} = \mathbf{M}_R \vee \mathbf{M}_S$$

24. Let R be a relation on a set A that has n elements. Show that $\mathbf{M}_{R^*} = \mathbf{M}_{R^\infty} \vee \mathbf{I}_n$, where $\mathbf{I}_n$ is the $n \times n$ identity matrix.

In Exercises 25–28, let R be the relation whose digraph is given in Fig. 8.

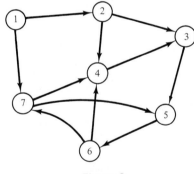

Figure 8

25. If $\pi_1 = 1, 2, 4, 3$ and $\pi_2 = 3, 5, 6, 4$, find the composition $\pi_1 \circ \pi_2$.
26. If $\pi_1 = 1, 7, 5$ and $\pi_2 = 5, 6, 7, 4, 3$, find the composition $\pi_1 \circ \pi_2$.
27. Find the inverse of the path $\pi = 5, 6, 7, 4, 3$.
28. Find the inverse of the path $\pi = 1, 2, 3, 5, 6, 7$.
29. Let π_1 be a path in a relation R from a to b and let π_2 be a path in R from b to c. Prove that

$$(\pi_1 \circ \pi_2)^{-1} = \pi_2^{-1} \circ \pi_1^{-1}$$

2.4 Properties of Relations

In many of the applications of relations to computer science and applied mathematics, we deal with relations on a set A rather than relations from A to B. Moreover, these relations often satisfy certain properties that will be discussed in this section.

Reflexive and Irreflexive Relations

A relation R on a set A is **reflexive** if $(a, a) \in R$ for every $a \in A$, that is, if $a \, R \, a$ for all $a \in A$. A relation R on a set A is **irreflexive** if $a \, \not{R} \, a$ for every $a \in A$.

Thus R is reflexive if every element $a \in A$ is related to itself and it is irreflexive if no element is related to itself.

Example 1

(a) Let $\Delta = \{(a, a) \mid a \in A\}$, so that Δ is the relation of **equality** on the set A. Then Δ is reflexive, since $(a, a) \in \Delta$ for all $a \in A$.

(b) Let $R = \{(a, b) \in A \times A \mid a \neq b\}$, so that R is the relation of **inequality** on the set A. Then R is irreflexive, since $(a, a) \notin R$ for all $x \in A$.

(c) Let $A = \{1, 2, 3\}$, and let $R = \{(1, 1), (1, 2)\}$. Then R is not reflexive since $(2, 2) \notin R$ and $(3, 3) \notin R$. Also, R is not irreflexive, since $(1, 1) \in R$.

(d) Let A be a nonempty set. Let $R = \varnothing \subseteq A \times A$, the **empty relation**. Then R is not reflexive, since $(a, a) \notin R$ for all $a \in A$ (the empty set has no elements). However, R is irreflexive.

We can characterize a reflexive or irreflexive relation by its matrix as follows. The matrix of a reflexive relation must have all 1's on its main diagonal, while the matrix of an irreflexive relation must have all 0's on its main diagonal.

Similarly, we can characterize a reflexive or irreflexive relation by its digraph as follows. A reflexive relation has a cycle of length 1 at every vertex, while an irreflexive relation has no cycles of length 1. Another useful way of saying the same thing uses the equality relation Δ on a set A: R is reflexive if and only if $\Delta \subseteq R$, and R is irreflexive if and only if $\Delta \cap R = \varnothing$.

Finally, we may note that if R is reflexive on a set A, then Dom $(R) =$ Ran $(R) = A$.

Symmetric, Asymmetric, and Antisymmetric Relations

A relation R on a set A is **symmetric** if whenever $a \, R \, b$, then $b \, R \, a$. It then follows that R is not symmetric if we have a and $b \in A$ with $a \, R \, b$, but $b \, \not{R} \, a$. A relation R on a set A is **asymmetric** if whenever $a \, R \, b$, then $b \, \not{R} \, a$. It then follows that R is not asymmetric if we have a and $b \in A$ with both $a \, R \, b$ and $b \, R \, a$.

A relation R on a set A is **antisymmetric** if whenever $a \, R \, b$ and $b \, R \, a$, then $a = b$. A restatement of this definition is that R is antisymmetric if whenever $a \neq b$, we have $a \, \not{R} \, b$ or $b \, \not{R} \, a$. It then follows that R is not antisymmetric if we have a and b in A, $a \neq b$, and both $a \, R \, b$ and $b \, R \, a$.

Given a relation R, we shall want to determine which properties hold for R. Keep in mind the following remarks. A property fails to hold in general if we can find one situation where the property does not hold. If there is no situation where the property fails to hold, we must conclude that the property holds at all times.

Example 2   Let $A = Z$, the set of integers and let

$$R = \{(a, b) \in A \times A \mid a < b\}$$

so that R is the relation **less than**. Is R symmetric, asymmetric, or antisymmetric?

   Solution.

Symmetry: If $a < b$, then it is not true that $b < a$, so R is not symmetric.
Asymmetry: If $a < b$, then $b \not< a$ (b is not less than a), so R is asymmetric.
Antisymmetry: If $a \neq b$, then either $a \not< b$ or $b \not< a$, so R is antisymmetric.

Example 3   Let A be a set of people and let

$$R = \{(x, y) \in A \times A \mid x \text{ is a cousin of } y\}$$

Then R is a symmetric relation (verify).

Example 4   Let $A = \{1, 2, 3, 4\}$ and let

$$R = \{(1, 2), (2, 2), (3, 4), (4, 1)\}$$

Then R is not symmetric, since $(1, 2) \in R$ but $(2, 1) \notin R$. Also, R is not asymmetric, since $(2, 2) \in R$. Finally, R is antisymmetric, since if $a \neq b$, either $(a, b) \notin R$ or $(b, a) \notin R$.

Example 5   Let $A = Z^+$, the set of positive integers, and let

$$R = \{(a, b) \in A \times A \mid a \text{ divides } b\}$$

Is R symmetric, asymmetric, or antisymmetric?

Solution.

If $a \mid b$, it does not follow that $b \mid a$, so R is not symmetric.
If $a = b = 3$, say, then $a \, R \, b$ and $b \, R \, a$, so R is not asymmetric.
If $a \mid b$ and $b \mid a$, then $a = b$, so R is antisymmetric.

We now characterize symmetric, asymmetric, or antisymmetric properties of a relation by properties of its matrix. The matrix $\mathbf{M}_R = [m_{ij}]$ of a symmetric relation satisfies the property that

$$\text{if } m_{ij} = 1, \qquad \text{then } m_{ji} = 1$$

Moreover, if $m_{ji} = 0$, then $m_{ij} = 0$. Thus $\mathbf{M}_R$ is a matrix such that each pair of locations, symmetrically placed about the main diagonal, are either both 0 or both 1. It follows that $\mathbf{M}_R = \mathbf{M}_R^T$, so that $\mathbf{M}_R$ is a symmetric matrix (see Section 1.8).

The matrix $\mathbf{M}_R = [m_{ij}]$ of an asymmetric relation R satisfies the property that

$$\text{if } m_{ij} = 1, \qquad \text{then} \qquad m_{ji} = 0.$$

If R is asymmetric, it follows that $m_{ii} = 0$ for all i; that is, the main diagonal of the matrix $\mathbf{M}_R$ consists entirely of 0's. This must be true since the asymmetric property implies that if $m_{ii} = 1$, then $m_{ii} = 0$, which is a contradiction.

Finally, the matrix $\mathbf{M}_R = [m_{ij}]$ of an antisymmetric relation R satisfies the property that if $i \neq j$, then $m_{ij} = 0$ or $m_{ji} = 0$.

Example 6   Consider the matrices in Fig. 1, each of which is the matrix of a relation, as indicated.

Relations R_1 and R_2 are symmetric since the matrices $\mathbf{M}_{R_1}$ and $\mathbf{M}_{R_2}$ are symmetric matrices. Relation R_3 is antisymmetric, since no symmetrically situated, off-diagonal positions of $\mathbf{M}_{R_3}$ can both contain 1's. Such positions may both have 0's, however, and the diagonal elements are unrestricted. The relation R_3 is not asymmetric because $\mathbf{M}_{R_3}$ has diagonal 1's.

Relation R_4 has none of the three properties; $\mathbf{M}_{R_4}$ is not symmetric. The presence of the 1 in position 4, 1 of $\mathbf{M}_{R_4}$ violates both asymmetry and antisymmetry.

Finally, R_5 is antisymmetric but not asymmetric, and R_6 is both asymmetric and antisymmetric.

We now consider the digraphs of these three types of relations. If R is an asymmetric relation, then the digraph of R cannot simultaneously have an edge from vertex i to vertex j, and an edge from vertex j to vertex i. This is true for any i and j, and in particular if i equals j, so there can be no cycles of length 1.

If R is an antisymmetric relation, then for different vertices i and j, there cannot be an edge from vertex i to vertex j, and an edge from vertex j to vertex i.

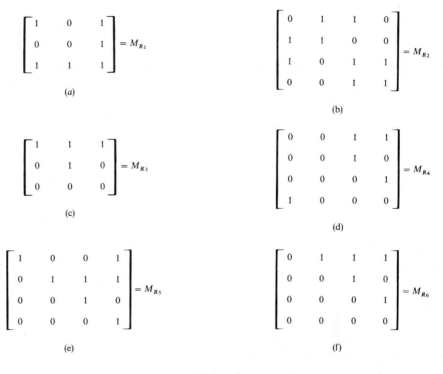

$$\begin{bmatrix} 1 & 0 & 1 \\ 0 & 0 & 1 \\ 1 & 1 & 1 \end{bmatrix} = M_{R_1}$$

(a)

$$\begin{bmatrix} 0 & 1 & 1 & 0 \\ 1 & 1 & 0 & 0 \\ 1 & 0 & 1 & 1 \\ 0 & 0 & 1 & 1 \end{bmatrix} = M_{R_2}$$

(b)

$$\begin{bmatrix} 1 & 1 & 1 \\ 0 & 1 & 0 \\ 0 & 0 & 0 \end{bmatrix} = M_{R_3}$$

(c)

$$\begin{bmatrix} 0 & 0 & 1 & 1 \\ 0 & 0 & 1 & 0 \\ 0 & 0 & 0 & 1 \\ 1 & 0 & 0 & 0 \end{bmatrix} = M_{R_4}$$

(d)

$$\begin{bmatrix} 1 & 0 & 0 & 1 \\ 0 & 1 & 1 & 1 \\ 0 & 0 & 1 & 0 \\ 0 & 0 & 0 & 1 \end{bmatrix} = M_{R_5}$$

(e)

$$\begin{bmatrix} 0 & 1 & 1 & 1 \\ 0 & 0 & 1 & 0 \\ 0 & 0 & 0 & 1 \\ 0 & 0 & 0 & 0 \end{bmatrix} = M_{R_6}$$

(f)

Figure 1

When $i = j$, no condition is imposed, so there may be cycles of length 1.

We consider the digraphs of symmetric relations in more detail.

The digraph of a symmetric relation R has the property that if there is an edge from vertex i to vertex j, then there is an edge from vertex j to vertex i. Thus if two vertices are connected by an edge, they must always be connected in both directions. Because of this, it is possible, and quite useful, to give a different representation of a symmetric relation. We keep the vertices as they appear in the digraph, but if two vertices a and b are connected by edges in each direction, we replace these two edges with one undirected edge, or a "two-way street." This undirected edge is just a single line without arrows, and connects a and b. The resulting diagram will be called the **graph** of the symmetric relation.

Example 7   Let $A = \{a, b, c, d, e\}$ and let R be the symmetric relation given by

$$R = \{(a, b), (b, a), (a, c), (c, a), (b, c), (c, b), (b, e), (e, b), (e, d), (d, e), (c, d), (d, c)\}$$

The usual digraph of R is shown in Fig. 2(a), while Fig. 2(b) shows the graph of R. Note that each undirected edge corresponds to two ordered pairs in the relation R.

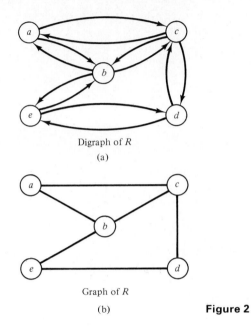

Digraph of R

(a)

Graph of R

(b)

Figure 2

An undirected edge between a and b, in the graph of a symmetric relation R, corresponds to a set $\{a, b\}$ such that $(a, b) \in R$ and $(b, a) \in R$. Sometimes we will also refer to such a set $\{a, b\}$ as an **undirected edge** of the relation R.

A symmetric relation R on a set A will be called **connected** if there is a path from any element of A to any other element of A. This simply means that the graph of R is all in one piece. In Fig. 3 we show the graphs of two symmetric relations. The graph in Fig. 3(a) is connected, whereas that in Fig. 3(b) is not connected.

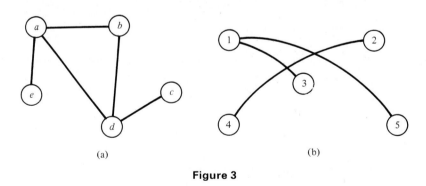

(a)   (b)

Figure 3

Transitive Relations

We say that a relation R on a set A is **transitive** if whenever $a\ R\ b$ and $b\ R\ c$, then $a\ R\ c$. It follows that R is not transitive if and only if we can find elements a, b, and c in A such that $a\ R\ b$ and $b\ R\ c$, but $a\ \not{R}\ c$.

Example 8   Let $A = Z$, the set of integers, and let R be the relation considered in Example 2. To see whether R is transitive, we assume that $a\ R\ b$ and $b\ R\ c$. Thus $a < b$ and $b < c$. It then follows that $a < c$, so $a\ R\ c$. Hence R is transitive.

Example 9   Let $A = Z^+$ and let R be the relation considered in Example 5. Is R transitive?

   Solution. Suppose that $a\ R\ b$ and $b\ R\ c$, so that $a\,|\,b$ and $b\,|\,c$. It then does follow that $a\,|\,c$. [See Section 1.7, Theorem 2, part (c).] Thus R is transitive.

Example 10   Let $A = \{1, 2, 3, 4\}$ and let

$$R = \{(1, 2), (1, 3), (4, 2)\}$$

Is R transitive?

   Solution. Since we cannot find elements a, b, and c in A such that $a\ R\ b$ and $b\ R\ c$, we conclude that R is transitive.

   We can characterize a transitive relation by its matrix $\mathbf{M}_R = [m_{ij}]$ as follows:

$$\text{if } m_{ij} = 1 \text{ and } m_{jk} = 1, \qquad \text{then } m_{ik} = 1$$

   To see what transitivity means for the digraph of a relation, we translate the definition of transitivity into geometric terms.
   If we consider particular vertices a and c, the conditions $a\ R\ b$ and $b\ R\ c$ can occur if and only if there is a path of length two from a to c, that is, if and only if $a\ R^2\ c$. We may rephrase the definition of transitivity as follows: If $a\ R^2\ c$, then $a\ R\ c$; that is, $R^2 \subseteq R$ (as subsets of $A \times A$).
   We can slightly generalize the foregoing geometric characterization of transitivity as follows.

Theorem 1   A relation R is transitive if and only if it satisfies the following property: If there is a path of length greater than 1 from vertex a to vertex b, there is a path of length 1 from a to b (that is, a is related to b). Algebraically stated, R is transitive if and only if $R^n \subseteq R$ for all $n \geq 1$.

   Proof. The proof is left as an exercise.

Equivalence Relations

A relation R on a set A is called an **equivalence relation** if it is reflexive, symmetric, and transitive.

Example 11   Let A be the set of all triangles in the plane and let R be the relation on A defined as follows.

$$R = \{(a, b) \in A \times A \mid a \text{ is congruent to } b\}$$

It is easy to see that R is an equivalence relation.

Example 12 Let $A = \{1, 2, 3, 4\}$ and let

$$R = \{(1, 1), (1, 2), (2, 1), (2, 2), (3, 4), (4, 3), (3, 3), (4, 4)\}$$

It is easy to verify that R is an equivalence relation.

Example 13 Let $A = Z$, the set of integers, and let

$$R = \{(a, b) \in A \times A \mid a \leq b\}$$

Is R an equivalence relation?

Solution. Since $a \leq a$, R is reflexive. If $a \leq b$, it need not follow that $b \leq a$, so R is not symmetric. Incidentally, R is transitive, since $a \leq b$ and $b \leq c$ imply that $a \leq c$. We see that R is not an equivalence relation.

Example 14 Let $A = Z$ and let

$$R = \{(a, b) \in A \times A \mid 2 \text{ divides } a - b\}$$

We write $a\ R\ b$ as

$$a \equiv b \pmod{2}$$

read "a is congruent to b mod 2." Show that congruence mod 2 is an equivalence relation.

Solution. First,

$$a \equiv a \pmod{2}$$

since $2 \mid (a - a)$.

Second, if $a \equiv b \pmod{2}$, then $2 \mid (a - b)$, or $a - b = 2k$ for some $k \in Z$. Then

$$b - a = 2(-k)$$

so $2 \mid (b - a)$ and $b \equiv a \pmod{2}$.

Finally, suppose that $a \equiv b \pmod{2}$ and $b \equiv c \pmod{2}$. Then $2 \mid (a - b)$, so

$$a - b = 2k_1 \tag{1}$$

where $k_1 \in Z$. Also, $2 \,|\, (b - c)$, so

$$b - c = 2k_2 \qquad (2)$$

where $k_2 \in Z$. Adding equations (1) and (2), we obtain

$$a - c = 2(k_1 + k_2)$$

so $2 \,|\, (a - c)$, which means that

$$a \equiv c \pmod 2$$

Hence congruence mod 2 is an equivalence relation.

Example 15   Let $A = Z$ and let $n \in Z^+$. We generalize the relation defined in Example 14 as follows. Let

$$R = \{(a, b) \in A \times A \,|\, n \text{ divides } a - b\}$$

and we write $a \; R \; b$ as

$$a \equiv b \pmod n$$

read "a is congruent to b mod n." Proceeding exactly as in Example 14, we can show that congruence mod n is an equivalence relation.

Equivalence Relations and Partitions

We shall now show that an equivalence relation on a set gives a partition of the set, and that conversely, a partition of a set determines an equivalence relation on the set.

Let R be an equivalence relation on the set A. If $a \in A$, then

$$[a] = \{x \in A \,|\, x \; R \; a\}$$

is called the **equivalence class** of a.

Observe that the equivalence class $[a]$ is never empty, since the reflexive property of R implies that $a \in [a]$.

Example 16   Let R be the equivalence relation defined in Example 12. Then

$$[1] = \{1, 2\} \qquad \text{and} \qquad [2] = \{1, 2\}$$

since

$$(1, 1), (1, 2), (2, 1), \text{ and } (2, 2) \in R$$

Lemma 1*   Let R be an equivalence relation on a set A. Then:

   (i) $[a] = [b]$ if and only if $a \, R \, b$.
   (ii) If $[a] \neq [b]$, then $[a] \cap [b] = \varnothing$.

   Proof.
   (i) Suppose that $[a] = [b]$. Since $a \in [a]$, we must have $a \in [b]$, so $a \, R \, b$. Conversely, suppose that $a \, R \, b$. We must show that $[a] = [b]$. Let $x \in [a]$. Then $x \, R \, a$. Since $a \, R \, b$, we conclude by transitivity that $x \, R \, b$, so $x \in [b]$. Hence $[a] \subseteq [b]$. Now let $y \in [b]$. Then $y \, R \, b$. If $a \, R \, b$, then $b \, R \, a$ by symmetry, therefore by transitivity we conclude that $y \, R \, a$, so $y \in [a]$. Hence $[b] \subseteq [a]$, and thus $[a] = [b]$.
   (ii) We prove the following equivalent statement: If $[a] \cap [b] \neq \varnothing$, then $[a] = [b]$. Thus let $c \in [a] \cap [b]$. Then $c \, R \, a$ and $c \, R \, b$. Since $c \, R \, a$, we have $a \, R \, c$ by symmetry. Then $a \, R \, c$ and $c \, R \, b$ imply by transitivity that $a \, R \, b$ and by (i) we conclude that $[a] = [b]$.

Theorem 2   If R is an equivalence relation on a set A, then the collection of all equivalence classes of the elements of A gives a partition of A, denoted by A/R.

   Proof. First, each element a belongs to the equivalence class of a. Second, by Lemma 1, if $[a] \neq [b]$, then $[a] \cap [b] = \varnothing$. Hence the set of all equivalence classes of the elements of A provides a partition of A.

Example 17   Let R be the relation defined in Example 12. Determine A/R.

   Solution. From Example 12, we have

$$[1] = \{1, 2\}$$

Similarly, $[3] = \{3, 4\}$. Hence $A/R = \{\{1, 2\}, \{3, 4\}\}$.

Example 18   Let R be the equivalence relation defined in Example 14. Determine A/R.

   Solution. First,

$$[0] = \{\ldots, -6, -4, -2, 0, 2, 4, 6, 8, \ldots\}$$

the set of even integers, since 2 divides the difference between two even integers.

---

*A lemma is a theorem whose main purpose is to aid in proving some other theorem.

Also,

$$[1] = \{\ldots, -5, -3, -1, 1, 3, 5, 7, \ldots\}$$

the set of odd integers, since 2 divides the difference between two odd integers. Hence A/R consists of the set of even integers and the set of odd integers.

Theorem 3   Given a partition of a set A, we can define an equivalence relation R on A so that the corresponding equivalence classes are the given blocks of the partition of A.

Proof. Suppose that $\{A_1, A_2, \ldots,\}$ is a partition of A. Define the relation R on A as follows:

   $a \, R \, b$ if and only if a and b are members of the same block

We now show that R is an equivalence relation. Of course, $a \, R \, a$. Next, if $a \, R \, b$, then a and b belong to the same block, so $b \, R \, a$. Finally, if $a \, R \, b$ and $b \, R \, c$, then a and b belong to the same block, say a and $b \in A_i$. Similarly, b and c belong to the same block, say b and $c \in A_j$. Thus $b \in A_i \cap A_j$. From the definition of a partition, it follows that $A_i = A_j$, so a and c are members of the same block and $a \, R \, c$. Hence R is an equivalence relation. From the manner in which R was constructed, it follows that the equivalence classes of the elements of A are the given blocks of the partition of A.

We have proved in Theorem 3 that if we have a partition on a set A, we can define an equivalence relation R on A so that the corresponding equivalence classes are the given blocks of the partition. Thus the partition of A corresponding to R, as described in Theorem 2, coincides with the original partition of A.

Example 19   Let $A = \{1, 2, 3, 4\}$, and consider the partition

$$\{\{1, 2, 3\}, \{4\}\}$$

of A. Determine the corresponding equivalence relation R on A.

Solution. Since the equivalence classes of the elements of A are the given blocks of the partition, we have

$$[1] = \{1, 2, 3\}$$
$$[4] = \{4\}$$

From the definition of $[1]$ and the fact that R is an equivalence relation, we see that

$$(1, 1), (1, 2), (2, 1), (2, 2), (1, 3), (3, 1), (2, 3), (3, 2), (3, 3) \in R$$

Similarly, $(4, 4) \in R$. Hence

$$R = \{(1, 1), (1, 2), (2, 1), (2, 2), (1, 3), (3, 1), (2, 3), (3, 2), (3, 3), (4, 4)\}$$

EXERCISE SET 2.4

In Exercises 1–8, let $A = \{1, 2, 3, 4\}$. Determine whether the relation is reflexive, irreflexive, symmetric, asymmetric, antisymmetric, or transitive.

1. $R = \{(1, 1), (1, 2), (2, 1), (2, 2), (3, 3), (3, 4), (4, 3), (4, 4)\}$

2. $R = \{(1, 2), (1, 3), (1, 4), (2, 3), (2, 4), (3, 4)\}$

3. $R = \{(1, 3), (1, 1), (3, 1), (1, 2), (3, 3), (4, 4)\}$

4. $R = \{(1, 1), (2, 2), (3, 3)\}$

5. $R = \varnothing$

6. $R = A \times A$

7. $R = \{(1, 2), (1, 3), (3, 1), (1, 1), (3, 3), (3, 2), (1, 4), (4, 2), (3, 4)\}$

8. $R = \{(1, 3), (4, 2), (2, 4), (3, 1), (2, 2)\}$

In Exercises 9 and 10, let $A = \{1, 2, 3, 4, 5\}$. Determine whether the relation R whose digraph is given is reflexive, irreflexive, symmetric, asymmetric, antisymmetric, or transitive.

9.   10.

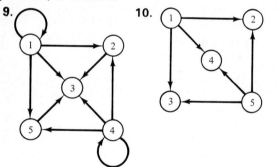

In Exercises 11 and 12, let $A = \{1, 2, 3, 4\}$. Determine whether the relation R whose matrix $\mathbf{M}_R$ is given is reflexive, irreflexive, symmetric, asymmetric, antisymmetric, or transitive.

11. $\begin{bmatrix} 0 & 1 & 0 & 1 \\ 1 & 0 & 1 & 1 \\ 0 & 1 & 0 & 0 \\ 1 & 1 & 0 & 0 \end{bmatrix}$ 12. $\begin{bmatrix} 1 & 1 & 0 & 0 \\ 1 & 1 & 0 & 0 \\ 0 & 0 & 1 & 0 \\ 0 & 0 & 0 & 1 \end{bmatrix}$

In Exercises 13–20, determine whether the rela-

tion R on the set A is reflexive, irreflexive, symmetric, asymmetric, antisymmetric, or transitive.

13. $A = Z$; $a\,R\,b$ if and only if $a \le b + 1$.

14. $A = Z^{+}$; $a\,R\,b$ if and only if $|a - b| \le 2$.

15. $A = Z^{+}$; $a\,R\,b$ if and only if $a = b^k$ for some $k \in Z^{+}$.

16. $A = Z$; $a\,R\,b$ if and only if $a + b$ is even.

17. $A = Z$; $a\,R\,b$ if and only if $|a - b| = 2$.

18. $A =$ the set of all real numbers; $a\,R\,b$ if and only if $a^2 + b^2 = 4$.

19. $A = Z^{+}$; $a\,R\,b$ if and only if GCD $(a, b) = 1$, that is, if and only if 1 is the only factor that a and b have in common. In this case, we say that a and b are **relatively prime**.

20. $A =$ the set of all ordered pairs of real numbers; $(a, b)\,R\,(c, d)$ if and only if $a = c$.

21. Let R be the following symmetric relation on the set $A = \{1, 2, 3, 4, 5\}$.

$$R = \{(1, 2), (2, 1), (3, 4), (4, 3), (3, 5), (5, 3), (4, 5), (5, 4), (5, 5)\}$$

Draw the graph of R.

22. Let $A = \{a, b, c, d\}$ and let R be the symmetric relation

$$R = \{(a, b), (b, a), (a, c), (c, a), (a, d), (d, a)\}$$

Draw the graph of R.

23. Consider the following graph of a symmetric relation R on $A = \{1, 2, 3, 4, 5, 6, 7\}$. Determine R (list all pairs).

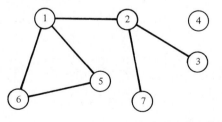

24. Consider the following graph of a symmetric relation R on $A = \{a, b, c, d, e\}$. Determine R (list all pairs).

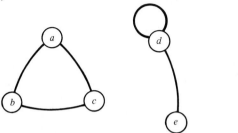

In Exercises 25 and 26, let $A = \{a, b, c\}$. Determine whether the relation R whose matrix $\mathbf{M}_R$ is given is an equivalence relation.

25. $\mathbf{M}_R = \begin{bmatrix} 1 & 0 & 0 \\ 0 & 1 & 1 \\ 0 & 1 & 1 \end{bmatrix}$ **26.** $\mathbf{M}_R = \begin{bmatrix} 1 & 0 & 1 \\ 0 & 1 & 0 \\ 0 & 0 & 1 \end{bmatrix}$

In Exercises 27 and 28, determine whether the relation R whose digraph is given is an equivalence relation.

27. **28.**

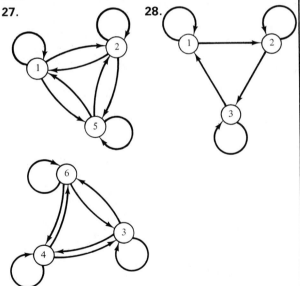

In Exercises 29–35, determine whether the relation R on the set A is an equivalence relation.

29. $A = \{a, b, c, d\}$, $R = \{(a, a), (b, a), (b, b), (c, c), (d, d), (d, c)\}$

30. $A = \{1, 2, 3, 4, 5\}$, $R = \{(1, 1), (1, 2), (1, 3), (2, 1), (2, 2), (3, 1), (2, 3), (3, 3), (4, 4), (3, 2), (5, 5)\}$

31. $A = \{1, 2, 3, 4\}$, $R = \{(1, 1), (1, 2), (2, 1), (2, 2), (3, 1), (3, 3), (1, 3), (4, 1), (4, 4)\}$

32. $A =$ the set of all members of the Software-of-the-Month Club; $a \, R \, b$ if and only if a and b buy the same number of programs.

33. $A =$ the set of all members of the Software-of-the-Month Club; $a \, R \, b$ if and only if a and b buy the same programs.

34. $A =$ the set of all people in the Social Security data base; $a \, R \, b$ if and only if a and b have the same last name.

35. $A =$ the set of all triangles in the plane; $a \, R \, b$ if and only if a is similar to b.

36. If $\{\{a, c, e\}, \{b, d, f\}\}$ is a partition of the set $A = \{a, b, c, d, e, f\}$, determine the corresponding equivalence relation R.

37. Let $S = \{1, 2, 3, 4, 5\}$ and let $A = S \times S$. Define the following relation R on A: $(a, b) \, R \, (a', b')$ if and only if $ab' = a'b$.

(a) Show that R is an equivalence relation.

(b) Compute A/R.

38. Let $S = \{1, 2, 3, 4\}$ and let $A = S \times S$. Define the following relation R on A: $(a, b) \, R \, (a', b')$ if and only if $a + b = a' + b'$.

(a) Show that R is an equivalence relation.

(b) Compute A/R.

39. If $\{\{1, 3, 5\}, \{2, 4\}\}$ is a partition of the set $A = \{1, 2, 3, 4, 5\}$, determine the corresponding equivalence relation R.

40. A relation R on a set A is called **circular** if $a \, R \, b$ and $b \, R \, c$ imply $c \, R \, a$. Show that R is reflexive and circular if and only if it is an equivalence relation.

41. Show that if a relation on a set A is transitive and irreflexive, then it is asymmetric.

42. Prove by induction that if a relation R on a set A is symmetric, then R^n is symmetric, for $n \geq 1$.

43. Let R be a nonempty relation on a set A. Suppose R is symmetric and transitive. Show that R is not irreflexive.

44. Show that if R_1 and R_2 are equivalence relations on A, then $R_1 \cap R_2$ is an equivalence relation on A.

2.5 Computer Representation of Relations and Digraphs

The most straightforward method of storing data items is to place them in a linear list or array. This generally corresponds to putting consecutive data items in consecutively numbered storage locations in a computer memory. Figure 1(a) illustrates this method for five data items $D_1, \ldots, D_5$. The method is an efficient use of space and provides, at least at the level of most programming languages, random access to the data. Thus the linear array might be A and the data would be in locations $A[1]$, $A[2]$, $A[3]$, $A[4]$, $A[5]$, and we would have access to any data item D_i by simply supplying its index i.

The main problem with this storage method is that one cannot insert new data between existing data without moving a possibly large number of items. Thus, to add another item E to the list in Fig. 1(a), and place E between D_2 and D_3, we would have to move D_3 to $A[4]$, D_4 to $A[5]$, and D_5 to $A[6]$, if room exists, then assign E to $A[3]$.

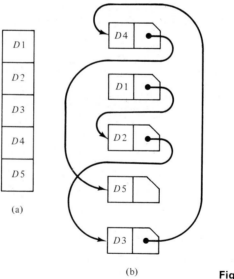

(a)

(b)

Figure 1

An alternative method of representing a sequence is by a **linked list**, shown in schematic fashion in Fig. 1(b). The basic unit of information storage is the **storage cell**. We imagine such cells to have room for two information items. The first can be "data" (numbers or symbols), and the second item is a **pointer**, that is, a number which tells us (points to) the location of the "next" cell to be considered. Thus cells may be arranged sequentially, but the data items that they represent are not assumed to be in the same sequence. Instead, we discover the proper data sequence by following the pointers from each item to the next.

As shown in Fig. 1(b), we represent the storage cell as a partitioned box $\boxed{\text{DATA} \ \bullet}$, with a dot in the right-hand side representing a pointer. A line is drawn from each such dot to the cell that the corresponding pointer designates as next. The symbol $\bullet\!\!-\!\!\perp$ means that data has ended, and that no further pointers need be followed.

In practice, the concept of a linked list may be implemented using two linear arrays, a data array A and a pointer array P as shown in Fig. 2. Note that once we have accessed the data in location $A[i]$, then the number in location $P[i]$ gives, or points to, the index of A containing the next data item.

Thus if we were at location $A[3]$, accessing data item D_2, then location $P[3]$ would contain 5, since the next data item, D_3, is located in $A[5]$. A zero in some location of P signifies that no more data items exist. In Fig. 2, $P[4]$ is zero because $A[4]$ contains D_5, the last data item. In this scheme, we need two arrays for the data that we previously represented in a single array, and we have only sequential access. Thus we cannot locate D_2 directly, but must go through the links until we come to it. The big advantage of this method, however, is that the actual physical order of the data does not have to be the same as the so-called logical, or natural order. In the example above, the natural order is $D_1 D_2 D_3 D_4 D_5$, but the data are not stored this way. The links allow us to pass naturally through the data, no matter how they are stored. Thus it is easy to add new items anywhere. If we want to insert item E between D_2 and D_3, we adjoin E to the end of the array A, change one pointer, and adjoin another pointer as shown in Fig. 3. This approach can be used no matter how long the list is. We should have one additional variable START holding the index of the first data item. In Figs. 2 and 3 START would contain 2 since D_1 is in $A[2]$.

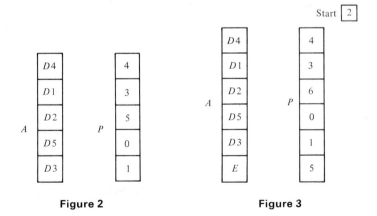

Figure 2   Figure 3

It does not matter how large the data item is, within computer constraints, so A might actually be a two-dimensional array, or matrix. The first row would hold several numbers describing the first data item, the second row would describe the next data item, and so on.

The problem of storing information to represent a relation or its digraph also has two solutions similar to those presented above for simple data. In the first place, we know from Section 2.2 that a relation R on A can be represented by an $n \times n$ matrix $\mathbf{M}_R$, if A has n elements. The matrix $\mathbf{M}_R$ has entries which are 0 or 1. Then a straightforward way of representing R in a computer would be by an $n \times n$ array having 0's and 1's stored in each location. Thus if $A = \{1, 2\}$ and $R = \{(1, 1), (1, 2), (2, 2)\}$, then

$$\mathbf{M}_R = \begin{bmatrix} 1 & 1 \\ 0 & 1 \end{bmatrix}$$

and these data could be represented by a two-dimensional array MAT where MAT[1, 1] = 1, MAT[1, 2] = 1, MAT[2, 1] = 0, and MAT[2, 2] = 1.

A second method of storing data for relations and graphs uses the linked list idea described above. For clarity, we use a graphical language. A linked list will be constructed which contains all the edges of the digraph, that is, the ordered pairs of numbers that determine those edges. The data can be represented by two arrays HEAD and TAIL, giving the end vertex and beginning vertex, respectively, for all "arrows." If we wish to make these edge data into a linked list, we will also need an array NEXT of pointers from each edge to the next edge.

Consider the relation whose digraph is shown in Fig. 4. The vertices are the integers 1 through 6 and we arbitrarily number the edges as shown.

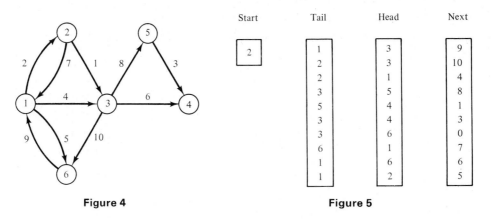

Figure 4   Figure 5

If we wish to store the digraph in linked-list form so that the logical order coincides with the numbering of edges, we can use a scheme such as that illustrated in Fig. 5. START contains 2, the index of the first data item, the edge (2, 3) (this edge is labeled with a 1 in Fig. 4). This edge is stored in the second entries of TAIL and HEAD, respectively. Since NEXT[2] contains 10, the next edge is the one located in position 10 of TAIL and HEAD, namely (1, 2) (labeled edge 2 in Fig. 4).

NEXT[10] contains 5, so we go next to data position 5, which contains the edge (5, 4). This process continues until we reach edge (3, 6) in data position 7. This is the last edge, and that fact is indicated by having NEXT[7] contain 0. We use 0 as a pointer, indicating the absence of any more data.

If we trace through this process, we will see that we encounter the edges in exactly the order corresponding to their numbering. We can arrange, in a similar way, to pass through the edges in any desired order.

This scheme, and the numerous equivalent variations of it, have important disadvantages. In many algorithms, it is efficient to locate a vertex and then immediately begin to investigate the edges which begin or end with that vertex. This is not possible in general with the storage mechanism shown in Fig. 5, so we now give a modification of it. We use an additional linear array VERT having one position for each vertex in the digraph. For each vertex I, VERT[I] is the index, in TAIL and HEAD, of the first edge we wish to consider leaving vertex I (in the digraph of Fig. 4, the first edge could be taken to be the edge with the smallest number labeling it). Thus VERT, like NEXT, contains pointers to edges. For each vertex I, we must arrange the pointers in NEXT so that they link together all edges leaving I, starting with the edge pointed to by VERT[I]. The last of these edges is made to point to zero in each case. In a sense, the data arrays TAIL and HEAD really contain several linked lists of edges, one list for each vertex. This method is shown in Fig. 6 for the digraph of Fig. 4.

Here VERT[1] contains 10, so the first edge leaving vertex 1 must be stored in the tenth data position. This is edge (1, 3).

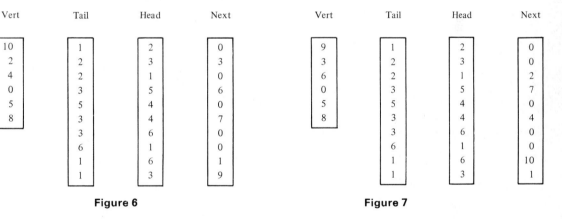

Vert		Tail	Head	Next		Vert		Tail	Head	Next
10		1	2	0		9		1	2	0
2		2	3	3		3		2	3	0
4		2	1	0		6		2	1	2
0		3	5	6		0		3	5	7
5		5	4	0		5		5	4	0
8		3	4	7		8		3	4	4
		3	6	0				3	6	0
		6	1	0				6	1	0
		1	6	1				1	6	10
		1	3	9				1	3	1

Figure 6   **Figure 7**

Since NEXT[10] = 9, the next edge leaving vertex 1 is (1, 6) located in data position 9. Again NEXT[9] = 1, which points us to edge (1, 2) in data position 1. Since NEXT[1] = 0, we have come to the end of those edges that begin at vertex 1. The order of the edges chosen here differs from the numbering in Fig. 4.

We then proceed to VERT[2] and get a pointer to position 2 in the data. This contains the first edge leaving vertex 2, namely (2, 3), and we can follow the pointers to visit all edges coming from vertex 2. In a similar way, we can trace through the edges (if any) coming from each vertex. Note that VERT[4] = 0, signifying that there are no edges beginning at vertex 4. Figure 7 shows an alternative to Fig. 6 for describing the digraph. The reader should check the accuracy of the method described in Fig. 7. We remind the reader again that the ordering of the edges leaving each vertex can be chosen arbitrarily.

We see then that we have (at least) two methods for storing the data for a relation or digraph; one using the matrix of the relation, and one using linked lists. A number of factors determine the choice of method to be used for storage. The total number of elements n in the set A, the number of ordered pairs in R or the ratio of this number to n^2 (the maximum possible number of ordered pairs), and the possible information that is to be extracted from R are all considerations. An analysis of such factors will determine which one of the storage methods is superior. We will consider two cases.

Suppose that $A = \{1, 2, \ldots, N\}$, and let R be a relation on A, whose matrix M_R is represented by the array MAT. Suppose that R contains P ordered pairs, so that MAT contains exactly P ones. First, we will consider the problem of adding a pair (I, J) to R, and second, the problem of testing R for transitivity.

Adding (I, J) to R is accomplished by the statement

$$\text{MAT}[I, J] \leftarrow 1$$

This is extremely simple with the matrix storage method.

Now, consider the following algorithm, which assigns RESULT the value T or F, depending on whether R is or is not transitive.

ALGORITHM TRANS
1. RESULT $\leftarrow T$
2. FOR $I = 1$ THRU N
  a. FOR $J = 1$ THRU N
    1. IF $(\text{MAT}[I, J] = 1)$ THEN
      a. FOR $K = 1$ THRU N
        1. IF $(\text{MAT}[J, K] = 1$ and $\text{MAT}[I, K] = 0)$ THEN
          a. RESULT $\leftarrow F$
END OF ALGORITHM TRANS

Here RESULT is originally set to T, and it is changed only if a situation is found where $(I, J) \in R$ and $(J, K) \in R$, but $(I, K) \notin R$ (a situation that violates transitivity).

We now provide a count of the number of steps required by algorithm TRANS. Observe that I and J each run from 1 to N. If (I, J) is not in R, we only perform the one test "IF $\text{MAT}[I, J] = 1$," which will be false and the rest of the algorithm will not be executed. Since $N^2 - P$ ordered pairs do not belong to R, we have $N^2 - P$ steps that must be executed for such elements. If $(I, J) \in R$, then the test "IF $\text{MAT}[I, J] = 1$" will be true and an additional loop

a. FOR $K = 1$ THRU N
  1. IF $(\text{MAT}[J, K] = 1$ and $\text{MAT}[I, K] = 0)$ THEN
    a. RESULT $\leftarrow F$

of N steps will be executed. Since R contains P ordered pairs, we have PN steps for such elements. Thus the total number of steps required by algorithm TRANS is

$$T_A = PN + (N^2 - P)$$

Suppose that $P = kN^2$, where $0 \leq k \leq 1$, since P must be between 0 and N^2. Then algorithm TRANS tests for transitivity in

$$T_A = kN^3 + (1 - k)N^2$$

steps.

Now consider the same digraph represented by our linked list scheme using VERT, TAIL, HEAD, and NEXT. First we deal with the problem of adding an edge (I, J). We assume that TAIL, HEAD, and NEXT have additional unused positions available, and that the total number of edges is counted by a variable P. Then the following algorithm adds an edge (I, J) to the relation R.

ALGORITHM ADDEDGE
1. $P \leftarrow P + 1$
2. TAIL$[P] \leftarrow I$
3. HEAD$[P] \leftarrow J$
4. NEXT$[P] \leftarrow$ VERT$[I]$
5. VERT$[I] \leftarrow P$
END OF ALGORITHM ADDEDGE

Figure 8 shows the situation diagrammatically in pointer form, both before and after the addition of edge (I, J). VERT$[I]$ now points to the new edge, and the pointer from that edge goes to the edge previously pointed to by VERT$[I]$, namely (I, J'). This method is not too involved, but clearly the matrix storage method has the advantage for the task of adding an edge.

We next consider the transitivity problem. To keep matters simple, we assume that there is available to us a function EDGE(I, J) which has the value T (true) if (I, J) is in R, otherwise has value F (false). The reader will be asked to construct such a function in the exercises. The algorithm below tests for transitivity of R with this storage method. Again RESULT will have value T if R is transitive, and otherwise will have value F.

ALGORITHM NEWTRANS
1. RESULT $\leftarrow$ T
2. FOR $I = 1$ THRU N
   a. $X \leftarrow$ VERT $[I]$
   b. WHILE $(X \neq 0)$
      1. $J \leftarrow$ HEAD$[X]$

2. $Y \leftarrow \text{VERT}[J]$
3. WHILE $(Y \neq 0)$
    a. $K \leftarrow \text{HEAD}[Y]$
    b. $\text{TEST} \leftarrow \text{EDGE}[I, K]$
    c. IF (TEST) THEN
       1. $Y \leftarrow \text{NEXT}[Y]$
    d. ELSE
       1. $\text{RESULT} \leftarrow F$
       2. $Y \leftarrow \text{NEXT}[Y]$
4. $X \leftarrow \text{NEXT}[X]$
END OF ALGORITHM NEWTRANS

The reader should follow the steps of this algorithm with several simple examples. For each vertex I, it searches through all paths of length 2 beginning at I, and checks these for transitivity. Thus it eventually checks each path of length 2 to see if

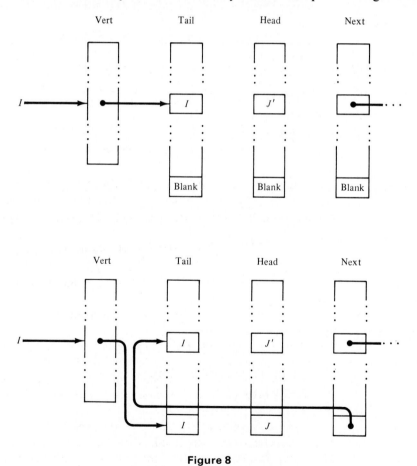

Figure 8

there is an equivalent direct path. Algorithm NEWTRANS is somewhat longer than algorithm TRANS, which corresponds to the matrix method of storage, and NEW-TRANS also uses the function EDGE, but it is much more like the human method of determining transitivity of R. Moreover, NEWTRANS may be more efficient.

Let us analyze the average number of steps that algorithm NEWTRANS takes to test for transitivity. Each of the P edges begins at a unique vertex, so, on the average, $P/N = D$ edges begin at a vertex. It is not hard to see that a function EDGE, such as needed above, can be made to take an average of about D steps, since it must check all edges beginning at a particular vertex. The main FOR loop of NEWTRANS will be executed N times, and each subordinate WHILE statement will average about D executions. Since the last WHILE calls EDGE each time, we see that the entire algorithm will average about ND^3 execution steps. As before, we suppose that $P = kN^2$ with $0 \le k \le 1$. Then NEWTRANS averages about

$$T_L = N\left(\frac{kN^2}{N}\right)^3 = k^3 N^4 \text{ steps}$$

Recall that algorithm TRANS, using matrix storage, required about $T_A = kN^3 + (1-k)N^2$ steps.

Consider now the ratio T_L/T_A of steps needed with linked storage versus steps needed with matrix storage, to test R for transitivity. Thus

$$\frac{T_L}{T_A} = \frac{k^3 N^4}{kN^3 + (1-k)N^2} = \frac{k^2 N}{1 + \left(\dfrac{1}{k} - 1\right)\dfrac{1}{N}}$$

When k is close to 1, that is, when there are many edges, then T_L/T_A is nearly N, so $T_L = T_A N$, and the linked list method averages N times as many steps as the matrix storage method. Thus the matrix storage method is N times faster than the linked list method.

On the other hand, if k is very small, then T_L/T_A may be nearly zero. This means that if the number of edges is small compared with N^2, it is considerably more efficient to test for transitivity in a linked list storage method than with adjacency matrix storage.

We have, of course, made some oversimplifications. All steps do not take the same time to execute and each algorithm to test for transitivity may be shortened by aborting the search when the first counterexample to transitivity is discovered. In spite of this, the conclusions remain true and illustrate the important point that the choice of a data structure to represent objects such as sets, relations, and digraphs has an important effect on the efficiency with which information about the objects may be extracted.

Virtually all relations and digraphs, that are of practical importance, are too large to be explored by hand. Thus the computer storage of relations and the algorithmic implementation of methods for exploring them are of great importance.

EXERCISE SET 2.5

1. Verify that the linked-list arrangement of Fig. 7 correctly describes the digraph of Fig. 4.

2. Construct a function EDGE(I, J) (in pseudocode), which has the value T (true) if the pair (i, j) is in R and F (false) otherwise. Assume that the relation R is given by arrays VERT, TAIL, HEAD, and NEXT, as described in this section.

3. Show that the function EDGE of Exercise 2 runs in an average of D steps, where $D = P/N$, P is the number of edges of R, and N is the number of vertices of R. (*Hint*: Let P_{ij} be the number of edges running from vertex i to vertex j. Express the total number of steps executed by EDGE for each pair of vertices, then average. Use the fact that $\sum_{i=1, j=1} P_{ij} = P$.)

4. Let NUM be a linear array holding N positive integers, and let NEXT be a linear array of the same length. Suppose that START is a pointer to a "first" integer in NUM, and for each I, NEXT[I] points to the "next" integer in NUM to be considered. If NEXT[I] = 0, the list ends.

Write a function LOOK(NUM, NEXT, START, N, K) in pseudocode to search NUM using the pointers in NEXT, for an integer K. If K is found, the position of K in NUM is returned. If not, LOOK prints "NOT FOUND." You may use a function PRINT, as in the text, for messages.

5. Let $A = \{1, 2, 3, 4\}$ and let $R = \{(1, 1), (1, 2), (1, 3), (2, 3), (2, 4), (3, 1), (3, 4), (4, 2)\}$ be a relation on A. Compute both the matrix $\mathbf{M}_R$ giving the representation of R, and the values of arrays VERT, TAIL, HEAD, and NEXT describing R as a linked list. You may link in any reasonable way.

6. Let $A = \{1, 2, 3, 4\}$ and let R be the relation whose digraph is shown in Fig. 9. Describe arrays VERT, TAIL, HEAD, and NEXT, setting up a linked list representation of R, so that the edges out of each vertex are reached in the list in increasing order (relative to their numbering in Fig. 9).

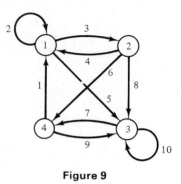

Figure 9

7. Consider the following arrays.

$$\text{VERT} = [1, 2, 6, 4]$$
$$\text{TAIL} = [1, 2, 2, 4, 4, 3, 4, 1]$$
$$\text{HEAD} = [2, 2, 3, 3, 4, 4, 1, 3]$$
$$\text{NEXT} = [8, 3, 0, 5, 7, 0, 0, 0]$$

These describe a relation R on the set $A = \{1, 2, 3, 4\}$. Compute both the digraph of R and the matrix $\mathbf{M}_R$.

8. Let $A = \{1, 2, 3, 4, 5\}$ and let R be a relation on A such that

$$\mathbf{M}_R = \begin{bmatrix} 1 & 0 & 0 & 1 & 0 \\ 0 & 1 & 1 & 0 & 0 \\ 0 & 0 & 0 & 1 & 0 \\ 1 & 0 & 1 & 0 & 1 \\ 0 & 1 & 0 & 0 & 1 \end{bmatrix}$$

Construct a linked list representation, VERT, TAIL, HEAD, NEXT, for the relation R.

2.6 Manipulation of Relations

Just as we can manipulate numbers and formulas using the rules of algebra, we can also define operations that allow us to manipulate relations. With these operations we can change, combine, and refine existing relations to produce new ones.

Let R and S be relations from a set A to a set B. Then if we remember that R and S are simply subsets of $A \times B$, we can use set operations on R and S. For example, the complement of R, $\bar{R}$ is referred to as the **complementary relation**. It is, of course, a relation from A to B which can be expressed simply in terms of R:

$$a \, \bar{R} \, b \quad \text{if and only if} \quad a \, \mathcal{R} \, b$$

We can also form the intersection $R \cap S$ and the union $R \cup S$ of the relations R and S. In relational terms, we see that $a \, (R \cap S) \, b$ means that $a \, R \, b$ and $a \, S \, b$, $a \, (R \cup S) \, b$ means that $a \, R \, b$ or $a \, S \, b$. All of our set-theoretic operations can be used in this way to produce new relations. The reader should try to give a relational description of the relation $R \oplus S$ (see Section 1.3).

A different type of operation on a relation R from A to B is the formation of the **inverse**, usually written R^{-1}. The relation R^{-1} is a relation from B to A (reverse order from R), defined by

$$b \, R^{-1} \, a \quad \text{if and only if} \quad a \, R \, b$$

It is not hard to see that $\text{Dom}\,(R^{-1}) = \text{Ran}\,(R)$ and $\text{Ran}\,(R^{-1}) = \text{Dom}\,(R)$. We leave these simple facts as exercises.

Observe that the relation R^{-1} consists of all ordered pairs in R, written in reverse order. It is clear from this observation that $(R^{-1})^{-1} = R$.

Example 1   Let $A = \{1, 2, 3, 4\}$ and $B = \{a, b, c\}$. Let

$$R = \{(1, a), (1, b), (2, b) \, (2, c), (3, b), (4, a)\}$$
$$S = \{(1, b), (2, c), (3, b), (4, b)\}$$

Compute: (a) $\bar{R}$; (b) $R \cap S$; (c) $R \cup S$; (d) R^{-1}.

Solution.
(a) We first find

$$A \times B = \{(1, a), (1, b), (1, c), (2, a), (2, b), (2, c), (3, a), (3, b), (3, c), (4, a), (4, b), (4, c)\}$$

Then the complement of R in $A \times B$ is

$$\bar{R} = \{(1, c), (2, a), (3, a), (3, c), (4, b), (4, c)\}$$

(b) We have

$$R \cap S = \{(1, b), (3, b), (2, c)\}$$

(c) We have

$$R \cup S = \{(1, a), (1, b), (2, b), (2, c), (3, b), (4, a), (4, b)\}$$

(d) Since $(x, y) \in R^{-1}$ if and only if $(y, x) \in R$, we have

$$R^{-1} = \{(a, 1), (b, 1), (b, 2), (c, 2), (b, 3), (a, 4)\}$$

Example 2   Let $A = B = \mathbb{R}$. Let R be the relation " $\leq$ " and let S be " $\geq$." Then the complement of R is the relation " $>$," since $a \nleq b$ means that $a > b$. Similarly, the complement of S is " $<$." On the other hand, $R^{-1} = S$, since for any numbers a and b,

$$a \, R^{-1} \, b \quad \text{if and only if} \quad b \, R \, a \quad \text{if and only if} \quad b \leq a \quad \text{if and only if} \quad a \geq b$$

Similarly, we have $S^{-1} = R$. Also, we note that $R \cap S$ is the relation of equality, since $a \, (R \cap S) \, b$ if and only if $a \leq b$ and $b \leq a$ if and only if $a = b$. Since, for any a and b, $a \leq b$ or $b \leq a$ must hold, we see that $R \cup S = A \times B$, that is $R \cup S$ is the "universal" relation in which any a is related to any b.

Example 3   Let $A = \{a, b, c, d, e\}$ and let R and S be two relations on A whose corresponding digraphs are shown in Fig. 1. Then the reader can verify the following facts:

$$\bar{R} = \{(a, a), (b, b), (a, c), (b, a), (c, b), (c, d), (c, e), (c, a), (d, b), (d, a),$$
$$(d, e), (e, b), (e, a), (e, d), (e, c)\}$$
$$R^{-1} = \{(b, a), (e, b), (c, c), (c, d), (d, d), (d, b), (c, b), (d, a), (e, e), (e, a)\}$$
$$R \cap S = \{(a, b), (b, e), (c, c)\}$$

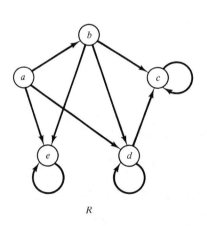

R

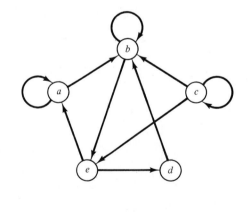

S

Figure 1

Example 4   Let $A = \{1, 2, 3\}$ and let R and S be relations on A. Suppose that the matrices of R and S are

$$\mathbf{M}_R = \begin{bmatrix} 1 & 0 & 1 \\ 0 & 1 & 1 \\ 0 & 0 & 0 \end{bmatrix} \quad \text{and} \quad \mathbf{M}_S = \begin{bmatrix} 0 & 1 & 1 \\ 1 & 1 & 0 \\ 0 & 1 & 0 \end{bmatrix}$$

Then we can verify that

$$\mathbf{M}_{\bar{R}} = \begin{bmatrix} 0 & 1 & 0 \\ 1 & 0 & 0 \\ 1 & 1 & 1 \end{bmatrix}, \quad \mathbf{M}_{R^{-1}} = \begin{bmatrix} 1 & 0 & 0 \\ 0 & 1 & 0 \\ 1 & 1 & 0 \end{bmatrix}$$

$$\mathbf{M}_{R \cap S} = \begin{bmatrix} 0 & 0 & 1 \\ 0 & 1 & 0 \\ 0 & 0 & 0 \end{bmatrix}, \quad \mathbf{M}_{R \cup S} = \begin{bmatrix} 1 & 1 & 1 \\ 1 & 1 & 1 \\ 0 & 1 & 0 \end{bmatrix}$$

Example 4 illustrates some general facts. Recalling the operations on Boolean matrices from Section 1.8, we can show (Exercise 36) that if R and S are relations on set A, then

$$\mathbf{M}_{R \cap S} = \mathbf{M}_R \wedge \mathbf{M}_S$$
$$\mathbf{M}_{R \cup S} = \mathbf{M}_R \vee \mathbf{M}_S$$
$$\mathbf{M}_{R^{-1}} = (\mathbf{M}_R)^T$$

Moreover, if $\mathbf{M}$ is a Boolean matrix, we define the **complement** $\overline{\mathbf{M}}$ of $\mathbf{M}$ as the matrix obtained from $\mathbf{M}$ by replacing every 1 in $\mathbf{M}$ by a 0, and every 0 by a 1. Thus if

$$\mathbf{M} = \begin{bmatrix} 1 & 0 & 0 \\ 0 & 1 & 1 \\ 1 & 0 & 0 \end{bmatrix}$$

then

$$\overline{\mathbf{M}} = \begin{bmatrix} 0 & 1 & 1 \\ 1 & 0 & 0 \\ 0 & 1 & 1 \end{bmatrix}$$

We can also show (Exercise 36) that if R is a relation on a set A, then

$$\mathbf{M}_{\bar{R}} = \overline{\mathbf{M}}_R$$

We know that a symmetric relation is a relation R such that $\mathbf{M}_R = (\mathbf{M}_R)^T$, and since $(\mathbf{M}_R)^T = \mathbf{M}_{R^{-1}}$, we see that R is symmetric if and only if $R = R^{-1}$.

We now prove a few useful properties about combinations of relations.

Theorem 1   Suppose that R and S are relations from A to B.

(a) If $R \subseteq S$, then $R^{-1} \subseteq S^{-1}$.

(b) If $R \subseteq S$, then $\bar{S} \subseteq \bar{R}$.

(c) $(R \cap S)^{-1} = R^{-1} \cap S^{-1}$ and $(R \cup S)^{-1} = R^{-1} \cup S^{-1}$

(d) $\overline{(R \cap S)} = \bar{R} \cup \bar{S}$ and $\overline{(R \cup S)} = \bar{R} \cap \bar{S}$

Proof.

(b) and (d) are special cases of general set properties proved in Section 1.3.

We now prove (a). Suppose that $R \subseteq S$ and let $(a, b) \in R^{-1}$. Then $(b, a) \in R$, so $(b, a) \in S$. This, in turn, implies that $(a, b) \in S^{-1}$. Since each element of R^{-1} is in S^{-1}, we are done.

We next prove (c). For the first part, suppose that $(a, b) \in (R \cap S)^{-1}$. Then $(b, a) \in R \cap S$, so $(b, a) \in R$ and $(b, a) \in S$. This means that $(a, b) \in R^{-1}$ and $(a, b) \in S^{-1}$, so $(a, b) \in R^{-1} \cap S^{-1}$. The converse containment can be proved by reversing the steps. A similar argument works to show that $(R \cup S)^{-1} = R^{-1} \cup S^{-1}$.

We will now see what effect our relational operations have on some of the properties of relations which we presented in Section 2.4.

Theorem 2   Let R and S be relations on a set A.

(a) If R is reflexive, so is R^{-1}.

(b) R is reflexive if and only if $\bar{R}$ is irreflexive.

(c) If R and S are reflexive, then so are $R \cap S$ and $R \cup S$.

Proof. Let Δ be the equality relation on A. We know that R is reflexive if and only if $\Delta \subseteq R$. Clearly, $\Delta = \Delta^{-1}$, so if $\Delta \subseteq R$, then $\Delta = \Delta^{-1} \subseteq R^{-1}$ by Theorem 1, so R^{-1} is also reflexive. This proves (a). To prove (c), we note that if $\Delta \subseteq R$ and $\Delta \subseteq S$, then $\Delta \subseteq R \cap S$ and $\Delta \subseteq R \cup S$. To show (b), we note that a relation S is irreflexive if and only if $S \cap \Delta = \varnothing$. Then R is reflexive if and only if $\Delta \subseteq R$ if and only if $\Delta \cap \bar{R} = \varnothing$ if and only if $\bar{R}$ is irreflexive.

Example 5   Let $A = \{1, 2, 3\}$ and let

$$R = \{(1, 1), (1, 2), (1, 3), (2, 2), (3, 3)\}$$

$$S = \{(1, 1), (1, 2), (2, 2), (3, 2), (3, 3)\}$$

Then
    (a) $R^{-1} = \{(1, 1), (2, 1), (3, 1), (2, 2), (3, 3)\}$, so R and R^{-1} are both reflexive.
    (b) $\bar{R} = \{(2, 1), (2, 3), (3, 1), (3, 2)\}$, which is irreflexive while R is reflexive.
    (c) $R \cap S = \{(1, 1), (1, 2), (2, 2), (3, 3)\}$ and $R \cup S = \{(1, 1), (1, 2), (1, 3), (2, 2), (3, 2), (3, 3)\}$, which are both reflexive.

Theorem 3   Let R be a relation on a set A. Then

    (a) R is symmetric if and only if $R = R^{-1}$.
    (b) R is antisymmetric if and only if $R \cap R^{-1} \subseteq \Delta$.
    (c) R is asymmetric if and only if $R \cap R^{-1} = \varnothing$.

    Proof. This theorem is straightforward, and the proof is left as an exercise.

Theorem 4   Let R and S be relations on A.

    (a) If R is symmetric, so are R^{-1} and $\bar{R}$.
    (b) If R and S are symmetric, so are $R \cap S$ and $R \cup S$.

    Proof. If R is symmetric, $R = R^{-1}$ and thus $(R^{-1})^{-1} = R = R^{-1}$, which means that R^{-1} is also symmetric. Also, $(a, b) \in (\bar{R})^{-1}$ if and only if $(b, a) \in \bar{R}$ if and only if $(b, a) \notin R$ if and only if $(a, b) \notin R^{-1} = R$ if and only if $(a, b) \in \bar{R}$, so $\bar{R}$ is symmetric and (a) is proved. The proof of (b) follows immediately from property (c) of Theorem 1.

Example 6   Let $A = \{1, 2, 3\}$ and consider the symmetric relations

$$R = \{(1, 1), (1, 2), (2, 1), (1, 3), (3, 1)\}$$
$$S = \{(1, 1), (1, 2), (2, 1), (2, 2), (3, 3)\}$$

Then
    (a) $R^{-1} = \{(1, 1), (2, 1), (1, 2), (3, 1), (1, 3)\}$, $\bar{R} = \{(2, 2), (2, 3), (3, 2), (3, 3)\}$, so R^{-1} and $\bar{R}$ are symmetric.
    (b) $R \cap S = \{(1, 1), (1, 2), (2, 1)\}$ and $R \cup S = \{(1, 1), (1, 2), (1, 3), (2, 1), (2, 2), (3, 1), (3, 3)\}$, which are both symmetric.

Theorem 5   Let R and S be relations on A.

    (a) $(R \cap S)^2 \subseteq R^2 \cap S^2$
    (b) If R and S are transitive, so is $R \cap S$.
    (c) If R and S are equivalence relations, so is $R \cap S$.

Proof. We prove (a) geometrically. We have $a \, (R \cap S)^2 \, b$ if and only if there is a path of length 2 from a to b in $R \cap S$. Then both edges of this path lie in R and in S, so $a \, R^2 \, b$ and $a \, S^2 \, b$, which implies that $a \, (R^2 \cap S^2) \, b$. To show (b), recall from Section 2.4 that a relation T is transitive if and only if $T^2 \subseteq T$. If R and S are transitive, then $R^2 \subseteq R$, $S^2 \subseteq S$, so $(R \cap S)^2 \subseteq R^2 \cap S^2$ [by part (a)] $\subseteq R \cap S$ so $R \cap S$ is transitive. We next prove (c). Relations R and S are each reflexive, symmetric, and transitive. The same properties hold for $R \cap S$ from Theorems 2(c), 4(b), and 5(b), respectively. Hence $R \cap S$ is an equivalence relation.

Example 7   Let R and S be equivalence relations on a finite set A, and let π_R and π_S be the corresponding partitions. We describe the partition corresponding to $R \cap S$ as follows. Let $\pi_R = \{A_1, A_2, \ldots, A_n\}$, $\pi_S = \{B_1, B_2, \ldots, B_p\}$, and $\pi_{R \cap S} = \{C_1, C_2, \ldots, C_m\}$. Suppose that a and b are two elements in a fixed set C_k. Then $a \, (R \cap S) \, b$, so $a \, R \, b$ and $a \, S \, b$. This means that a and b belong to the same set A_i and the same set B_j, thus a and b belong to $A_i \cap B_j$. Since the steps in this argument are reversible, we have $A_i \cap B_j = C_k$. Thus we get the partition corresponding to $R \cap S$ by taking all possible intersections of members of π_R with members of π_S. That is, $\pi_{R \cap S} = \{A_i \cap B_j \mid i = 1, \ldots, n, \quad j = 1, \ldots, p\}$.

We can generalize this example to infinite sets, and the results are analogous.

Closures

If R is a relation on a set A, it may well happen that R lacks some of the important relational properties discussed in Section 2.4, especially reflexivity, symmetry, and transitivity. If R does not possess a particular property, we may wish to add related pairs to R until we get a relation that *does* have the required property. Naturally we want to add as few new pairs as possible, so what we need to find is the *smallest* relation R_1 on A that contains R and possesses the property we desire. Sometimes R_1 does not exist. If a relation such as R_1 does exist, we call it the **closure** of R with respect to the property in question.

Example 8   Suppose that R is a relation on a set A, and R is not reflexive. This can only occur because some pairs of the diagonal relation Δ are not in R. Thus $R_1 = R \cup \Delta$ is the smallest reflexive relation on A, containing R; that is, the **reflexive closure** of R is $R \cup \Delta$.

Example 9   Suppose now that R is a relation on A which is not symmetric. Then there must exist pairs (x, y) in R such that (y, x) is not in R. Of course, $(y, x) \in R^{-1}$, so if R is to be symmetric we must add all pairs from R^{-1}; that is, we must enlarge R to

$R \cup R^{-1}$. Clearly, $(R \cup R^{-1})^{-1} = R \cup R^{-1}$, so $R \cup R^{-1}$ is the smallest symmetric relation containing R; that is, $R \cup R^{-1}$ is the **symmetric closure** of R.

If $A = \{a, b, c, d\}$ and $R = \{(a, b), (b, c), (a, c), (c, d)\}$, then $R^{-1} = \{(b, a), (c, b), (c, a), (d, c)\}$, so the symmetric closure of R is

$$R \cup R^{-1} = \{(a, b), (b, a), (b, c), (c, b), (a, c), (c, a), (c, d), (d, c)\}$$

The symmetric closure of a relation R is very easy to visualize geometrically. All edges in the digraph of R become "two-way streets" in $R \cup R^{-1}$. Thus the graph of the symmetric closure of R is simply the digraph of R with all edges made bidirectional. We show in Fig. 2(a) the digraph of the relation R of Example 9. Figure 2(b) shows the graph of the symmetric closure $R \cup R^{-1}$.

The transitive closure of a relation will be discussed in the next section.

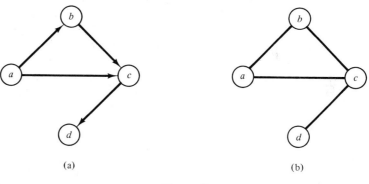

(a)   (b)

Figure 2

Composition

Now suppose that A, B, and C are sets, R is a relation from A to B, and S is a relation from B to C. We can then define a new relation, the **composition** of R and S, written $R \circ S$. The relation $R \circ S$ is a relation from A to C, and is defined as follows. If a is in A and c is in C, then $a(R \circ S)c$ if and only if for some b in B, we have $a R b$ and $b S c$. In other words, a is related to c by $R \circ S$ if we can get from a to c in two stages; first to an intermediate vertex b by relation R, and then from b to c by relation S.

Example 10   Let $A = \{1, 2, 3, 4\}$, $R = \{(1, 2), (1, 1), (1, 3), (2, 4), (3, 2)\}$, and $S = \{(1, 4), (1, 3), (2, 3), (3, 1), (4, 1)\}$. Since $(1, 2) \in R$ and $(2, 3) \in S$, we must have $(1, 3) \in R \circ S$. Similarly, since $(1, 1) \in R$ and $(1, 4) \in S$, we see that $(1, 4) \in R \circ S$. Proceeding in this way, we find that $R \circ S = \{(1, 4), (1, 3), (1, 1), (2, 1), (3, 3)\}$.

Example 11   Let $A = \{a, b, c\}$, and let R and S be relations on A whose matrices are

$$\mathbf{M}_R = \begin{bmatrix} 1 & 0 & 1 \\ 1 & 1 & 1 \\ 0 & 1 & 0 \end{bmatrix}, \quad \mathbf{M}_S = \begin{bmatrix} 1 & 0 & 0 \\ 0 & 1 & 1 \\ 1 & 0 & 1 \end{bmatrix}$$

We see from the matrices that

$$(a, a) \in R \quad \text{and} \quad (a, a) \in S, \quad \text{so} \quad (a, a) \in R \circ S$$
$$(a, c) \in R \quad \text{and} \quad (c, a) \in S, \quad \text{so} \quad (a, a) \in R \circ S$$
$$(a, c) \in R \quad \text{and} \quad (c, c) \in S, \quad \text{so} \quad (a, c) \in R \circ S$$

It is easily seen that $(a, b) \notin R \circ S$, since if we had $(a, x) \in R$ and $(x, b) \in S$, matrix $\mathbf{M}_R$ tells us that x would have to be a or c, but matrix $\mathbf{M}_S$ tells us that neither (a, b) nor (c, b) is an element of S.

We see that the first row of $\mathbf{M}_{R \circ S}$ is 1   0   1. The reader may show by similar analysis that

$$\mathbf{M}_{R \circ S} = \begin{bmatrix} 1 & 0 & 1 \\ 1 & 1 & 1 \\ 0 & 1 & 1 \end{bmatrix}$$

We note that $\mathbf{M}_{R \circ S} = \mathbf{M}_R \odot \mathbf{M}_S$ (verify).

Example 11 illustrates a general and useful fact. Let A, B, and C be finite sets with n, p, and m elements, respectively, let R be a relation from A to B, and let S be a relation from B to C. Then R and S have Boolean matrices $\mathbf{M}_R$ and $\mathbf{M}_S$ with respective sizes $n \times p$ and $p \times m$. Then $\mathbf{M}_R \odot \mathbf{M}_S$ can be computed, and it equals $\mathbf{M}_{R \circ S}$.

To see this, let $A = \{a_1, \ldots, a_n\}$, $B = \{b_1, \ldots, b_p\}$, and $C = \{c_1, \ldots, c_m\}$. Also suppose that $\mathbf{M}_R = [r_{ij}]$, $\mathbf{M}_S = [s_{ij}]$, and $\mathbf{M}_{R \circ S} = [t_{ij}]$. Then $t_{ij} = 1$ if and only if $(a_i, a_j) \in R \circ S$, which means that for some k, $(a_i, a_k) \in R$ and $(a_k, a_j) \in S$. In other words, $r_{ik} = 1$ and $s_{kj} = 1$ for some k between 1 and p. This condition is identical to the condition needed for $\mathbf{M}_R \odot \mathbf{M}_S$ to have a 1 in position i, j, and thus $\mathbf{M}_{R \circ S}$ and $\mathbf{M}_R \odot \mathbf{M}_S$ are equal.

In the special case where R and S are equal, we have $R \circ S = R^2$ and $\mathbf{M}_{R \circ S} = \mathbf{M}_{R^2} = \mathbf{M}_R \odot \mathbf{M}_R$, as was shown in Section 2.3.

Example 12   Let us redo Example 10 using matrices. We see that

$$\mathbf{M}_R = \begin{bmatrix} 1 & 1 & 1 & 0 \\ 0 & 0 & 0 & 1 \\ 0 & 1 & 0 & 0 \\ 0 & 0 & 0 & 0 \end{bmatrix} \quad \text{and} \quad \mathbf{M}_S = \begin{bmatrix} 0 & 0 & 1 & 1 \\ 0 & 0 & 1 & 0 \\ 1 & 0 & 0 & 0 \\ 1 & 0 & 0 & 0 \end{bmatrix}$$

Then

$$\mathbf{M}_R \odot \mathbf{M}_S = \begin{bmatrix} 1 & 0 & 1 & 1 \\ 1 & 0 & 0 & 0 \\ 0 & 0 & 1 & 0 \\ 0 & 0 & 0 & 0 \end{bmatrix}$$

so

$$R \circ S = \{(1,\,1),\,(1,\,3),\,(1,\,4),\,(2,\,1),\,(3,\,3)\}$$

as we found before. In cases where the number of pairs in R and S is large, the matrix method is much more reliable and systematic.

Theorem 6   Let A, B, C, and D be sets, R a relation from A to B, S a relation from B to C, and T a relation from C to D. Then

$$(R \circ S) \circ T = R \circ (S \circ T)$$

Proof. The relations R, S, and T are determined by their Boolean matrices $\mathbf{M}_R$, $\mathbf{M}_S$, and $\mathbf{M}_T$, respectively. As we showed after Example 11, the matrix of the composition is the Boolean matrix product, that is, $\mathbf{M}_{R \circ S} = \mathbf{M}_R \odot \mathbf{M}_S$. Thus

$$\mathbf{M}_{(R \circ S) \circ T} = \mathbf{M}_{R \circ S} \odot \mathbf{M}_T = (\mathbf{M}_R \odot \mathbf{M}_S) \odot \mathbf{M}_T$$

Similarly,

$$\mathbf{M}_{R \circ (S \circ T)} = \mathbf{M}_R \odot (\mathbf{M}_S \odot \mathbf{M}_T)$$

Since matrix multiplication is associative [see Exercise 33(c) in Section 1.8] we must have

$$(\mathbf{M}_R \odot \mathbf{M}_S) \odot \mathbf{M}_T = \mathbf{M}_R \odot (\mathbf{M}_S \odot \mathbf{M}_T)$$

and therefore

$$\mathbf{M}_{(R \circ S) \circ T} = \mathbf{M}_{R \circ (S \circ T)}$$

Then

$$(R \circ S) \circ T = R \circ (S \circ T)$$

since these relations have the same matrices.

In general, $R \circ S \neq S \circ R$, as shown in the following example.

Example 13   Let $A = \{a, b\}$, $R = \{(a, a), (b, a), (b, b)\}$, and $S = \{(a, b), (b, a), (b, b)\}$. Then $R \circ S = \{(a, b), (b, a), (b, b)\}$ while $S \circ R = \{(a, a), (a, b), (b, a), (b, b)\}$.

Theorem 7   Let A, B, and C be sets, R a relation from A to B, and S a relation from B to C. Then $(R \circ S)^{-1} = S^{-1} \circ R^{-1}$.

Proof. Let $c \in C$ and $a \in A$. Then $(c, a) \in (R \circ S)^{-1}$ if and only if $(a, c) \in R \circ S$, that is, if and only if there is a $b \in B$ with $(a, b) \in R$ and $(b, c) \in S$. Finally, this is equivalent with the statement that $(c, b) \in S^{-1}$ and $(b, a) \in R^{-1}$, that is, $(c, a) \in S^{-1} \circ R^{-1}$.

EXERCISE SET 2.6

In Exercises 1–4, let R and S be the given relations from A to B. Compute (a) $\bar{R}$; (b) $R \cup S$; (c) $R \cap S$; (d) S^{-1}.

1. $A = \{1, 2, 3, 4\}$; $B = \{1, 2, 3\}$,
   $R = \{(1, 1), (1, 2), (2, 1), (2, 3), (3, 2), (4, 3)\}$;
   $S = \{(1, 3), (2, 3), (3, 2), (3, 3)\}$

2. $A = \{a, b, c\}$; $B = \{a, b\}$;
   $R = \{(a, a), (a, b), (b, a), (c, b)\}$; $S = \{(b, b), (c, a)\}$

3. $A = B = \{1, 2, 3\}$; $R = \{(1, 1), (1, 2), (2, 3), (3, 1)\}$;
   $S = \{(2, 1), (3, 1), (3, 2), (3, 3)\}$

4. $A = \{a, b, c\}$; $B = \{1, 2, 3\}$; $R = \{(a, 1), (b, 1), (c, 2), (c, 3)\}$; $S = \{(a, 1), (a, 2), (b, 1), (b, 2)\}$

In Exercises 5 and 6, let R and S be two relations whose corresponding digraphs are shown. Compute (a) $\bar{R}$; (b) $R \cap S$; (c) $R \cup S$; (d) S^{-1}.

5.

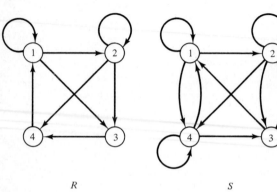

R $\qquad\qquad$ S

6.

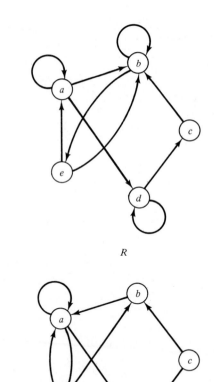

R

S

In Exercises 7 and 8, let $A = \{1, 2, 3\}$ and $B = \{1, 2, 3, 4\}$. Let R and S be the relations from A to B whose matrices are given. Compute (a) $\bar{S}$; (b) $R \cap S$; (c) $R \cup S$; (d) R^{-1}.

7. $\mathbf{M}_R = \begin{bmatrix} 1 & 1 & 0 & 1 \\ 0 & 0 & 0 & 1 \\ 1 & 1 & 1 & 0 \end{bmatrix}$, $\mathbf{M}_S = \begin{bmatrix} 0 & 1 & 1 & 0 \\ 1 & 0 & 0 & 1 \\ 1 & 1 & 0 & 0 \end{bmatrix}$

8. $\mathbf{M}_R = \begin{bmatrix} 1 & 0 & 1 & 0 \\ 0 & 0 & 0 & 1 \\ 1 & 1 & 1 & 0 \end{bmatrix}$, $\mathbf{M}_S = \begin{bmatrix} 1 & 1 & 1 & 1 \\ 0 & 0 & 0 & 1 \\ 0 & 1 & 0 & 1 \end{bmatrix}$

9. Let $A = B = \{1, 2, 3, 4\}$, and let R be the relation on A given by $R = \{(1, 3), (2, 1), (2, 3), (3, 1), (3, 2), (3, 4), (4, 2)\}$. Compute (a) R^{-1}; (b) the digraph of R^{-1}; (c) the matrix of $\bar{R}$.

In Exercises 10 and 11, let $A = \{1, 2, 3, 4\}$ and $B = \{1, 2, 3\}$. Given the matrices $\mathbf{M}_R$ and $\mathbf{M}_S$ of the relations R and S from A to B, compute (a) $\mathbf{M}_{R \cap S}$; (b) $\mathbf{M}_{R \cup S}$; (c) $\mathbf{M}_{R^{-1}}$; (d) $\mathbf{M}_{\bar{S}}$.

10. $\mathbf{M}_R = \begin{bmatrix} 1 & 0 & 1 \\ 0 & 1 & 1 \\ 0 & 1 & 0 \\ 1 & 0 & 1 \end{bmatrix}$, $\mathbf{M}_S = \begin{bmatrix} 0 & 1 & 0 \\ 1 & 0 & 1 \\ 1 & 0 & 1 \\ 1 & 1 & 1 \end{bmatrix}$

11. $\mathbf{M}_R = \begin{bmatrix} 0 & 1 & 0 \\ 0 & 1 & 1 \\ 0 & 0 & 1 \\ 1 & 1 & 1 \end{bmatrix}$, $\mathbf{M}_S = \begin{bmatrix} 1 & 0 & 1 \\ 1 & 0 & 1 \\ 0 & 1 & 0 \\ 0 & 1 & 0 \end{bmatrix}$

12. Let $A = B = \{1, 2, 3, 4\}$, $R = \{(1, 1), (1, 3), (2, 3), (3, 1), (4, 2), (4, 4)\}$, and $S = \{(1, 2), (2, 3), (3, 1), (3, 2), (4, 3)\}$. Compute (a) $\mathbf{M}_{R \cap S}$; (b) $\mathbf{M}_{R \cup S}$; (c) $\mathbf{M}_{R^{-1}}$; (d) $\mathbf{M}_{\bar{S}}$.

13. Let $A = B = \{1, 2, 3\}$, $R = \{(1, 2), (2, 3), (3, 2)\}$, $S = \{(1, 2), (2, 3), (3, 1), (3, 2), (3, 3)\}$, and $T = \{(2, 1), (2, 3), (3, 2), (3, 3)\}$.
   (a) Verify (a) of Theorem 1 with R and S.
   (b) Verify (b) of Theorem 1 with R and S.
   (c) Verify (c) of Theorem 1 with S and T.
   (d) Verify (d) of Theorem 1 with S and T.

14. Let $A = \{1, 2, 3\}$, $R = \{(1, 1), (1, 2), (2, 1), (2, 2), (2, 3), (3, 3)\}$, and $S = \{(1, 1), (1, 2), (2, 2), (3, 1), (3, 2), (3, 3)\}$. Verify Theorem 2.

15. Let $A = \{a, b, c\}$ and $R = \{(a, a), (a, b), (b, a), (a, c), (c, a)\}$. Verify (a) of Theorem 3.

16. Let $A = \{1, 2, 3\}$, and $R = \{(1, 1), (1, 2), (2, 1), (2, 2), (2, 3), (3, 1), (1, 3), (3, 2)\}$. Verify (a) of Theorem 4.

17. Let $A = \{1, 2, 3, 4, 5, 6\}$ and

$R = \{(1, 2), (1, 1), (2, 1), (2, 2), (3, 3), (4, 4), (5, 5), (5, 6), (6, 5), (6, 6)\}$

$S = \{(1, 1), (1, 2), (1, 3), (2, 1), (2, 2), (2, 3), (3, 1), (3, 2), (3, 3), (4, 6), (4, 4), (6, 4), (6, 6), (5, 5)\}$

be equivalence relations on A. Compute the partition corresponding to $R \cap S$.

18. Let $A = \{1, 2, 3, 4, 5, 6\}$ and let R and S be the equivalence relations of Exercise 17. Compute the matrix $\mathbf{M}_{R \cup S}$.

19. Let $A = \{a, b, c, d\}$, $B = \{1, 2, 3\}$, and $C = \{\square, \triangle, \diamondsuit\}$. Let R and S be the following relations from A to B and from B to C, respectively.

$R = \{(a, 1), (a, 2), (b, 2), (b, 3), (c, 1), (d, 3), (d, 2)\}$

$S = \{(1, \square), (2, \triangle), (3, \triangle), (1, \diamondsuit)\}$

   (a) Is $(b, \triangle) \in R \circ S$?
   (b) Is $(c, \triangle) \in R \circ S$?
   (c) Compute $R \circ S$.

20. $A = B =$ the set of all real numbers. Let R be the relation "$<$" and let S be the relation "$>$." Describe (a) $R \cap S$; (b) $R \cup S$; (c) S^{-1}.

21. Let $A =$ a set of people. Let $a \, R \, b$ if and only if a and b are brothers; let $a \, S \, b$ if and only if a and b are sisters. Describe $R \cup S$.

22. Let $A =$ a set of people. Let $a \, R \, b$ if and only if a is older than b; let $a \, S \, b$ if and only if a is a brother of b. Describe $R \cap S$.

23. Let $A =$ the set of all people in the Social Security data base. Let $a \, R \, b$ if and only if a and b receive the same benefits; let $a \, S \, b$ if and only if a and b have the same last name. Describe $R \cap S$.

24. Let $A =$ a set of people. Let $a \, R \, b$ if and only if a is the father of b; let $a \, S \, b$ if and only if a is the mother of b. Describe $R \cup S$.

25. Let $A = \{2, 3, 6, 12\}$ and let R and S be the following relations on A: $x\,R\,y$ if and only if $2\,|\,(x - y)$; $x\,S\,y$ if and only if $3\,|\,(x - y)$. Compute (a) $\bar{R}$; (b) $R \cap S$; (c) $R \cup S$; (d) S^{-1}.

26. Let $A = B = C = $ the set of all real numbers. Let R and S be the following relations from A to B and from B to C, respectively:

$$R = \{(a, b)\,|\,a \leq 2b\}$$
$$S = \{(b, c)\,|\,b \leq 3c\}$$

(a) Is $(1, 5) \in R \circ S$?
(b) Is $(2, 3) \in R \circ S$?
(c) Describe $R \circ S$.

27. Let $A = \{1, 2, 3, 4\}$. Let

$$R = \{(1, 1), (1, 2), (2, 3), (2, 4), (3, 4), (4, 1), (4, 2)\}$$
$$S = \{(3, 1), (4, 4), (2, 3), (2, 4), (1, 1), (1, 4)\}$$

(a) Is $(1, 3) \in R \circ R$?
(b) Is $(4, 3) \in R \circ S$?
(c) Is $(1, 1) \in S \circ R$?
(d) Compute $R \circ R$.
(e) Compute $R \circ S$.
(f) Compute $S \circ R$.
(g) Compute $S \circ S$.

28. If R and S are equivalence relations on a set A, is $R \circ S$ an equivalence relation on A? Prove or disprove your statement.

29. Which properties of relations on a set A are preserved by composition? Prove or disprove your statement.

In Exercises 30 and 31, let $A = \{1, 2, 3, 4, 5\}$ and let $\mathbf{M}_R$ and $\mathbf{M}_S$ be the matrices of the relations R and S on A. Compute (a) $\mathbf{M}_{R \circ R}$; (b) $\mathbf{M}_{R \circ S}$; (c) $\mathbf{M}_{S \circ R}$; (d) $\mathbf{M}_{S \circ S}$.

30. $\mathbf{M}_R = \begin{bmatrix} 1 & 0 & 1 & 1 & 1 \\ 0 & 1 & 1 & 0 & 0 \\ 1 & 0 & 0 & 1 & 0 \\ 1 & 0 & 1 & 0 & 0 \\ 0 & 1 & 1 & 1 & 1 \end{bmatrix}$, $\mathbf{M}_S = \begin{bmatrix} 1 & 0 & 0 & 1 & 0 \\ 1 & 0 & 1 & 0 & 0 \\ 1 & 0 & 1 & 0 & 0 \\ 0 & 1 & 1 & 1 & 1 \\ 1 & 0 & 0 & 0 & 1 \end{bmatrix}$

31. $\mathbf{M}_R = \begin{bmatrix} 1 & 1 & 0 & 0 & 1 \\ 0 & 0 & 0 & 1 & 0 \\ 1 & 1 & 0 & 0 & 1 \\ 0 & 1 & 0 & 1 & 1 \\ 1 & 0 & 0 & 0 & 0 \end{bmatrix}$, $\mathbf{M}_S = \begin{bmatrix} 0 & 0 & 0 & 1 & 1 \\ 1 & 0 & 0 & 0 & 1 \\ 0 & 1 & 0 & 1 & 0 \\ 1 & 1 & 0 & 1 & 1 \\ 1 & 0 & 1 & 0 & 0 \end{bmatrix}$

32. Let R and S be relations on a set A. If R and S are asymmetric, either prove or disprove that $R \cap S$ and $R \cup S$ are asymmetric.

33. Let R and S be relations on a set A. If R and S are antisymmetric, either prove or disprove that $R \cap S$ and $R \cup S$ are antisymmetric.

34. Let R be a relation from A to B and let S and T be relations from B to C. Prove
(a) $R \circ (S \cup T) = (R \circ S) \cup (R \circ T)$
(b) $R \circ (S \cap T) = (R \circ S) \cap (R \circ T)$

35. Let R and S be relations from A to B and let T be a relation from B to C. Show that if $R \subseteq S$, then $R \circ T \subseteq S \circ T$.

36. Show that if R and S are relations on a set A, then
(a) $\mathbf{M}_{R \cap S} = \mathbf{M}_R \wedge \mathbf{M}_S$
(b) $\mathbf{M}_{R \cup S} = \mathbf{M}_R \vee \mathbf{M}_S$
(c) $\mathbf{M}_{R^{-1}} = (\mathbf{M}_R)^T$
(d) $\mathbf{M}_{\bar{R}} = \bar{\mathbf{M}}_R$

37. Let R and S be relations on a set A. Prove that $(R \cap S)^n \subseteq R^n \cap S^n$, for $n \geq 1$.

38. Let R be a relation from A to B. Prove
(a) Dom $(R^{-1}) = $ Ran (R)
(b) Ran $(R^{-1}) = $ Dom (R)

39. Prove Theorem 3.

2.7 Connectivity and Warshall's Algorithm

In this section we consider a construction that has several interpretations and many important applications. Suppose that R is a relation on a set A and that R is not transitive. We will show that the transitive closure of R (see Section 2.6) is just the connectivity relation R^∞, defined in Section 2.3.

Theorem 1   Let R be a relation on a set A. Then R^∞ is the transitive closure of R.

Proof. We recall that if a and b are in the set A, then $a \ R^\infty \ b$ if and only if there is a path in R from a to b. Now R^∞ is certainly transitive, since if $a \ R^\infty \ b$ and $b \ R^\infty \ c$, the composition of the paths from a to b and from b to c form a path from a to c in R, and so $a \ R^\infty \ c$. To show that R^∞ is the smallest transitive relation containing R, we must show that if S is any transitive relation on A, and $R \subseteq S$, then $R^\infty \subseteq S$. Theorem 1 of Section 2.4 tells us that if S is transitive, then $S^n \subseteq S$ for all n; that is, if a and b are connected by a path of length n, then $a \ S \ b$. It follows that $S^\infty = \bigcup_{n=1}^{\infty} S^n \subseteq S$. It is also true that if $R \subseteq S$, then $R^\infty \subseteq S^\infty$, since any path in R is also a path in S. Putting these facts together, we see that if $R \subseteq S$ and S is transitive on A, then $R^\infty \subseteq S^\infty \subseteq S$. This means that R^∞ is the smallest of all transitive relations on A that contain R.

We see that R^∞ has several interpretations. From a geometric point of view, it is called the connectivity relation, since it specifies which vertices are connected (by paths) to other vertices. If we include the relation Δ (see Section 2.4), then $R^\infty \cup \Delta$ is the reachability relation R^* (see Section 2.3) which is frequently more useful. On the other hand, from the algebraic point of view, R^∞ is the transitive closure of R, as we have shown in Theorem 1 above. In this form, it plays important roles in the theory of equivalence relations and in the theory of certain languages (see Section 5.3).

Example 1   Let $A = \{1, 2, 3, 4\}$, and let $R = \{(1, 2), (2, 3), (3, 4), (2, 1)\}$. Find the transitive closure of R.

Solution.
Method 1. The digraph of R is shown in Fig. 1. Since R^∞ is the transitive closure, we can proceed geometrically by computing all paths. We see that from vertex 1 we have paths to vertices 2, 3, 4, and 1. Note that the path from 1 to 1 proceeds from 1 to 2 to 1. Thus we see that the ordered pairs $(1, 1)$, $(1, 2)$, $(1, 3)$, and

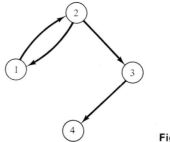

Figure 1

(1, 4) are in R^∞. Starting from vertex 2, we have paths to vertices 2, 1, 3, and 4, so the ordered pairs (2, 1), (2, 2), (2, 3), and (2, 4) are in R^∞. The only other path is from vertex 3 to vertex 4, so we have

$$R^\infty = \{(1, 1), (1, 2), (1, 3), (1, 4), (2, 1), (2, 2), (2, 3), (2, 4), (3, 4)\}$$

Method 2. The matrix of R is

$$\mathbf{M}_R = \begin{bmatrix} 0 & 1 & 0 & 0 \\ 1 & 0 & 1 & 0 \\ 0 & 0 & 0 & 1 \\ 0 & 0 & 0 & 0 \end{bmatrix}$$

We may proceed algebraically and compute the powers of $\mathbf{M}_R$. Thus

$$(\mathbf{M}_R)_\odot^2 = \begin{bmatrix} 1 & 0 & 1 & 0 \\ 0 & 1 & 0 & 1 \\ 0 & 0 & 0 & 0 \\ 0 & 0 & 0 & 0 \end{bmatrix}, \quad (\mathbf{M}_R)_\odot^3 = \begin{bmatrix} 0 & 1 & 0 & 1 \\ 1 & 0 & 1 & 0 \\ 0 & 0 & 0 & 0 \\ 0 & 0 & 0 & 0 \end{bmatrix},$$

$$(\mathbf{M}_R)_\odot^4 = \begin{bmatrix} 1 & 0 & 1 & 0 \\ 0 & 1 & 0 & 1 \\ 0 & 0 & 0 & 0 \\ 0 & 0 & 0 & 0 \end{bmatrix}$$

Continuing in this way, we can see that $(\mathbf{M}_R)_\odot^n$ equals $(\mathbf{M}_R)_\odot^2$ if n is even, and equals $(\mathbf{M}_R)_\odot^3$ if n is odd. Thus

$$\mathbf{M}_{R^\infty} = \mathbf{M}_R \vee (\mathbf{M}_R)_\odot^2 \vee (\mathbf{M}_R)_\odot^3 = \begin{bmatrix} 1 & 1 & 1 & 1 \\ 1 & 1 & 1 & 1 \\ 0 & 0 & 0 & 1 \\ 0 & 0 & 0 & 0 \end{bmatrix}$$

and this gives the same relation as with Method 1.

In Example 1 we did not need to consider all powers R^n, to obtain R^∞. This observation is true in all cases whenever the set A is finite, as we will now prove.

Theorem 2   Let A be a set with $|A| = n$, and let R be a relation on A. Then

$$R^\infty = R \cup R^2 \cup \cdots \cup R^n$$

In other words, powers of R greater than n are not needed to compute R^∞.

Proof. Let a and b be in A, and suppose that $a, x_1, x_2, \ldots, x_m, b$ is a path from a to b in R, that is, $(a, x_1), (x_1, x_2), \ldots, (x_m, b)$ are all in R. If x_i and x_j are equal, say $i < j$, then the path can be divided into three sections. First, a path from a to x_i, then a path from x_i to x_j, and finally, a path from x_j to b. The middle path is a cycle, since $x_i = x_j$, so we simply leave it out and put the first two paths together. This gives us a shorter path from a to b (see Fig. 2).

Now let $a, x_1, x_2, \ldots, x_k, b$ be the shortest path from a to b. If $a \neq b$, then all vertices $a, x_1, x_2, \ldots, x_k, b$ are distinct. Otherwise, the discussion above shows that we could find a shorter path. Thus the length of the path is at most $n - 1$ (since $|A| = n$). If $a = b$, then for similar reasons, the vertices $a, x_1, x_2, \ldots, x_k$ are distinct, so the length of the path is at most n. In other words, if $a\ R^\infty\ b$, then $a\ R^k\ b$ for some k from 1 to n. Thus $R^\infty = R \cup R^2 \cup \cdots \cup R^n$.

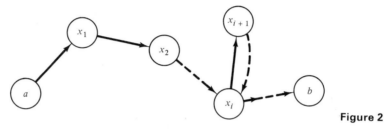

Figure 2

The methods used to solve Example 1 each have certain difficulties. The graphical method is impractical for large sets and relations, and is not systematic. The matrix method can be used in general, and is systematic enough to program for computer, but is inefficient and, for large matrices, can be prohibitively costly. Fortunately, a more efficient algorithm for computing transitive closure is available. It is known as Warshall's algorithm, after its creator, and we describe it next.

Warshall's Algorithm

Let R be a relation on a set $A = \{a_1, a_2, \ldots, a_n\}$. We introduce a sequence of matrices $\mathbf{W}_0, \mathbf{W}_1, \ldots, \mathbf{W}_n$ such that $\mathbf{W}_0 = \mathbf{M}_R$, $\mathbf{W}_n = \mathbf{M}_{R^\infty}$, and each matrix is easily computed from the previous one. These are different from the powers of $\mathbf{M}_R$, and this difference results in considerable savings of computation. In a path $x_1, x_2, \ldots, x_m$, any vertices other than the first and last (in this case $x_2, x_3, \ldots, x_{m-1}$) are called **interior** vertices. For any fixed integer k between 1 and n, we define $\mathbf{W}_k$ to be the matrix that has a 1 in position i, j if and only if there is a path in R from a_i to a_j whose interior vertices, if any, come from the set $\{a_1, a_2, \ldots, a_k\}$. In other words, no interior vertices can have subscript greater than k. Define $\mathbf{W}_0$ to be $\mathbf{M}_R$. We see that $\mathbf{W}_n = \mathbf{M}_{R^\infty}$ since any path must have interior vertices coming from $\{a_1, a_2, \ldots, a_n\} = A$. We now show how to compute each $\mathbf{W}_k$ from the matrices $\mathbf{W}_0, \mathbf{W}_1, \ldots, \mathbf{W}_{k-1}$. Let $\mathbf{W}_k = [w_{ij}^{(k)}]$; that is, $w_{ij}^{(k)}$ is the entry in the i, j position of matrix $\mathbf{W}_k$. Then we have the following result.

Theorem 3   For any $i, j, k \quad 0 \leq i, j, k \leq n$.

$$w_{ij}^{(k)} = w_{ij}^{(k-1)} \vee (w_{ik}^{(k-1)} \wedge w_{kj}^{(k-1)}) \tag{1}$$

Proof. In more detail, the theorem states that $w_{ij}^{(k)} = 1$ if and only if either $w_{ij}^{(k-1)} = 1$ or both $w_{ik}^{(k-1)}$ and $w_{kj}^{(k-1)}$ are 1.

To see this, consider a path which goes from a_i to a_j and has all interior vertices from the set $\{a_1, \ldots, a_k\}$. The existence of such a path is equivalent to the statement that $w_{ij}^{(k)} = 1$. Also, we may assume, as in the proof of Theorem 2, that all vertices of the path are distinct, except possibly the first and last if $a_i = a_j$. Only two possibilities exist. Either this path does not have a_k as an interior vertex or it does. If it does not, the interior vertices are from the set $\{a_1, \ldots, a_{k-1}\}$, so we must have $w_{ij}^{(k-1)} = 1$. If a_k is an interior vertex, the situation is as shown in Fig. 3. Subpath 1 goes from a_i to a_k and the interior vertices for subpath 1 come from the set $\{a_1, \ldots, a_{k-1}\}$. This is so because they are also interior vertices for the original path and so come from the set $\{a_1, \ldots, a_k\}$, but a_k appears only once, as shown. The existence of subpath 1 shows that $w_{ik}^{(k-1)} = 1$, and similarly, the existence of subpath 2 tells us that $w_{kj}^{(k-1)} = 1$.

Conversely, if $w_{ij}^{(k-1)} = 1$ or both $w_{ik}^{(k-1)}$ and $w_{kj}^{(k-1)}$ are 1, then $w_{ij}^{(k)} = 1$. This completes the proof.

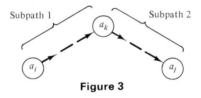

Figure 3

Theorem 3 is the basis for Warshall's algorithm. We begin with $\mathbf{W}_0 = \mathbf{M}_R$, and use equation (1) to compute each $\mathbf{W}_k$ from $\mathbf{W}_{k-1}$, until we reach $\mathbf{W}_n = \mathbf{M}_{R^\infty}$. Equation (1) shows that if $w_{ij}^{(k-1)} = 1$, then $w_{ij}^{(k)} = 1$. This means that every entry that is a 1 in matrix $\mathbf{W}_{k-1}$ remains a 1 in matrix $\mathbf{W}_k$. Equation (1) also tells us that we get a new 1 in position i, j of $\mathbf{W}_k$ (that is, an entry of 1 that was not in $\mathbf{W}_{k-1}$) only if there were 1's in positions i, k and k, j of $\mathbf{W}_{k-1}$. Thus we may examine column k and row k of $\mathbf{W}_{k-1}$, and if there is a 1 in location i of column k and a 1 in location j of row k, a 1 will be entered in position i, j of $\mathbf{W}_k$. This observation leads to the following procedure for computing $\mathbf{W}_k$ from $\mathbf{W}_{k-1}$.

Step 1. First transfer to $\mathbf{W}_k$ all 1's in $\mathbf{W}_{k-1}$.

Step 2. List the locations $p_1, p_2, \ldots$, in column k of $\mathbf{W}_{k-1}$, where the entry is 1, and the locations $q_1, q_2, \ldots$, in row k of $\mathbf{W}_{k-1}$, where the entry is 1.

Step 3. Put 1's in all the positions p_i, q_j of $\mathbf{W}_k$ (if they are not already there.)

Example 2   Consider the relation R defined in Example 1. Then

$$\mathbf{W}_0 = \mathbf{M}_R = \begin{bmatrix} 0 & 1 & 0 & 0 \\ 1 & 0 & 1 & 0 \\ 0 & 0 & 0 & 1 \\ 0 & 0 & 0 & 0 \end{bmatrix}$$

and $n = 4$.

First we find $\mathbf{W}_1$, so that $k = 1$. $\mathbf{W}_0$ has 1's in location 2 of column 1 and location 2 of row 1. Thus $\mathbf{W}_1$ is just $\mathbf{W}_0$ with a new 1 in position 2,2.

$$\mathbf{W}_1 = \begin{bmatrix} 0 & 1 & 0 & 0 \\ 1 & 1 & 1 & 0 \\ 0 & 0 & 0 & 1 \\ 0 & 0 & 0 & 0 \end{bmatrix}$$

Now we compute $\mathbf{W}_2$, so that $k = 2$. We must consult column 2 and row 2 of $\mathbf{W}_1$. Matrix $\mathbf{W}_1$ has 1's in locations 1 and 2 of column 2 and locations 1, 2, and 3 of row 2.

Thus to obtain $\mathbf{W}_2$, we must put 1's in positions 1,1, 1,2, 1,3, 2,1, 2,2, and 2,3 of matrix $\mathbf{W}_1$ (if 1's are not already there). We see that

$$\mathbf{W}_2 = \begin{bmatrix} 1 & 1 & 1 & 0 \\ 1 & 1 & 1 & 0 \\ 0 & 0 & 0 & 1 \\ 0 & 0 & 0 & 0 \end{bmatrix}$$

Proceeding, we see that column 3 of $\mathbf{W}_2$ has 1's in locations 1 and 2, and row 3 of $\mathbf{W}_2$ has a 1 in location 4. To obtain $\mathbf{W}_3$ we must put 1's in positions 1,4 and 2,4 of $\mathbf{W}_2$, so

$$\mathbf{W}_3 = \begin{bmatrix} 1 & 1 & 1 & 1 \\ 1 & 1 & 1 & 1 \\ 0 & 0 & 0 & 1 \\ 0 & 0 & 0 & 0 \end{bmatrix}$$

Finally, $\mathbf{W}_3$ has 1's in locations 1, 2, 3 of column 4 and no 1's in row 4, so no new 1's are added and $\mathbf{M}_{R^\infty} = \mathbf{W}_4 = \mathbf{W}_3$. Thus we have obtained the same result as in Example 1.

The procedure illustrated in Example 2 results in the following algorithm for computing the matrix CLOSURE of the transitive closure of the $N \times N$ matrix MAT representing a relation R.

ALGORITHM WARSHALL
1. CLOSURE $\leftarrow$ MAT
2. FOR $K = 1$ THRU N
   a. FOR $I = 1$ THRU N
      1. FOR $J = 1$ THRU N
         a. CLOSURE$[I, J] \leftarrow$ CLOSURE$[I, J]$
                    $\vee$ (CLOSURE$[I, K] \wedge$ CLOSURE$[K, J]$)
END OF ALGORITHM WARSHALL

The algorithm above was set up to proceed exactly as we have outlined previously. With some slight rearrangement of the steps, it can be made a little more efficient. If we think of the testing and assignment line as one step, then algorithm WARSHALL requires n^3 steps in all. The Boolean product of two $n \times n$ Boolean matrices $\mathbf{A}$ and $\mathbf{B}$ also requires n^3 steps, since we must compute n^2 entries, and each of these requires n comparisons. Thus in order to compute all products $(\mathbf{M}_R)^2_\odot$, $(\mathbf{M}_R)^3_\odot$, ..., $(\mathbf{M}_R)^n_\odot$, we require $n^3(n-1)$ steps, since we will need $n-1$ matrix multiplications. The formula

$$\mathbf{M}_{R^\infty} = \mathbf{M}_R \vee (\mathbf{M}_R)^2_\odot \vee \cdots \vee (\mathbf{M}_R)^n_\odot \tag{2}$$

if implemented directly, would require about n^4 steps without the final joins. Thus Warshall's algorithm is a significant improvement over direct computation of $\mathbf{M}_{R^\infty}$ using the formula given in (2).

An interesting application of the transitive closure is to equivalence relations. We showed in Section 2.6 that if R and S are equivalence relations on a set A, then $R \cap S$ is also an equivalence relation on A. The relation $R \cap S$ is the largest equivalence relation contained in both R and S, since it is the largest subset of $A \times A$ contained in both R and S. We would like to know the smallest equivalence relation that contains both R and S. The natural candidate is $R \cup S$, but this relation is not necessarily transitive. The solution is given in the next theorem.

Theorem 4   If R and S are equivalence relations on a set A, then the smallest equivalence relation containing both R and S is $(R \cup S)^\infty$.

Proof. Recall that Δ is the relation of equality on A, and that a relation is reflexive if and only if it contains Δ. Then $\Delta \subseteq R$, $\Delta \subseteq S$ since both are reflexive, so $\Delta \subseteq R \cup S \subseteq (R \cup S)^\infty$, and $(R \cup S)^\infty$ is also reflexive.

Since R and S are symmetric, $R = R^{-1}$ and $S = S^{-1}$, so $(R \cup S)^{-1} = R^{-1} \cup S^{-1} = R \cup S$, and $R \cup S$ is also symmetric. Because of this, all paths in $R \cup S$ are "two-way streets," and it follows from the definitions that $(R \cup S)^\infty$ must also be symmetric. Since we already know that $(R \cup S)^\infty$ is transitive, it is an equivalence relation containing $R \cup S$. It is the smallest one, because no smaller set containing $R \cup S$ can be transitive, by definition of the transitive closure.

Example 3   Let $A = \{1, 2, 3, 4, 5\}$, $R = \{1, 1), (1, 2), (2, 1), (2, 2), (3, 3), (3, 4), (4, 3), (4, 4), (5, 5)\}$, $S = \{(1, 1), (2, 2), (3, 3), (4, 4), (4, 5), (5, 4), (5, 5)\}$. The reader may verify that both R and S are equivalence relations. The partition A/R of A corresponding to R is $\{\{1, 2\}, \{3, 4\}, \{5\}\}$, and the partition A/S of A corresponding to S is $\{\{1\}, \{2\}, \{3\}, \{4, 5\}\}$. Find the smallest equivalence relation containing R and S, and compute the partition of A that it produces.

Solution. We have

$$\mathbf{M}_R = \begin{bmatrix} 1 & 1 & 0 & 0 & 0 \\ 1 & 1 & 0 & 0 & 0 \\ 0 & 0 & 1 & 1 & 0 \\ 0 & 0 & 1 & 1 & 0 \\ 0 & 0 & 0 & 0 & 1 \end{bmatrix} \quad \text{and} \quad \mathbf{M}_S = \begin{bmatrix} 1 & 0 & 0 & 0 & 0 \\ 0 & 1 & 0 & 0 & 0 \\ 0 & 0 & 1 & 0 & 0 \\ 0 & 0 & 0 & 1 & 1 \\ 0 & 0 & 0 & 1 & 1 \end{bmatrix}$$

so

$$\mathbf{M}_{R \cup S} = \mathbf{M}_R \vee \mathbf{M}_S = \begin{bmatrix} 1 & 1 & 0 & 0 & 0 \\ 1 & 1 & 0 & 0 & 0 \\ 0 & 0 & 1 & 1 & 0 \\ 0 & 0 & 1 & 1 & 1 \\ 0 & 0 & 0 & 1 & 1 \end{bmatrix}$$

We now compute $\mathbf{M}_{(R \cup S)^\infty}$ by Warshall's algorithm. First, $\mathbf{W}_0 = \mathbf{M}_{R \cup S}$. We next compute $\mathbf{W}_1$, so $k = 1$. Since $\mathbf{W}_0$ has 1's in locations 1 and 2 of column 1 and in locations 1 and 2 of row 1, we find that no new 1's must be adjoined to $\mathbf{W}_1$. Thus

$$\mathbf{W}_1 = \mathbf{W}_0$$

We now compute $\mathbf{W}_2$, so $k = 2$. Since $\mathbf{W}_1$ has 1's in locations 1 and 2 of column 2 and in locations 1 and 2 of row 2, we find that no new 1's must be added to $\mathbf{W}_1$. Thus

$$\mathbf{W}_2 = \mathbf{W}_1$$

We next compute $\mathbf{W}_3$, so $k = 3$. Since $\mathbf{W}_2$ has 1's in locations 3 and 4 of column 2 and in locations 3 and 4 of row 3, we find that no new 1's must be added to $\mathbf{W}_2$. Thus

$$\mathbf{W}_3 = \mathbf{W}_2$$

Things change when we now compute $\mathbf{W}_4$. Since $\mathbf{W}_3$ has 1's in locations 3, 4, and 5 of column 4 and in locations 3, 4, and 5 of row 4, we must add new 1's to $\mathbf{W}_3$ in positions 3,5 and 5,3. Thus

$$\mathbf{W}_4 = \begin{bmatrix} 1 & 1 & 0 & 0 & 0 \\ 1 & 1 & 0 & 0 & 0 \\ 0 & 0 & 1 & 1 & 1 \\ 0 & 0 & 1 & 1 & 1 \\ 0 & 0 & 1 & 1 & 1 \end{bmatrix}$$

The reader may verify that $\mathbf{W}_5 = \mathbf{W}_4$ and thus

$$(R \cup S)^\infty =$$
$$\{(1, 1), (1, 2), (2, 1), (2, 2), (3, 3), (3, 4), (3, 5), (4, 3), (4, 4), (4, 5), (5, 3), (5, 4), (5, 5)\}$$

The corresponding partition of A is then (verify) $\{\{1, 2\}, \{3, 4, 5\}\}$.

EXERCISE SET 2.7

1. Let $A = \{1, 2, 3\}$ and let $R = \{(1, 1), (1, 2), (2, 3),$ $(1, 3), (3, 1), (3, 2)\}$. Compute the matrix $\mathbf{M}_{R^\infty}$, of the transitive closure of R, by using the formula $\mathbf{M}_{R^\infty} = \mathbf{M}_R \vee (\mathbf{M}_R)_\odot^2 \vee (\mathbf{M}_R)_\odot^3$.

2. List the relation R^∞ whose matrix was computed in Exercise 1.

3. For the relation R of Exercise 1, compute the transitive closure R^∞, by using Warshall's algorithm.

4. Let $A = \{a_1, a_2, a_3, a_4, a_5\}$ and let R be a relation on A whose matrix is

$$\mathbf{M}_R = \begin{bmatrix} 1 & 0 & 0 & 1 & 0 \\ 0 & 1 & 0 & 0 & 0 \\ 0 & 0 & 0 & 1 & 1 \\ 1 & 0 & 0 & 0 & 0 \\ 0 & 1 & 0 & 0 & 1 \end{bmatrix} = \mathbf{W}_0$$

Compute $\mathbf{W}_1$, $\mathbf{W}_2$, and $\mathbf{W}_3$ as in Warshall's algorithm.

5. Prove that if R is reflexive and transitive, then $R^n = R$ for all n.

6. Let R be a relation on a set A, and let $S = R^2$. Prove that if $a, b \in A$, then $a \, S^\infty \, b$ if and only if there is a path in R from a to b having an even number of edges.

In Exercises 7–9, let $A = \{1, 2, 3, 4\}$. For the relation R whose matrix is given, find the matrix of the transitive closure by using Warshall's algorithm.

7. $\mathbf{M}_R = \begin{bmatrix} 1 & 0 & 0 & 1 \\ 1 & 1 & 0 & 0 \\ 0 & 0 & 1 & 0 \\ 0 & 0 & 0 & 1 \end{bmatrix}$

8. $\mathbf{M}_R = \begin{bmatrix} 1 & 1 & 0 & 0 \\ 1 & 0 & 0 & 0 \\ 0 & 0 & 0 & 0 \\ 0 & 0 & 1 & 0 \end{bmatrix}$

9. $\mathbf{M}_R = \begin{bmatrix} 1 & 0 & 0 & 1 \\ 0 & 1 & 1 & 0 \\ 0 & 1 & 1 & 0 \\ 1 & 0 & 0 & 1 \end{bmatrix}$

In Exercises 10 and 11, let $A = \{1, 2, 3, 4, 5\}$ and let R and S be the equivalence relations on A whose matrices are given. Compute the matrix of the smallest equivalence relation containing R and S, and list the elements of this relation.

10. $\mathbf{M}_R = \begin{bmatrix} 1 & 1 & 1 & 0 & 0 \\ 1 & 1 & 1 & 0 & 0 \\ 1 & 1 & 1 & 0 & 0 \\ 0 & 0 & 0 & 1 & 1 \\ 0 & 0 & 0 & 1 & 1 \end{bmatrix}$, $\mathbf{M}_S = \begin{bmatrix} 1 & 0 & 0 & 0 & 0 \\ 0 & 1 & 1 & 1 & 0 \\ 0 & 1 & 1 & 1 & 0 \\ 0 & 1 & 1 & 1 & 0 \\ 0 & 0 & 0 & 0 & 1 \end{bmatrix}$

11. $\mathbf{M}_R = \begin{bmatrix} 1 & 0 & 0 & 0 & 0 \\ 0 & 1 & 1 & 0 & 0 \\ 0 & 1 & 1 & 0 & 0 \\ 0 & 0 & 0 & 1 & 1 \\ 0 & 0 & 0 & 1 & 1 \end{bmatrix}$, $\mathbf{M}_S = \begin{bmatrix} 1 & 1 & 0 & 0 & 0 \\ 1 & 1 & 0 & 0 & 0 \\ 0 & 0 & 1 & 0 & 0 \\ 0 & 0 & 0 & 1 & 0 \\ 0 & 0 & 0 & 0 & 1 \end{bmatrix}$

KEY IDEAS FOR REVIEW

☐ $A \times B$ (product set or Cartesian product): $\{(a, b) \mid a \in A \text{ and } b \in B\}$.

☐ $|A \times B| = |A| \, |B|$.

☐ Partition or quotient set: see page 80.

☐ Relation from A to B: subset of $A \times B$.

☐ Domain and range of a relation: see pages 83 and 84.

☐ Matrix of a relation: see page 87.

☐ Digraph of a relation: pictorial representation of a relation; see page 88.

☐ Path of length n from a to b in a relation R: finite sequence $a, x_1, x_2, \ldots, x_{n-1}, b$ such that $a\ R\ x_1, x_1\ R\ x_2, \ldots, x_{n-1}\ R\ b$.

☐ $x\ R^n\ y$ (R a relation on A): There is a path of length n from x to y in R.

☐ $x\ R^\infty\ y$ (connectivity relation for R): Some path exists in R from x to y.

☐ $\mathbf{M}_{R^n} = \mathbf{M}_R \odot \mathbf{M}_R \odot \cdots \odot \mathbf{M}_R$: see page 95.

☐ Properties of relations on a set A.

Reflexive:	$(a, a) \in R$ for all $a \in A$
Irreflexive:	$(a, a) \notin R$ for all $a \in A$
Symmetric:	$(a, b) \in R$ implies that $(b, a) \in R$
Asymmetric:	$(a, b) \in R$ implies that $(b, a) \notin R$
Antisymmetric:	$(a, b) \in R$ and $(b, a) \in R$ imply that $a = b$
Transitive:	$(a, b) \in R$ and $(b, c) \in R$ imply that $(a, c) \in R$

☐ Graph of symmetric relation: see page 101.

☐ Equivalence relation: reflexive, symmetric, transitive; see page 103.

☐ Partition corresponding to an equivalence relation: see pages 106 and 107.

☐ Linked-list computer representation of a relation: see page 112.

☐ $a\ \bar{R}\ b$ (complement of R): $a\ \bar{R}\ b$ if and only if $a\ \not R\ b$.

☐ R^{-1}: $(x, y) \in R^{-1}$ if and only if $(y, x) \in R$.

☐ $R \cup S$, $R \cap S$: see page 119.

☐ $\mathbf{M}_{R \cap S} = \mathbf{M}_R \wedge \mathbf{M}_S$.

☐ $\mathbf{M}_{R \cup S} = \mathbf{M}_R \vee \mathbf{M}_S$.

☐ $\mathbf{M}_{R^{-1}} = (\mathbf{M}_R)^T$: see page 121.

☐ $\mathbf{M}_{\bar{R}} = \overline{\mathbf{M}_R}$: see page 121.

☐ If R and S are equivalence relations, so is $R \cap S$: see page 123.

☐ $\mathbf{M}_{R \circ S} = \mathbf{M}_R \odot \mathbf{M}_S$: see page 126.

☐ Theorem: R^∞ is the smallest transitive relation on A that contains R; see page 131.

☐ If $|A| = n$, $R^\infty = R \cup R^2 \cup \cdots \cup R^n$.

☐ Warshall's algorithm: computes $\mathbf{M}_{R^\infty}$ efficiently; see page 133.

☐ Theorem: If R and S are equivalence relations on A, $(R \cup S)^\infty$ is the smallest equivalence relation on A containing both A and B.

FURTHER READING

AHO, ALFRED V., JOHN E. HOPCROFT, and JEFFREY D. ULLMAN, *The Design and Analysis of Computer Algorithms*, Addison-Wesley, Reading, Mass., 1974.

BONDY, J. A. and U. S. R. MURTY, *Graph Theory with Applications*, Elsevier, New York, 1977.

BUSACKER, ROBERT G., and THOMAS L. SAATY, *Finite Graphs and Networks*, McGraw-Hill, New York, 1965.

EVEN, SHIMON, *Algorithmic Combinatorics*, Macmillan, New York, 1973.

HARARY, F., *Graph Theory*, Addison-Wesley, Reading, Mass., 1969.

KNUTH, DONALD E., *The Art of Computer Programming*, v. 1, 2nd ed., Addison-Wesley, Reading, Mass., 1973.

ORE, OYSTEIN, *Graphs and their Uses*, Random House, New York, 1963.

ORE, OYSTEIN, *Theory of Graphs*, American Mathematical Society, v. 38, Rhode Island, 1962.

WARSHALL, S. "A Theorem on Boolean Matrices," Journal of the Association of Computing Machinery, 11–12, 1962.

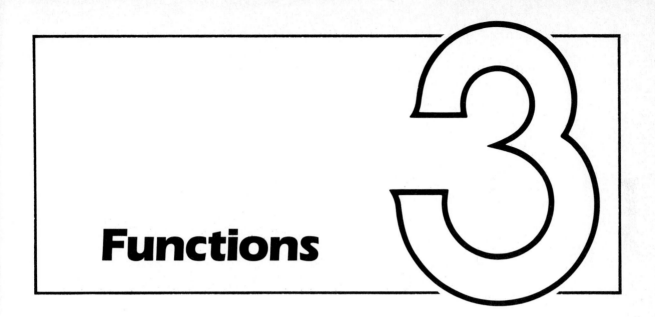

Functions

In this chapter we focus our attention on a special type of relation that plays an important role in mathematics, computer science, and many applications.

3.1 Functions

In this section we define the notion of a function, a special type of relation. We study its basic properties and then discuss several special types of functions. A number of important applications of functions will occur in later sections of the book, so it is essential to get a good grasp of the material in this section.

Let A and B be nonempty sets. A **function** f from A to B denoted $f: A \rightarrow B$ is a relation from A to B such that

(a) Dom $(f) = A$
(b) If (a, b) and (a, c) belong to f, then $b = c$.

Property (b) tells us that if $(a, b) \in f$, then b is uniquely determined by a. For this reason, we often write $b = f(a)$, and we list the relation f as $\{(a, f(a)) \mid a \in A\}$. Functions are also called **mappings** or **transformations**, since they can be geometrically viewed as rules that assign to each element $a \in A$, the unique element $f(a) \in B$ (see Fig. 1). The element a is called an **argument** of the function f, and $f(a)$ is called the **value** of the function for the argument a and is also referred to as the **image** of a under f. Figure 1 is a schematic or pictorial display of property (b) of the above definition of a function, and we will use several other similar diagrams. They should

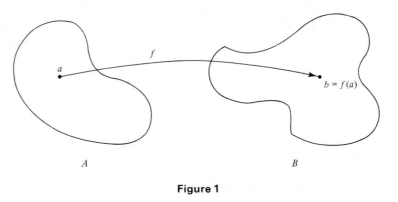

Figure 1

not be confused with the digraph of the relation f, which we will not generally display.

Example 1   Let $A = \{1, 2, 3, 4\}$ and $B = \{a, b, c, d\}$ and let

$$f = \{(1, a), (2, a), (3, d), (4, c)\}$$

Then f is a function, since no element of A appears as the first element of two different ordered pairs. Here we have

$$f(1) = a$$
$$f(2) = a$$
$$f(3) = d$$
$$f(4) = c$$

The range of f, Ran $(f) = \{a, d, c\}$. We will sometimes write the range of a function f as $f(A)$. Note that the element $a \in B$ appears as the second element of two different ordered pairs in f. This does not conflict with the definition of a function. Thus a function may take the same value at two different elements of A.

Example 2   Let $A = \{1, 2, 3\}$ and $B = \{x, y, z\}$. Consider the relations

$$R = \{(1, x), (2, x)\} \quad \text{and} \quad S = \{(1, x), (1, y), (2, z), (3, y)\}$$

Neither of these relations is a function from A to B, but for different reasons. The relation S is not a function since it contains the ordered pairs $(1, x)$ and $(1, y)$ in violation of property (b) of the definition of a function.

The relation R is not a function from A to B, since Dom $(R) \neq A$. This difficulty is not as serious as the difficulty that S has, since R is a function from the set $\{1, 2\}$ to B. This illustrates the general point that if a relation f from A to B satisfies property (b) of the definition of a function, then f is a function from Dom (f) to B.

Example 3   Let P be a computer program which accepts an integer as input and produces an integer as output. Let $A = B = Z$. Then P determines a relation f_P defined as follows:

$(m, n) \in f_P$ means that n is the output produced by program P when the input is m.

It is clear that f_P is a function, since any particular input corresponds to a unique output (computer results are reproducible, that is, they are the same each time the program is run).

Example 3 can be generalized to a program with any set A of possible inputs, and set B of corresponding outputs. In general, therefore, we may think of functions as **input-output** relations.

Example 4   Let $A = B =$ the set of all real numbers, and let $p(x) = a_0 + a_1 x + \ldots + a_n x^n$ be a real polynomial. Then this formula produces a relation p as follows:

$(x, y) \in p$ means that y is the value of the polynomial at the number x, or as it is usually written, $y = p(x)$.

The relation p is clearly a function, since p has a unique value at x.

In elementary mathematics, the *formula* (in the case of Example 4, the polynomial) is usually confused with the *function* it produces. This is not harmful, unless the student comes to expect a formula for every type of function.

Similar examples can be constructed with such elementary "functions" as e^x, $\sin x$, $\tan x$, $1/(1 + x^2)$, and so on.

Example 5   A **labeled digraph** is a digraph in which the vertices or the edges (or both) are labeled with information from a set. If V is the set of vertices and S is the set of labels of a labeled digraph, then the labeling of V can be specified to be a function $f: V \rightarrow S$, where for each $v \in V, f(v)$ is the label we wish to attach to v. Similarly, we can define a labeling of the edges E, as a function $g: E \rightarrow S$, where for each $e \in E, g(e)$ is the label we wish to attach to e. An example of a labeled digraph is a map where the vertices are labeled with the names of cities and the edges are labeled with the distances or travel times between the cities. Another example is a flowchart of a program, where the vertices are labeled with the steps that are to be performed at that point in the program; the edges indicate the flow from one part of the program to another part. Figure 2 shows two labeled digraphs.

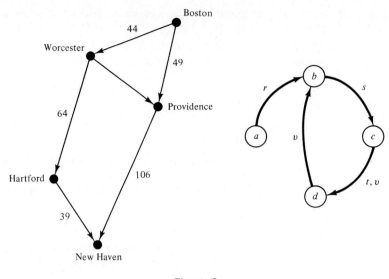

Figure 2

Example 6   Let $A = B = Z$ and let $f: A \rightarrow B$ be defined by

$$f(a) = a + 1 \qquad \text{for } a \in A$$

That is, f consists of all ordered pairs $(a, a + 1)$ for $a \in Z$. Then every $a \in A$ appears as a first element in some pair, so Dom $(f) = A$. Also, if $(a, b) \in f$ and $(a, c) \in f$, so that

$$b = f(a) = a + 1$$

and

$$c = f(a) = a + 1$$

then

$$b = c$$

Thus f is a function.

Example 7   Let $A = Z$ and let $B = \{0, 1\}$. Let $f: A \rightarrow B$ be defined by

$$f(a) = \begin{cases} 0 & \text{if } a \text{ is even} \\ 1 & \text{if } a \text{ is odd} \end{cases}$$

Then f is a function (verify).

Example 8   Let A be an arbitrary nonempty set. The **identity function** on A, denoted by 1_A, is defined by

$$1_A(a) = a$$

The reader may notice that 1_A is the relation we previously called Δ (see Section 2.4), which stands for the diagonal subset of $A \times A$. In the context of functions, the notation 1_A is preferred, since it emphasizes the input–output, or functional nature of the relation.

If $f: A \to B$ and $g: B \to C$ are functions, then the composition, $f \circ g$, of f and g (as relations) is also a function. To see this, we note that $(a, c) \in (f \circ g)$ if and only if $(a, b) \in f$ and $(b, c) \in g$ for some $b \in B$. Then $b = f(a)$ and $c = g(b)$. Hence

$$c = g(b) = g(f(a))$$

The second element of any ordered pair in $f \circ g$ is then determined by the first element, so $f \circ g$ is a function. As a bonus, we have a formula for computing the second element, namely

$$(f \circ g)(a) = g(f(a))$$

This means that to compute the value of $f \circ g$ at an argument a, we compute the value $f(a) = b$, and then compute the value of g at b. This is illustrated diagrammatically in Fig. 3.

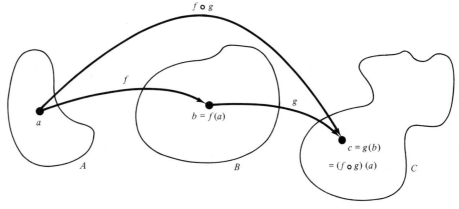

Figure 3

Example 9   Let $A = Z$, $B = Z$, and $C =$ the set of all even integers. Let $f: A \to B$ and $g: B \to C$ be defined by

$$f(a) = a + 1$$

$$g(b) = 2b$$

Find $f \circ g$.

Solution. We have

$$(f \circ g)(a) = g(f(a))$$
$$= g(a + 1)$$
$$= 2(a + 1)$$

Special Types of Functions

A function $f: A \to B$ is said to be **one to one** if for all a, a' in A,

$$a \neq a' \quad \text{implies that} \quad f(a) \neq f(a')$$

Alternatively, we may restate this last condition as

$$f(a) = f(a') \quad \text{implies that} \quad a = a'$$

Example 10   Consider the function f defined in Example 1. Since

$$f(1) = f(2) = a$$

we can conclude that f is not one to one.

Example 11   Consider the function f defined in Example 6. Is f one to one?

Solution. Suppose that

$$f(a) = f(a')$$

for a and a' in A. Then

$$a + 1 = a' + 1$$

so

$$a = a'$$

Hence f is one to one.

A function $f: A \to B$ is said to be **onto** if $f(A) = B$, that is, if Ran $(f) = B$. Thus f is onto if every element $b \in B$ is the second element in some ordered pair $(a, b) \in f$.

Alternatively, f is onto if for every $b \in B$ we can find some $a \in A$ such that $b = f(a)$. We note that Ran $(f) = B$ if and only if Dom $(f^{-1}) = B$.

Example 12   Consider the function f defined in Example 1. Since there is no element $x \in$ in A such that $f(x) = b$, we conclude that f is not onto.

Example 13   Consider the function f defined in Example 6. Is f onto?

Solution. Let b be an arbitrary element of B. Can we find an element $a \in A$ such that

$$f(a) = b$$

Since

$$f(a) = a + 1$$

we need an element a in A such that

$$a + 1 = b$$

Of course,

$$a = b - 1$$

will satisfy the desired equation since $b - 1$ is in A. Hence f is onto.

Example 14   Let $A = \{a_1, a_2, a_3\}$, $B = \{b_1, b_2, b_3\}$, $C = \{c_1, c_2\}$, and $D = \{d_1, d_2, d_3, d_4\}$. Consider the following four functions, from A to B, A to D, B to C, and A to B, respectively.

(a) $f_1 = \{(a_1, b_2), (a_2, b_3), (a_3, b_1)\}$

(b) $f_2 = \{(a_1, d_2), (a_2, d_1), (a_3, d_4)\}$

(c) $f_3 = \{(b_1, c_2), (b_2, c_2), (b_3, c_1)\}$

(d) $f_4 = \{(a_1, b_1), (a_2, b_1), (a_3, b_3)\}$

Determine whether or not each function is one to one, and whether each function is onto.

Solution.

(a) f_1 is one to one and onto.

(b) f_2 is one to one but not onto.

(c) f_3 is not one to one but is onto.

(d) f_4 is neither one to one nor onto.

When a function f is both one to one and onto, we say that f is a **bijection**, or a **one-to-one correspondence**.

Invertible Functions

A function $f: A \to B$ is said to be **invertible** if its inverse relation, f^{-1}, is also a function. The following example shows that a function is not necessarily invertible.

Example 15   Let f be the function of Example 1. Then

$$f^{-1} = \{(a, 1), (a, 2), (d, 3), (c, 4)\}$$

We see that f^{-1} is not a function from B to A, since both properties (a) and (b) in the definition of a function are violated. That is, $b \in B$, but b is not the first element of any ordered pair in f^{-1}. Also, both $(a, 1)$ and $(a, 2)$ belong to f^{-1}.

The following theorem is frequently used.

Theorem 1   Let $f: A \to B$ be a function.

(a) Then f^{-1} is a function from B to A if and only if f is both one to one and onto (that is, if and only if f is a bijection).

(b) If (a) holds, then the function f^{-1} is also a bijection, and $(f^{-1})^{-1} = f$.

Proof. (a) Suppose first that the function f is a bijection. We show that f^{-1} satisfies both properties (a) and (b) in the definition of a function. First establish (a) for f^{-1}. Choose any $b \in B$. Since f is onto, $b = f(a)$ for some $a \in A$; that is, $(a, b) \in f$. Thus $(b, a) \in f^{-1}$ and since b can be any element of B, Dom $(f^{-1}) = B$. Next we establish (b) for f^{-1}. Now suppose that $(b, a_1) \in f^{-1}$ and $(b, a_2) \in f^{-1}$. Then $(a_1, b) \in f$ and $(a_2, b) \in f$, so $f(a_1) = b = f(a_2)$. Since f is one-to-one, $a_1 = a_2$. Thus f^{-1} satisfies both parts of the definition of a function.

Conversely, suppose that the relation f^{-1} is a function. If $b \in B$, then $f^{-1}(b) = a$ for some $a \in A$, since Dom $(f^{-1}) = B$. Thus $(b, a) \in f^{-1}$, so $(a, b) \in f$ and hence $b \in$ Ran (f). Thus $B \subseteq$ Ran (f). Since we always have Ran $(f) \subseteq B$, we see that $B =$ Ran (f), so f is onto. If $f(a_1) = f(a_2) = b$, then $(a_1, b) \in f$ and $(a_2, b) \in f$, so $(b, a_1) \in f^{-1}$ and $(b, a_2) \in f^{-1}$. Since f^{-1} is a function, $a_1 = a_2$. We have then shown that f is one to one.

(b) We know from section 2.6 that $(f^{-1})^{-1} = f$. Since f is a function, part (a) shows that f^{-1} is also one to one and onto.

Example 16 Consider the function f defined in Example 6. Since f is both one to one and onto, it is invertible.

Example 17 Let $\mathbb{R}$ be the set of all real numbers, and let $f\colon \mathbb{R} \to \mathbb{R}$ be defined by $f(x) = x^2$. Is f invertible?

Solution. We must determine whether f is both one to one and onto. Since

$$f(2) = f(-2) = 4$$

we conclude that f is not one to one. Hence f is not invertible. Incidentally, f is also not onto, since there is no $x \in R$ such that

$$f(x) = -1$$

There are some simple and useful results concerning the composition of functions. We summarize these in the following theorem.

Theorem 2 Let $f\colon A \to B$ be any function. Then

(a) $f \circ 1_B = f$
(b) $1_A \circ f = f$

and if f is invertible,

(c) $f \circ f^{-1} = 1_A$
(d) $f^{-1} \circ f = 1_B$

Proof. All parts are easily proved. We illustrate by proving part (c) and leave the proofs of the other parts as exercises. Since $f\colon A \to B$ and $f^{-1}\colon B \to A$, the composition $f \circ f^{-1}$ is a relation on A. Then $(a_1, a_2) \in f \circ f^{-1}$ if and only if there is an element $b \in B$ such that $(a_1, b) \in f$ and $(b, a_2) \in f^{-1}$. Then $(a_1, b) \in f$ and $(a_2, b) \in f$. However, since we have assumed f to be invertible, then f is one to one by Theorem 1, so $a_1 = a_2$. We have thus shown that if $(a_1, a_2) \in f \circ f^{-1}$, then $(a_1, a_2) \in 1_A$. The converse follows easily, so $f \circ f^{-1} = 1_A$.

Theorem 3 (a) Let $f\colon A \to B$ and $g\colon B \to A$ be functions such that $f \circ g = 1_A$ and $g \circ f = 1_B$. Then f is invertible and $g = f^{-1}$.

(b) Let $f\colon A \to B$ and $g\colon B \to C$ be invertible. Then $f \circ g$ is invertible, and $(f \circ g)^{-1} = g^{-1} \circ f^{-1}$.

Proof.

(a) We show that the relations g and f^{-1} contain the same ordered pairs, so they are equal. Then f^{-1} is a function (since it equals g), so f is invertible.

Suppose that $(b, a) \in f^{-1}$. Then $(a, b) \in f$ and $(b, g(b)) \in g$, so $(a, g(b)) \in f \circ g = 1_A$. This means that $a = g(b)$, so $(b, a) \in g$.

Conversely, suppose that $(b, a) \in g$. We also have $(a, f(a)) \in f$, so $(b, f(a)) \in g \circ f = 1_B$ and $b = f(a)$. Thus $(a, b) \in f$, so $(b, a) \in f^{-1}$. Hence $f^{-1} = g$, and we are done.

(b) Since f and g are invertible, the relations f^{-1} and g^{-1} are functions. Then

$$(f \circ g) \circ (g^{-1} \circ f^{-1}) = f \circ (g \circ (g^{-1} \circ f^{-1})) = f \circ ((g \circ g^{-1}) \circ f^{-1}) = f \circ (1_B \circ f^{-1})$$

$$= f \circ f^{-1} = 1_A$$

A similar computation shows that

$$(g^{-1} \circ f^{-1}) \circ (f \circ g) = 1_C$$

Since $g^{-1} \circ f^{-1}$ is a function whose composition with $f \circ g$ gives identities on both sides, part (a) tells us that $f \circ g$ is invertible, and its inverse is $g^{-1} \circ f^{-1}$.

Example 18 Let $A = B = \mathbb{R}$, the set of all real numbers. Let $f : A \to B$ be given by the formula

$$f(x) = 2x^3 - 1$$

and let $g : B \to A$ be given by

$$g(y) = \sqrt[3]{\tfrac{1}{2}y + \tfrac{1}{2}}$$

Show that $g = f^{-1}$, so that both f and g are bijections.

Solution. Let $(x, y) \in f$. Then $y = 2x^3 - 1$, so $\tfrac{1}{2}(y + 1) = x^3$, and $x = \sqrt[3]{\tfrac{1}{2}y + \tfrac{1}{2}} = g(y)$. Thus $(y, x) \in g$. Conversely, if $(y, x) \in g$, then $x = \sqrt[3]{\tfrac{1}{2}y + \tfrac{1}{2}}$, so $x^3 = \tfrac{1}{2}(y + 1)$ and $y = 2x^3 - 1 = f(x)$, so $(x, y) \in f$. Since $(x, y) \in f$ if and only if $(y, x) \in g$, we have $g = f^{-1}$.

As Example 18 shows, it is often easier to show that a function, such as f, is one to one and onto, by constructing an inverse instead of proceeding directly.

Example 19 Let A be a nonempty set. Let $\mathscr{E}(A)$ be the set of all equivalence relations on A and let $\mathscr{P}(A)$ be the set of all partitions of A. In Section 2.4 we have already seen (see Theorems 2 and 3) that if $R \in \mathscr{E}(A)$, then we can determine a partition P_R, and if $P \in \mathscr{P}(A)$, then we can determine an equivalence relation R_P. We define the functions $f : \mathscr{E}(A) \to \mathscr{P}(A)$ and $g : \mathscr{P}(A) \to \mathscr{E}(A)$ by

$$f(R) = P_R \qquad \text{and} \qquad g(P) = R_P$$

We now show that

$$g(P_R) = R \qquad \text{and} \qquad f(R_P) = P$$

Consider the partition P_R. Then $g(P_R) = R'$ is the equivalence relation on A defined as follows: $a \ R' \ b$ if and only if a and b belong to the same block of P_R. Since the blocks of P_R are the equivalence classes of R, we conclude that $a \ R' \ b$ if and only if $a \ R \ b$. Hence $R = R'$.

Next, consider the equivalence relation R_P. Then $f(R_P) = P'$ is the partition of A consisting of all the equivalence classes of R_P. Since the equivalence classes of R_P are the blocks of P, we conclude that P' is the partition P. Hence $f(R_P) = P$. It thus follows that

$$g(f(R)) = R \qquad \text{and} \qquad f(g(P)) = P$$

or

$$(f \circ g)(R) = R \qquad \text{and} \qquad (g \circ f)(P) = P$$

which means that

$$f \circ g = 1_{\mathscr{E}(A)} \qquad \text{and} \qquad g \circ f = 1_{\mathscr{P}(A)}$$

It then follows from Theorems 1 and 3 that f is a bijection. The result just described is often stated as follows: There is a one-to-one correspondence between the set of all equivalence classes on A and the set of all partitions of A.

Finally, we discuss briefly some special results that hold when A and B are finite sets. Let $A = \{a_1, \ldots, a_n\}$ and $B = \{b_1, \ldots, b_n\}$, and let f be a function from A to B. If f is one to one, then $f(a_1), f(a_2), \ldots, f(a_n)$ are n different elements of B. Thus we must have all of B, so f is also onto. On the other hand, if f is onto, then $f(a_1), \ldots, f(a_n)$ form the entire set B, so they must all be different. Hence f is also one to one. We have therefore shown:

Theorem 4   Let A and B be two finite sets with the same number of elements, and let $f : A \to B$ be a function.

(a) If f is one to one, then f is onto.

(b) If f is onto, then f is one to one.

Thus for finite sets, A and B, with the same number of elements, and particularly if $A = B$, we need only prove that a function is one to one *or* onto to show that it is a bijection.

EXERCISE SET 3.1

In Exercises 1–4, let $A = \{a, b, c, d\}$ and $B = \{1, 2, 3\}$. Determine whether the relation R from A to B is a function. If it is a function, give its range.

1. $R = \{(a, 1), (b, 2), (c, 1), (d, 2)\}$

2. $R = \{(a, 1), (b, 2), (a, 2), (c, 1), (d, 2)\}$

3. $R = \{(a, 3), (b, 2), (c, 1)\}$

4. $R = \{(a, 1), (b, 1), (c, 1), (d, 1)\}$

In Exercises 5 and 6, determine whether the relation from A to B is a function.

5. $A =$ the set of all recipients of Medicare in the United States, $B = \{x \mid x$ is a nine-digit number$\}$, $a\ R\ b$ if b is a's Social Security number.

6. $A =$ a set of people in the United States, $B = \{x \mid x$ is a nine-digit number$\}$, $a\ R\ b$ if b is a's passport number.

In Exercises 7–10, verify that the formula yields a function from A to B.

7. $A = B = Z; f(a) = a^2$

8. $A = B = \mathbb{R}; f(a) = e^a$

9. $A = \mathbb{R}, B = \{0, 1\}$; let Z be the set of integers and note that $Z \subseteq \mathbb{R}$. Then for any real number a, let
$$f(a) = \begin{cases} 0 & \text{if } a \notin Z \\ 1 & \text{if } a \in Z \end{cases}$$

10. $A = \mathbb{R}, B = Z; f(a) =$ the greatest integer less than or equal to a.

11. Let $A = B = C = \mathbb{R}$, and let $f: A \to B$, $g: B \to C$ be defined by $f(a) = a - 1$ and $g(b) = b^2$. Find

  (a) $(g \circ f)(2)$ (b) $(f \circ g)(2)$

  (c) $(f \circ g)(x)$ (d) $(g \circ f)(x)$

  (e) $(f \circ f)(y)$ (f) $(g \circ g)(y)$

12. Let $A = B = C = \mathbb{R}$ and let $f: A \to B$, $g: B \to C$ be defined by $f(a) = a + 1$ and $g(b) = b^2 + 2$. Find

  (a) $(f \circ g)(-2)$ (b) $(g \circ f)(-2)$

  (c) $(f \circ g)(x)$ (d) $(g \circ f)(x)$

  (e) $(f \circ f)(y)$ (f) $(g \circ g)(y)$

13. Let $A = B = \{x \mid x$ is a real number and $x \neq 0, 1\}$. Consider the following six functions from A to B, each defined by its formula.

$$f_1(x) = x \qquad f_2(x) = 1 - x$$

$$f_3(x) = \frac{1}{x} \qquad f_4(x) = \frac{1}{1 - x}$$

$$f_5(x) = \frac{x}{x - 1} \qquad f_6(x) = \frac{x - 1}{x}$$

Show by substituting one formula into another that the composition of any two of these six functions is another one of the six.

14. In each part, sets A and B, and a function from A to B are given. Determine whether the function is one to one or onto (or both or neither).

  (a) $A = \{1, 2, 3, 4\} = B$;
  $f = \{(1, 1), (2, 3), (3, 4), (4, 2)\}$

  (b) $A = \{1, 2, 3\}; B = \{a, b, c, d\}$;
  $f = \{(1, a), (2, a), (3, c)\}$

  (c) $A = \{\frac{1}{2}, \frac{1}{3}, \frac{1}{4}\}; B = \{x, y, z, w\}$;
  $f = \{(\frac{1}{2}, x), (\frac{1}{4}, y), (\frac{1}{3}, w)\}$

  (d) $A = \{1.1, 7, 0.06\}; B = \{p, q\}$;
  $f = \{(1.1, p), (7, q), (0.06, p)\}$

In Exercises 15–21, let f be a function from A to B. Determine whether each function f is one to one and whether it is onto.

15. $A = B = Z; f(a) = a - 1$

16. $A = B = \mathbb{R}; f(a) = |a|$

17. $A = \mathbb{R}, B = \{x \mid x$ is real and $x \geq 0\}; f(a) = |a|$

18. $A = \mathbb{R} \times \mathbb{R}, B = \mathbb{R}; f((a, b)) = a$

19. Let $S = \{1, 2, 3\}$, $T = \{a, b\}$. Let $A = B = S \times T$ and let f be defined by $f(n, a) = (n, b), n = 1, 2, 3$, and $f(n, b) = (1, a), n = 1, 2, 3$.

20. $A = B = \mathbb{R} \times \mathbb{R}; f((a, b)) = (a + b, a - b)$

21. $A = \mathbb{R}, B = \{x \mid x$ is real and $x \geq 0\}; f(a) = a^2$

In Exercises 22–25, $f: A \to B$ and $g: B \to A$. Verify that $g = f^{-1}$.

22. $A = B = Z; f(a) = \dfrac{a + 1}{2}, g(b) = 2b - 1$

23. $A = \{x \mid x \text{ is real and } x \geq 0\}; B = \{y \mid y \text{ is real and } y \geq -1\}; f(a) = a^2 - 1, g(b) = \sqrt{b + 1}$

24. $A = B = P(S)$, where S is a set. If $X \in P(S)$, let $f(X) = \bar{X} = g(X)$.

25. $A = B = \{1, 2, 3, 4\}; f = \{(1, 4), (2, 1), (3, 2), (4, 3)\}$,
$g = \{(1, 2), (2, 3), (3, 4), (4, 1)\}$

In Exercises 26–29, f is a function from A to B. Find f^{-1}.

26. $A = \{x \mid x \text{ is real and } x \geq -1\}; B = \{x \mid x \text{ is real and } x \geq 0\}; f(a) = \sqrt{a + 1}$

27. $A = B = \mathbb{R}; f(a) = a^3 + 1$

28. $A = B = \mathbb{R}; f(a) = \dfrac{2a - 1}{3}$

29. $A = B = \{1, 2, 3, 4, 5\};$
$f = \{(1, 3), (2, 2), (3, 4), (4, 5), (5, 1)\}$

In Exercises 30 and 31, let f be a function from $A = \{1, 2, 3, 4\}$ to $B = \{a, b, c, d\}$. Determine whether f^{-1} is a function.

30. $f = \{(1, a), (2, a), (3, c), (4, d)\}$

31. $f = \{(1, a), (2, c), (3, b), (4, d)\}$

32. Let $A = B = C = \mathbb{R}$ and consider the functions $f: A \to B$ and $g: B \to C$ defined by $f(a) = 2a + 1$, $g(b) = b/3$. Verify (b) of Theorem 4: $(f \circ g)^{-1} = g^{-1} \circ f^{-1}$.

33. If a set A has n elements, how many functions are there from A to A?

34. If a set A has n elements, how many bijections are there from A to A?

35. If A has m elements and B has n elements, how many functions are there from A to B?

36. Prove that if $f: A \to B$ and $g: B \to C$ are one-to-one functions, then $f \circ g$ is one to one.

37. Prove that if $f: A \to B$ and $g: B \to C$ are onto functions, then $f \circ g$ is onto.

38. Let $f: A \to B$ and $g: B \to C$ be functions. Show that if $f \circ g$ is one to one, then f is one to one.

39. Let $f: A \to B$ and $g: B \to C$ be functions. Show that if $f \circ g$ is onto, then g is onto.

40. Let A be a set, and let $f: A \to A$ be a bijection. For any integer $k \geq 1$, let $f^k = f \circ f \cdots \circ f$ (composition k times), and let $f^{-k} = f^{-1} \circ f^{-1} \circ \cdots \circ f^{-1}$ (k times). Define f^0 to be 1_A. Then f^n is defined for all $n \in Z$. For any $a \in A$, let $O(a, f) = \{f^n(a) \mid n \in Z\}$.
Prove that if $a_1, a_2 \in A$ and $O(a_1, f) \cap O(a_2, f) \neq \emptyset$, then $O(a_1, f) = O(a_2, f)$.

3.2 Permutations

In this section we discuss bijections from a set A to itself. Of special importance is the case when A is finite. Bijections on a finite set occur in a wide variety of applications in mathematics, computer science, and physics.

A bijection from a set A to itself is called a **permutation** of A.

Example 1   Let $A = \mathbb{R}$ and let $f: A \to A$ be defined by $f(a) = 2a + 1$. Since f is one to one and onto (verify), it follows that f is a permutation of A.

If $A = \{a_1, a_2, \ldots, a_n\}$ is a finite set and p is a bijection on A, we list the elements of A, and the corresponding function values $p(a_1), p(a_2), \ldots, p(a_n)$ in the following form:

$$\begin{pmatrix} a_1 & a_2 & \cdots & a_n \\ p(a_1) & p(a_2) & \cdots & p(a_n) \end{pmatrix} \tag{1}$$

Observe that (1) completely describes p since it gives the value of p for every element of A. We often write

$$p = \begin{pmatrix} a_1 & a_2 & \cdots & a_n \\ p(a_1) & p(a_2) & \cdots & p(a_n) \end{pmatrix}$$

Thus if p is a permutation of a finite set $A = \{a_1, a_2, \ldots, a_n\}$, then the sequence $p(a_1), p(a_2), \ldots, p(a_n)$ is just a rearrangement of the elements of A, and so corresponds exactly to a permutation of A in the sense of Sec. 1.4.

Example 2   Let $A = \{1, 2, 3\}$. Then all the permutations of A are

$$1_A = \begin{pmatrix} 1 & 2 & 3 \\ 1 & 2 & 3 \end{pmatrix}, \quad p_1 = \begin{pmatrix} 1 & 2 & 3 \\ 1 & 3 & 2 \end{pmatrix}, \quad p_2 = \begin{pmatrix} 1 & 2 & 3 \\ 2 & 1 & 3 \end{pmatrix},$$

$$p_3 = \begin{pmatrix} 1 & 2 & 3 \\ 2 & 3 & 1 \end{pmatrix}, \quad p_4 = \begin{pmatrix} 1 & 2 & 3 \\ 3 & 1 & 2 \end{pmatrix}, \quad p_5 = \begin{pmatrix} 1 & 2 & 3 \\ 3 & 2 & 1 \end{pmatrix}$$

Example 3   Using the permutations of Example 2, compute (a) p_4^{-1}; (b) $p_2 \circ p_3$.

Solution.
(a) Viewing p_4 as a function, we have

$$p_4 = \{(1, 3), (2, 1), (3, 2)\}$$

Then

$$p_4^{-1} = \{(3, 1), (1, 2), (2, 3)\}$$

or when written in increasing order of the first component of each ordered pair, we have

$$p_4^{-1} = \{(1, 2), (2, 3), (3, 1)\}$$

Thus

$$p_4^{-1} = \begin{pmatrix} 1 & 2 & 3 \\ 2 & 3 & 1 \end{pmatrix} = p_3$$

(b) The function p_2 takes 1 to 2 and p_3 takes 2 to 3, so $p_2 \circ p_3$ takes 1 to 3. Also, p_2 takes 2 to 1 and p_3 takes 1 to 2, so $p_2 \circ p_3$ takes 2 to 2. Finally, p_2 takes 3 to 3 and p_3 takes 3 to 1, so $p_2 \circ p_3$ takes 3 to 1. Thus

$$p_2 \circ p_3 = \begin{pmatrix} 1 & 2 & 3 \\ 3 & 2 & 1 \end{pmatrix}$$

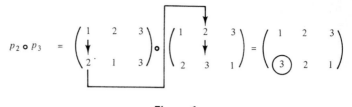

$$p_2 \circ p_3 \quad = \quad \begin{pmatrix} 1 & 2 & 3 \\ 2 & 1 & 3 \end{pmatrix} \circ \begin{pmatrix} 1 & 2 & 3 \\ 2 & 3 & 1 \end{pmatrix} = \begin{pmatrix} 1 & 2 & 3 \\ 3 & 2 & 1 \end{pmatrix}$$

Figure 1

We may view the process of forming $p_2 \circ p_3$ as shown in Fig. 1. Observe that $p_2 \circ p_3 = p_5$.

The composition of two permutations is another permutation usually referred to as the **product** of these permutations. In the remainder of this chapter, we will follow this convention.

Theorem 1 If $A = \{a_1, a_2, \ldots, a_n\}$ is a set containing n elements, then there are $n! = n \cdot (n - 1) \ldots 2 \cdot 1$ permutations of A.

Proof. This result follows from Theorem 3 of Section 1.4 by letting $r = n$.

Let $b_1, b_2, \ldots, b_r$ be r distinct elements of the set $A = \{a_1, a_2, \ldots, a_n\}$. The permutation $p : A \to A$ defined by

$$p(b_1) = b_2$$
$$p(b_2) = b_3$$
$$\vdots$$
$$p(b_{r-1}) = b_r$$
$$p(b_r) = b_1$$
$$p(x) = x \quad \text{if} \quad x \in A, \quad x \notin \{b_1, b_2, \ldots, b_r\}$$

is called a **cyclic permutation** of length r, or simply a **cycle** of length r, and will be denoted by $(b_1, b_2, \ldots, b_r)$. Do not confuse this terminology with that used for cycles in a digraph (Section 2.3). The two concepts are different and we use slightly different notations. If the elements $b_1, b_2, \ldots, b_r$ are arranged uniformly on a circle, as shown in Fig. 2, then a cycle p of length r moves these elements in a clockwise direction so that b_1 is sent to b_2, b_2 to $b_3, \ldots, b_{r-1}$ to b_r, and b_r to b_1. All the other elements of A are left fixed by p.

Example 4 Let $A = \{1, 2, 3, 4, 5\}$. The cycle $(1, 3, 5)$ denotes the permutation

$$\begin{pmatrix} 1 & 2 & 3 & 4 & 5 \\ 3 & 2 & 5 & 4 & 1 \end{pmatrix}$$

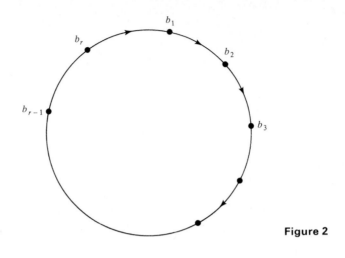

Figure 2

Observe that if $p = (b_1, b_2, \cdots, b_r)$ is a cycle of length r, then we can also write p by starting with any b_i, $1 \le i \le r$, and moving in a clockwise direction as shown in Fig. 2. Thus

$$(3, 5, 8, 2) = (5, 8, 2, 3) = (8, 2, 3, 5) = (2, 3, 5, 8)$$

Note also that the notation for a cycle does not indicate the number of elements in the set A. Thus the cycle $(3, 2, 1, 4)$ could be a permutation of the set $\{1, 2, 3, 4\}$ or of $\{1, 2, 3, 4, 5, 6, 7, 8\}$. We need to be told explicitly the set on which a cycle is defined. It follows from the definition that a cycle on a set A is of length 1 if and only if it is the identity permutation, 1_A.

Since cycles are permutations, we can form their product. However, as we show in the following example, the product of two cycles need not be a cycle.

Example 5 Let $A = \{1, 2, 3, 4, 5, 6\}$. Compute $(4, 1, 3, 5) \circ (5, 6, 3)$ and $(5, 6, 3) \circ (4, 1, 3, 5)$.

Solution. We have

$$(4, 1, 3, 5) = \begin{pmatrix} 1 & 2 & 3 & 4 & 5 & 6 \\ 3 & 2 & 5 & 1 & 4 & 6 \end{pmatrix}$$

and

$$(5, 6, 3) = \begin{pmatrix} 1 & 2 & 3 & 4 & 5 & 6 \\ 1 & 2 & 5 & 4 & 6 & 3 \end{pmatrix}$$

Then

$$(4, 1, 3, 5) \circ (5, 6, 3) = \begin{pmatrix} 1 & 2 & 3 & 4 & 5 & 6 \\ 3 & 2 & 5 & 1 & 4 & 6 \end{pmatrix} \circ \begin{pmatrix} 1 & 2 & 3 & 4 & 5 & 6 \\ 1 & 2 & 5 & 4 & 6 & 3 \end{pmatrix}$$

$$= \begin{pmatrix} 1 & 2 & 3 & 4 & 5 & 6 \\ 5 & 2 & 6 & 1 & 4 & 3 \end{pmatrix}$$

and

$$(5, 6, 3) \circ (4, 1, 3, 5) = \begin{pmatrix} 1 & 2 & 3 & 4 & 5 & 6 \\ 1 & 2 & 5 & 4 & 6 & 3 \end{pmatrix} \circ \begin{pmatrix} 1 & 2 & 3 & 4 & 5 & 6 \\ 3 & 2 & 5 & 1 & 4 & 6 \end{pmatrix}$$

$$= \begin{pmatrix} 1 & 2 & 3 & 4 & 5 & 6 \\ 3 & 2 & 4 & 1 & 6 & 5 \end{pmatrix}$$

Observe that

$$(4, 1, 3, 5) \circ (5, 6, 3) \neq (5, 6, 3) \circ (4, 1, 3, 5)$$

Two cycles of a set A are said to be **disjoint** if no element of A appears in both cycles.

Example 6   Let $A = \{1, 2, 3, 4, 5, 6\}$. Then the cycles $(1, 2, 5)$ and $(3, 4, 6)$ are disjoint, whereas the cycles $(1, 2, 5)$ and $(2, 4, 6)$ are not.

It is not difficult to show that if $p_1 = (a_1, a_2, \ldots, a_r)$ and $p_2 = (b_1, b_2, \ldots, b_s)$ are disjoint cycles of A, then $p_1 \circ p_2 = p_2 \circ p_1$. This can be seen by observing that p_1 affects only the a's, while p_2 affects only the b's.

We shall now present a fundamental theorem, and instead of giving its proof, we shall do an example which completely imitates the proof.

Theorem 2   A permutation of a finite set that is not the identity or a cycle can be written as a product of disjoint cycles of length ≥ 2.

Example 7   Write the permutation

$$p = \begin{pmatrix} 1 & 2 & 3 & 4 & 5 & 6 & 7 & 8 \\ 3 & 4 & 6 & 5 & 2 & 1 & 8 & 7 \end{pmatrix}$$

of the set $A = \{1, 2, 3, 4, 5, 6, 7, 8\}$ as a product of disjoint cycles.

Solution. We start with 1 and find that $p(1) = 3$, $p(3) = 6$, and $p(6) = 1$, so we have the cycle (1, 3, 6). Next, we choose the first element of A that has not appeared in a previous cycle. We choose 2, and we have $p(2) = 4$, $p(4) = 5$, and $p(5) = 2$, so we obtain the cycle (2, 4, 5). We now choose 7, the first element of A that has not appeared in a previous cycle. Since $p(7) = 8$ and $p(8) = 7$, we obtain the cycle (7, 8). We can then write p as a product of disjoint cycles as

$$p = (1, 3, 6) \circ (2, 4, 5) \circ (7, 8)$$

It is not difficult to show that in Theorem 2, when a permutation is written as a product of disjoint cycles, the product is unique except for the order of the cycles.

Even and Odd Permutations

A cycle of length 2 is called a **transposition**. That is, a transposition is a cycle $p = (a_i, a_j)$, where $p(a_i) = a_j$ and $p(a_j) = a_i$.

Observe that if $p = (a_i, a_j)$ is a transposition of A, then $p \circ p = 1_A$, the identity permutation of A.

Every cycle can be written as a product of transpositions. In fact,

$$(b_1, b_2, \ldots, b_r) = (b_1, b_2) \circ (b_1, b_3) \circ \cdots \circ (b_1, b_{r-1}) \circ (b_1, b_r)$$

This can be verified by induction on r, as follows.

Basis Step. If $r = 2$, then the cycle is just (b_1, b_2), which already has the proper form.

Induction Step. Assume true for $r = n$, and let $(b_1, b_2, \ldots, b_n, b_{n+1})$ be a cycle of length $n + 1$. Then $(b_1, b_2, \ldots, b_n, b_{n+1}) = (b_1, b_2, \ldots, b_n) \circ (b_1, b_{n+1})$, as may be verified by computing the composition. By the induction assumption, $(b_1, b_2, \ldots, b_n) = (b_1, b_2) \circ (b_1, b_3) \circ \cdots \circ (b_1, b_n)$. Thus, by substitution, $(b_1, b_2, \ldots, b_{n+1}) = (b_1, b_2) \circ (b_1, b_3) \circ \cdots \circ (b_1, b_n) \circ (b_1, b_{n+1})$. This completes the induction step. For example,

$$(1, 2, 3, 4, 5) = (1, 2) \circ (1, 3) \circ (1, 4) \circ (1, 5)$$

We now obtain the following corollary of Theorem 2.

Corollary 1 Every permutation of a finite set with at least two elements can be written as a product of transpositions.

Observe that the transpositions in Corollary 1 need not be disjoint.

Example 8 Write the permutation p of Example 7 as a product of transpositions.

Solution. We have

$$p = (1, 3, 6) \circ (2, 4, 5) \circ (7, 8)$$

Since we can write

$$(1, 3, 6) = (1, 3) \circ (1, 6)$$

$$(2, 4, 5) = (2, 4) \circ (2, 5)$$

we have

$$p = (1, 3) \circ (1, 6) \circ (2, 4) \circ (2, 5) \circ (7, 8)$$

We have observed above that every cycle can be written as a product of transpositions. However, this can be done in many different ways. For example,

$$(1, 2, 3) = (1, 2) \circ (1, 3)$$
$$= (2, 3) \circ (2, 1)$$
$$= (2, 3) \circ (3, 2) \circ (1, 2) \circ (1, 3) \circ (3, 1) \circ (1, 3)$$

It then follows that every permutation on a set of two or more elements can be written as a product of transpositions in many ways. However, the following theorem, whose proof we omit, brings some order to the situation.

Theorem 3   If a permutation of a finite set can be written as a product of an even number of transpositions, then it can never be written as a product of an odd number of transpositions, and conversely.

A permutation of a finite set is called **even** if it can be written as a product of an even number of transpositions, and it is called **odd** if it can be written as a product of an odd number of transpositions.

Example 9   Is the permutation

$$p = \begin{pmatrix} 1 & 2 & 3 & 4 & 5 & 6 & 7 \\ 2 & 4 & 5 & 7 & 6 & 3 & 1 \end{pmatrix}$$

even or odd?

Solution. We first write p as a product of disjoint cycles, obtaining

$$p = (1, 2, 4, 7) \circ (3, 5, 6)$$

Next, we write each of the cycles as a product of transpositions:

$$(1, 2, 4, 7) = (1, 2) \circ (1, 4) \circ (1, 7)$$

$$(3, 5, 6) = (3, 5) \circ (3, 6)$$

Then

$$p = (1, 2) \circ (1, 4) \circ (1, 7) \circ (3, 5) \circ (3, 6)$$

Since p is a product of an odd number of transpositions, it is an odd permutation.

From the definition of even and odd permutations, it follows (Exercise 23) that

(a) The product of two even permutations is even.
(b) The product of two odd permutations is even.
(c) The product of an even and an odd permutation is odd.

Theorem 4   Let $A = \{a_1, a_2, \ldots, a_n\}$ be a finite set with n elements, $n \geq 2$. There are $n!/2$ even permutations and $n!/2$ odd permutations.

Proof. Let A_n be the set of all even permutations of A and let B_n be the set of all odd permutations. We shall define a function $f: A_n \rightarrow B_n$ which we show is one to one and onto, and this will show that A_n and B_n have the same number of elements.

Since $n \geq 2$, we can choose a particular transposition q_0 of A. Say that $q_0 = (a_{n-1}, a_n)$. We now define the function $f: A_n \rightarrow B_n$ by

$$f(p) = p \circ q_0, \quad p \in A_n$$

Observe that if $p \in A_n$, then p is an even permutation, so $p \circ q_0$ is an odd permutation and thus $f(p) \in B_n$. Suppose now that p_1 and p_2 are in A_n and

$$f(p_1) = f(p_2)$$

Then

$$p_1 \circ q_0 = p_2 \circ q_0 \tag{2}$$

We now compose each side of equation (2) with q_0:

$$(p_1 \circ q_0) \circ q_0 = (p_2 \circ q_0) \circ q_0$$

so by the associative property

$$p_1 \circ (q_0 \circ q_0) = p_2 \circ (q_0 \circ q_0)$$

or since $q_0 \circ q_0 = 1_A$,

$$p_1 \circ 1_A = p_2 \circ 1_A$$
$$p_1 = p_2$$

Thus f is one to one.

Now let $q \in B$. Then $q \circ q_0 \in A_n$, and

$$f(q \circ q_0) = (q \circ q_0) \circ q_0$$
$$= q \circ (q_0 \circ q_0)$$
$$= q \circ 1_A$$
$$= q$$

which means that f is an onto function. Since $f: A_n \to B_n$ is one to one and onto, we conclude that A_n and B_n have the same number of elements. Note that $A_n \cap B_n = \varnothing$, since no permutation can be both even and odd. Also, by Theorem 1, $|A_n \cup B_n| = n!$. Thus by Theorem 2 of Section 1.3,

$$n! = |A_n \cup B_n| = |A_n| + |B_n| - |A_n \cap B_n| = 2|A_n|$$

We then have

$$|A_n| = |B_n| = \frac{n!}{2}$$

EXERCISE SET 3.2

1. Which of the following functions $f: \mathbb{R} \to \mathbb{R}$ are permutations of $\mathbb{R}$?

  (a) f is defined by $f(a) = a - 1$.
  (b) f is defined by $f(a) = a^2$.
  (c) f is defined by $f(a) = a^3$.
  (d) f is defined by $f(a) = e^a$.

2. Which of the following functions $f: Z \to Z$ are permutations of Z?

  (a) f is defined by $f(a) = a + 1$.
  (b) f is defined by $f(a) = (a - 1)^2$.
  (c) f is defined by $f(a) = a^2 + 1$.
  (d) f is defined by $f(a) = a^3 - 3$.

In Exercises 3 and 4, let $A = \{1, 2, 3, 4, 5, 6\}$ and

$$p_1 = \begin{pmatrix} 1 & 2 & 3 & 4 & 5 & 6 \\ 3 & 4 & 1 & 2 & 6 & 5 \end{pmatrix},$$

$$p_2 = \begin{pmatrix} 1 & 2 & 3 & 4 & 5 & 6 \\ 2 & 3 & 1 & 5 & 4 & 6 \end{pmatrix},$$

$$p_3 = \begin{pmatrix} 1 & 2 & 3 & 4 & 5 & 6 \\ 6 & 3 & 2 & 5 & 4 & 1 \end{pmatrix}$$

3. Compute

  (a) p_1^{-1} (b) $p_1 \circ p_3$
  (c) $p_2 \circ (p_1 \circ p_2)$ (d) $(p_2^{-1} \circ p_3) \circ p_1$

4. Compute

  (a) p_3^{-1} (b) $p_2^{-1} \circ p_1^{-1}$
  (c) $p_1 \circ (p_2 \circ p_3)$ (d) $(p_1 \circ p_2)^{-1} \circ p_3$

In Exercises 5–8, let $A = \{1, 2, 3, 4, 5, 6, 7, 8\}$. Compute the products.

5. $(1, 3, 2) \circ (3, 5, 7, 8)$

6. $(2, 5, 3, 4) \circ (3, 5, 7, 8) \circ (2, 6)$

7. $(1, 4, 6, 7) \circ (2, 4, 5, 6) \circ (1, 4)$

8. $(3, 5, 6, 7) \circ (1, 2, 3, 4) \circ (5, 8)$

In Exercises 9 and 10, let $A = \{a, b, c, d, e, f, g\}$. Compute the products.

9. $(b, c, d, e) \circ (a, f, g)$

10. $(a, b, c) \circ (b, c, f) \circ (f, g)$

In Exercises 11–14, let $A = \{1, 2, 3, 4, 5, 6, 7, 8\}$. Write each permutation as the product of disjoint cycles.

11. $\begin{pmatrix} 1 & 2 & 3 & 4 & 5 & 6 & 7 & 8 \\ 4 & 3 & 2 & 5 & 1 & 8 & 7 & 6 \end{pmatrix}$

12. $\begin{pmatrix} 1 & 2 & 3 & 4 & 5 & 6 & 7 & 8 \\ 2 & 3 & 4 & 1 & 7 & 5 & 8 & 6 \end{pmatrix}$

13. $\begin{pmatrix} 1 & 2 & 3 & 4 & 5 & 6 & 7 & 8 \\ 6 & 5 & 7 & 8 & 4 & 3 & 2 & 1 \end{pmatrix}$

14. $\begin{pmatrix} 1 & 2 & 3 & 4 & 5 & 6 & 7 & 8 \\ 2 & 3 & 1 & 4 & 6 & 7 & 8 & 5 \end{pmatrix}$

In Exercises 15 and 16, let $A = \{a, b, c, d, e, f, g\}$. Write each permutation as the product of disjoint cycles.

15. $\begin{pmatrix} a & b & c & d & e & f & g \\ g & d & b & a & c & f & e \end{pmatrix}$

16. $\begin{pmatrix} a & b & c & d & e & f & g \\ d & e & a & b & g & f & c \end{pmatrix}$

In Exercises 17 and 18, let $A = \{1, 2, 3, 4, 5, 6, 7, 8\}$. Write each permutation as a product of transpositions.

17. $(2, 1, 4, 5, 8, 6)$ **18.** $(4, 8, 2, 5) \circ (3, 1, 6)$

In Exercises 19–22, let $A = \{1, 2, 3, 4, 5, 6, 7, 8\}$. Determine whether the permutation is even or odd.

19. $\begin{pmatrix} 1 & 2 & 3 & 4 & 5 & 6 & 7 & 8 \\ 4 & 2 & 1 & 6 & 5 & 8 & 7 & 3 \end{pmatrix}$

20. $\begin{pmatrix} 1 & 2 & 3 & 4 & 5 & 6 & 7 & 8 \\ 7 & 3 & 4 & 2 & 1 & 8 & 6 & 5 \end{pmatrix}$

21. $(6, 4, 2, 1, 5)$

22. $(2, 4, 7, 1) \circ (3, 5, 2, 1) \circ (4, 8)$

23. Prove that

(a) The product of two even permutations is even.

(b) The product of two odd permutations is even.

(c) The product of an even and an odd permutation is odd.

24. Show that if p is a permutation of a finite set A, then $p^2 = p \circ p$ is a permutation of A.

25. (a) Use mathematical induction to show that if p is a permutation of a finite set A, then $p^n = p \circ p \circ \cdots \circ p$ is a permutation of A for $n \in Z^+$.

(b) If A is a finite set and p is a permutation of A, show that $p^m = 1_A$ for some $m \in Z^+$.

26. Let p be a permutation of a set A. Define the following relation R on A: $a \, R \, b$ if and only if $p^n(a) = b$ for some $n \in Z$. [p^0 is defined as the identity permutation and p^{-n} is defined as $(p^{-1})^n$.] Show that R is an equivalence relation, and describe the equivalence classes.

KEY IDEAS FOR REVIEW

☐ Function: see page 141.

☐ Function defined by a formula: see page 143.

☐ Identity function, 1_A: $1_A(a) = a$.

☐ One-to-one function f from A to B: $a \neq a'$ implies $f(a) \neq f(a')$.

☐ Onto function f from A to B: Ran $(f) = B$.

☐ Bijection: one-to-one and onto function.

☐ If f is a function from A to B, $f \circ 1_B = f$; $1_A \circ f = f$.

☐ If f is an invertible function from A to B, $f \circ f^{-1} = 1_A$; $f^{-1} \circ f = 1_B$.

☐ Theorem: A function $f: A \to B$ is invertible if and only if f is one to one and onto.

☐ $(f \circ g)^{-1} = g^{-1} \circ f^{-1}$

☐ Permutation: a bijection from a set A to itself.

☐ Theorem: If A is a set that has n elements, then there are $n!$ permutations of A.

☐   Cycle of length r: $(b_1, b_2, \ldots, b_r)$; see page 155.

☐   Theorem: A permutation ot a finite set that is not the identity or a cycle can be written as a product of disjoint cycles.

☐   Transposition: a cycle of length 2.

☐   Corollary: Every permutation of a finite set with at least two elements can be written as a product of transpositions.

☐   Even (odd) permutation: one that can be written as a product of an even (odd) number of transpositions.

☐   Theorem: If a permutation of a finite set can be written as a product of an even number of transpositions, then it can never be written as a product of an odd number of transpositions, and conversely.

☐   The product of:

    (a)   Two even permutations is even.

    (b)   Two odd permutations is even.

    (c)   An even and an odd permutation is odd.

☐   Theorem: If A is a set that has n elements, then there are $n!/2$ even permutations and $n!/2$ odd permutations of A.

FURTHER READING

FRALEIGH, JOHN B., *A First Course in Abstract Algebra*, 3rd ed., Addison-Wesley, Reading, Mass., 1982.

Order, Relations and Structures

Prerequisites: Chapters 1 and 2.

In this chapter we study partially ordered sets, including lattices and Boolean algebras. The structures are useful in set theory, algebra, sorting and searching, and, especially in the case of Boolean algebras, in the construction of logical representations for computer circuits.

4.1 Partially Ordered Sets (Posets)

A relation R on a set A is called a **partial order** if R is reflexive, antisymmetric, and transitive. The set A together with the partial order R is called a **partially ordered set**, or simply a **poset**, and we will denote this poset by (A, R). If there is no possibility of confusion about the partial order, we may refer to the poset simply as A, rather than (A, R).

Example 1   Let A be a collection of subsets of a set S. The relation $\subseteq$ of set inclusion is a partial order on A, so $(A, \subseteq)$ is a poset.

Example 2   Let Z^+ be the set of all positive integers. The usual relation $\leq$ (less than or equal to) is a partial order on Z^+, as is $\geq$ (greater than or equal to).

Example 3   The relation of divisibility ($a \; R \; b$ if and only if $a \mid b$) is a partial order on Z^+.

Example 4 Let W be the set of all equivalence relations on a set A. Since W consists of subsets of $A \times A$, W is a partially ordered set under the partial order of set containment. If R and S are equivalence relations on A, the same property may be expressed in relational notation as follows.

$$R \subseteq S \text{ if and only if } x \, R \, y \quad \text{implies} \quad x \, S \, y \text{ for all } x, y \text{ in } A$$

Example 5 The relation $<$ on Z^+ is not a partial order since it is not reflexive.

Example 6 Let R be a partial order on a set A, and let R^{-1} be the inverse relation of R. Then R^{-1} is also a partial order. To see this we recall the characterizations of reflexive, antisymmetric, and transitive given in Section 2.6. If R has these three properties, then $\Delta \subseteq R$, $R \cap R^{-1} \subseteq \Delta$, and $R^2 \subseteq R$. By taking inverses, we have $\Delta = \Delta^{-1} \subseteq R^{-1}$, $R^{-1} \cap (R^{-1})^{-1} = R^{-1} \cap R \subseteq \Delta$, and $(R^{-1})^2 \subseteq R^{-1}$, so by Section 2.6, R^{-1} is reflexive, antisymmetric, and transitive. Thus R^{-1} is also a partial order. The poset (A, R^{-1}) is called the **dual** of the poset (A, R), and the partial order R^{-1} is called the **dual** of the partial order R.

The most familiar partial orders are the relations $\leq$ and $\geq$ on Z and $\mathbb{R}$. For this reason, when speaking in general of a partial order R on a set A, we shall often use the symbols $\leq$ or $\geq$ for R. This makes the properties of R more familiar and easier to remember. Thus the reader may see the symbol $\leq$ used for many different partial orders, on different sets. Do not mistake this to mean that these relations are all the same, or that they have anything to do with the familiar relation $\leq$ on Z or $\mathbb{R}$. If it becomes absolutely necessary to distinguish partial orders from one another, we may also use symbols such as $\leq_1$, $\leq'$, $\geq_1$, $\geq'$, and so on, to denote partial orders.

We will observe the following convention. Whenever $(A, \leq)$ is a poset, we will always use the symbol $\geq$ for the partial order $\leq^{-1}$, and thus $(A, \geq)$ will be the dual poset. Similarly, the dual of poset $(A, \leq_1)$ will be denoted by $(A, \geq_1)$ and the dual of the poset $(B, \leq')$ will be denoted by $(B, \geq')$. Again, this convention is to remind us of the familiar dual posets $(Z, \leq)$ and $(Z, \geq)$, as well as the posets $(\mathbb{R}, \leq)$ and $(\mathbb{R}, \geq)$.

If $(A, \leq)$ is a poset, the elements a and b of A are said to be **comparable** if

$$a \leq b \quad \text{or} \quad b \leq a$$

Observe that in a partially ordered set every pair of elements need not be comparable. For example, consider the poset in Example 3. The elements 2 and 7 are not comparable, since $2 \not| 7$ and $7 \not| 2$. Thus the word "partial" in partially ordered set means that some elements may not be comparable. If every pair of elements in a poset A are comparable, we say that A is a **linearly ordered** set and the partial order is called a **linear order**. We also say that A is a **chain**.

Example 7   The poset of Example 2 is linearly ordered.

The following theorem is sometimes useful, since it shows how to construct a new poset from given posets.

Theorem 1   If $(A, \le)$ and $(B, \le)$ are posets, then $(A \times B, \le)$ is a poset, with partial order $\le$ defined by

$$(a, b) \le (a', b') \qquad \text{if } a \le a' \text{ in } A \text{ and } b \le b' \text{ in } B$$

Note that the symbol $\le$ is being used to denote three distinct partial orders. The reader should find it easy to determine which of the three is meant at any time.

Proof. If $(a, b) \in A \times B$, then $(a, b) \le (a, b)$ since $a \le a$ in A and $b \le b$ in B, so $\le$ satisfies the reflexive property in $A \times B$. Now, suppose that $(a, b) \le (a', b')$ and $(a', b') \le (a, b)$, where a and $a' \in A$, and b and $b' \in B$. Then

$$a \le a' \qquad \text{and} \qquad a' \le a \qquad \text{in } A$$

and

$$b \le b' \qquad \text{and} \qquad b' \le b \qquad \text{in } B$$

Since A and B are posets, the antisymmetry of the partial orders in A and B implies that

$$a = a' \qquad \text{and} \qquad b = b'$$

Hence $\le$ satisfies the antisymmetric property in $A \times B$.
    Finally, suppose

$$(a, b) \le (a', b') \qquad \text{and} \qquad (a', b') \le (a'', b'')$$

where $a, a', a'' \in A$, and $b, b', b'' \in B$. Then

$$a \le a' \qquad \text{and} \qquad a' \le a''$$

so $a \le a''$, by the transitive property of the partial order in A. Similarly,

$$b \le b' \qquad \text{and} \qquad b' \le b''$$

so $b \le b''$, by the transitive property of the partial order in B. Hence

$$(a, b) \le (a'', b'')$$

Consequently, the transitive property holds for the partial order in $A \times B$, and we conclude that $A \times B$ is a poset.

The partial order $\le$ defined on the Cartesian product $A \times B$ as above is called the **product partial order**.

If $(A, \le)$ is a poset, we say that $a < b$ if $a \le b$ but $a \ne b$. Suppose now that $(A, \le)$ and $(B, \le)$ are posets. In Theorem 1 we have defined the product partial order on $A \times B$. Another useful partial order on $A \times B$, denoted by $\prec$, is defined as follows:

$$(a, b) \prec (a', b') \quad \text{if } a < a' \quad \text{or} \quad \text{if } a = a' \text{ and } b \le b'$$

This ordering is called **lexicographic**, or "dictionary" order. The ordering of the elements in the first coordinate dominates, except in case of "ties," when attention passes on to the second coordinate. If $(A, \le)$ and $(B, \le)$ are linearly ordered sets, then the lexicographic order $\prec$ on $A \times B$ is also a linear order.

Example 8   Let $A = \mathbb{R}$, with the usual ordering $\le$. Then the plane $\mathbb{R}^2 = \mathbb{R} \times \mathbb{R}$ may be given lexicographic order. This is illustrated in Fig. 1. We see that the plane is linearly ordered by lexicographic order. Each vertical line has the usual order and points on one line are less than any points on a line farther to the right. Thus in Fig. 1 $p_1 \prec p_2$, $p_1 \prec p_3$, and $p_2 \prec p_3$.

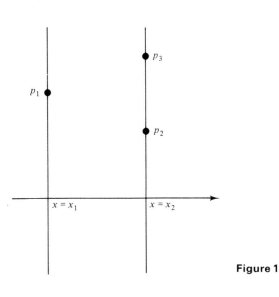

Figure 1

Lexicographic ordering is easily extended to Cartesian products $A_1 \times A_2 \times \cdots \times A_n$ as follows:

$$(a_1, a_2, \ldots, a_n) \prec (a'_1, a'_2, \ldots, a'_n) \qquad \text{if and only if}$$

$$a_1 < a'_1 \text{ or}$$

$$a_1 = a'_1 \text{ and } a_2 < a'_2 \text{ or}$$

$$a_1 = a'_1, a_2 = a'_2, \text{ and } a_3 < a'_3 \text{ or } \ldots$$

$$a_1 = a'_1, a_2 = a'_2, \ldots, a_{n-1} = a'_{n-1} \text{ and } a_n \leq a'_n$$

Thus the first coordinate dominates except for equality, in which case we consider the second coordinate. If equality holds again, we pass to the next coordinate, and so on.

Example 9   Let $S = \{a, b, \ldots, z\}$ be the ordinary alphabet, linearly ordered in the usual way ($a \leq b, b \leq c, \ldots, y \leq z$). Then S^n (see Section 2.1) can be identified with the set of all words having length n. Lexicographic order on S^n has the property that if $w_1 \prec w_2$ ($w_1, w_2 \in S^n$), then w_1 would precede w_2 in a dictionary listing. This fact accounts for the name of the ordering.

Thus *park* $\prec$ *part*, *help* $\prec$ *hind*, *jump* $\prec$ *mump*. The third is true since $j < m$, the second since $h = h$, but $e < i$, and the first is true since $p = p, a = a, r = r, k < t$.

If S is a poset we can extend lexicographic order to S^* (see Section 1.2) in the following way.

If $x = a_1 a_2 \cdots a_n$ and $y = b_1 b_2 \cdots b_k$ are in S^* with $n \leq k$, we say that $x \prec y$ if $(a_1, \ldots, a_n) \prec (b_1, \ldots, b_n)$ in S^n under lexicographic ordering of S^n. In other words, we chop off to the length of the shortest word, and then compare.

In the previous paragraph, we use the fact that the n-tuple $(a_1, a_2, \ldots, a_n) \in S^n$ and the string $a_1 a_2 \ldots a_n \in S^*$ are really the same sequence of length n, written in two different notations. The notations differ for historical reasons, and we will use them interchangeably, depending on context.

Example 10   Let $S = \{a, b, \ldots, z\}$, ordered as usual. Then S^* is the set of all possible "words" of any length, whether such words are meaningful or not.

Thus we have

$$help \prec helping$$

in S^* since

$$help \prec help$$

in S^4. Similarly, we have

$$helper \prec helping$$

since

$$helper \prec helpin$$

in S^6. As the example

$$help \prec helping$$

shows, this order includes "prefix order"; that is, any word is greater than all of its prefixes (beginning parts). This is also the way that words occur in the dictionary.

Thus we have dictionary ordering again, but this time for words of any finite length.

Since a partial order is a relation, we can look at the digraph of the partial order. We shall find that the digraphs of partial orders can be represented in a simpler manner than those of general relations. The following theorem provides the first result in this direction.

Theorem 2   The digraph of a partial order has no cycle of length greater than 1.

Proof. Suppose that the digraph of the partial order $\leq$ on the set A contains a cycle of length $n \geq 2$. Then there exist distinct elements $a_1, a_2, \ldots, a_n \in A$ such that

$$a_1 \leq a_2, a_2 \leq a_3, \ldots, a_{n-1} \leq a_n, a_n \leq a_1$$

By the transitivity of the partial order, used $n - 1$ times, $a_1 \leq a_n$. By antisymmetry, $a_n \leq a_1$ and $a_1 \leq a_n$ imply that $a_n = a_1$, a contradiction to the assumption that $a_1, a_2, \ldots, a_n$ are distinct.

Hasse Diagrams

Theorem 2 has shown that the digraph of a partial order has only cycles of length 1. Indeed, since a partial order is reflexive, every vertex in the digraph of the partial order is contained in a cycle of length 1. To simplify matters, we shall delete all such cycles from the digraph. Thus the digraph shown in Fig. 2(a) would be drawn as shown in Fig. 2(b).

We shall also eliminate all edges that are implied by the transitive property. Thus, if $a \leq b$, and $b \leq c$, it follows that $a \leq c$. In this case, we omit the edge from a to c; however, we do draw the edges from a to b and from b to c. For example, the digraph shown in Fig. 3(a) would be drawn as shown in Fig. 3(b). We also agree to draw the digraph of a partial order with all edges pointing upward, so that arrows

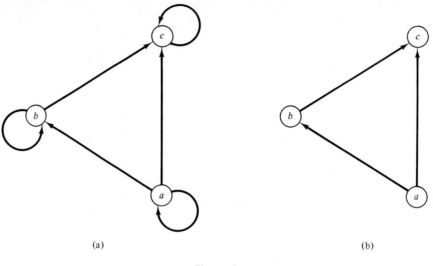

(a)   (b)

Figure 2

may be omitted from the edges. Finally, we replace the circles representing the vertices by dots. Thus the diagram shown in Fig. 4 gives the final form of the digraph shown in Fig. 2(a). The resulting diagram of a partial order, much simpler than its digraph, is called the **Hasse diagram** of the partial order or of the poset. Since the Hasse diagram completely describes the associated partial order, we shall find it to be a very useful tool.

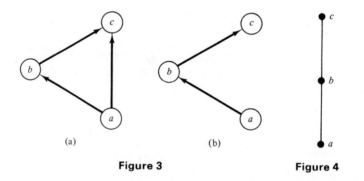

(a)   (b)

Figure 3   **Figure 4**

Example 11   Let $A = \{1, 2, 3, 4, 12\}$. Consider the partial order of divisibility on A. That is, if a and $b \in A$, $a \leq b$ if and only if $a \mid b$. Draw the Hasse diagram of the poset $(A, \leq)$.

Solution. The Hasse diagram is shown in Fig. 5.

To emphasize the simplicity of the Hasse diagram, we show in Fig. 6 the digraph of the poset in Fig. 5.

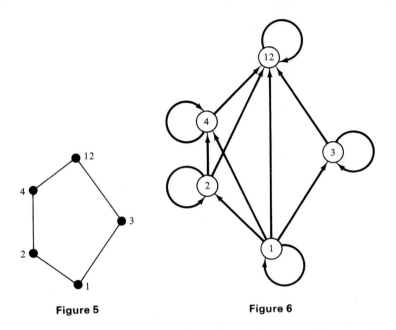

12

4

2

3

1

Figure 5

Figure 6

Example 12   Let $S = \{a, b, c\}$ and let $A = P(S)$. Draw the Hasse diagram of the poset A with the partial order $\subseteq$ (set inclusion).

Solution. We first determine A, obtaining

$$A = \{\varnothing, \{a\}, \{b\}, \{c\}, \{a, b\}, \{a, c\}, \{b, c\}, \{a, b, c\}\}$$

The Hasse diagram can then be drawn as shown in Fig. 7.

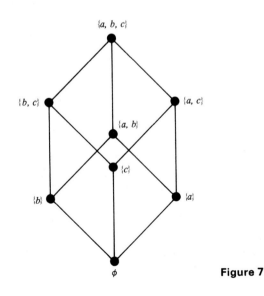

Figure 7

Observe that the Hasse diagram of a linearly ordered set is always of the form shown in Fig. 8.

It is easily seen that if $(A, \leq)$ is a poset and $(A, \geq)$ is the dual poset, the Hasse diagram of $(A, \geq)$ is just the Hasse diagram of $(A, \leq)$, turned upside down.

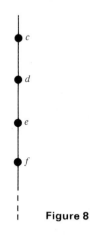

Figure 8

Example 13   Figure 9(a) shows the Hasse diagram of a poset $(A, \leq)$, where $A = \{a, b, c, d, e, f\}$. Figure 9(b) shows the Hasse diagram of the dual poset $(A, \geq)$. Notice that, as mentioned above, each of these diagrams can be constructed by turning the other upside down.

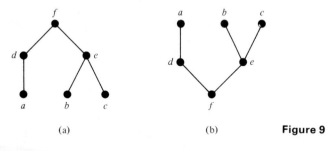

(a)   (b)   **Figure 9**

Topological Sorting

If A is a poset with partial order $\leq$, we sometimes need to find a linear order $\prec$ for the set A that will merely be an extension of the given partial order in the sense that if $a \leq b$, then $a \prec b$. The process of constructing a linear order such as $\prec$ is called **topological sorting.** This problem might arise when we have to enter a finite poset A into a computer. The elements of A must be entered in sequential order, and we might want them entered so that the partial order is preserved. That is, if $a \leq b$, then a is entered before b. A topological sorting $\prec$ will give an order of entry of the elements which meets this condition.

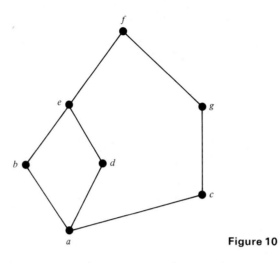

Figure 10

Example 14   Give a topological sorting for the poset whose Hasse diagram is shown in Fig. 10.

Solution. The partial order $\prec$ whose Hasse diagram is shown in Fig. 11(a), is clearly a linear order. It is easy to see that every pair in $\leq$ is also in the order $\prec$, so $\prec$ is a topological sorting of the partial order $\leq$. Figure 11(b) and (c) show two other solutions to this problem.

As Example 14 shows, there are many ways of topologically sorting a given poset. An algorithm for generating topological sortings will be given in Section 4.2.

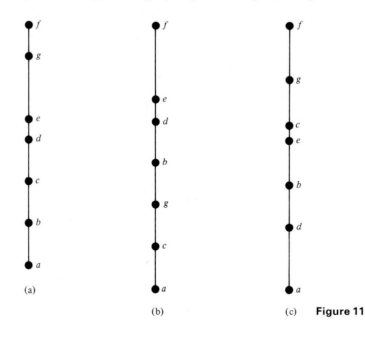

Figure 11

EXERCISE SET 4.1

In Exercises 1–4, determine whether the relation R is a partial order on the set A.

1. $A = Z$, and $a \ R \ b$ if and only if $a = 2b$.
2. $A = Z$, and $a \ R \ b$ if and only if $b^2 \mid a$.
3. $A = Z$, and $a \ R \ b$ if and only if $a = b^k$ for some $k \in Z^+$.
4. $A = \mathbb{R}$, and $a \ R \ b$ if and only if $a \leq b$.

In Exercises 5–8, determine whether the relation R is a linear order on the set A.

5. $A = \mathbb{R}$, and $a \ R \ b$ if and only if $a \leq b$.
6. $A = \mathbb{R}$, and $a \ R \ b$ if and only if $a \geq b$.
7. $A = P(S)$, where S is a set. The relation R is set inclusion.
8. $A = \mathbb{R} \times \mathbb{R}$, and $(a, b)R(a', b')$ if and only if $a \leq a'$ and $b \leq b'$, where $\leq$ is the usual partial order on $\mathbb{R}$.

9. On the set $A = \{a, b, c\}$, find all partial orders $\leq$ in which $a \leq b$.

10. What can you say about the relation R on a set A if R is a partial order and an equivalence relation?

In Exercises 11 and 12, determine the Hasse diagram of the relation R.

11. $A = \{1, 2, 3, 4\}$, $R = \{(1, 1), (1, 2), (2, 2), (2, 4), (1, 3), (3, 3), (3, 4), (1, 4), (4, 4)\}$
12. $A = \{a, b, c, d, e\}$, $R = \{(a, a), (b, b), (c, c), (a, c), (c, d), (c, e), (a, d), (d, d), (a, e), (b, c), (b, d), (b, e), (e, e)\}$

In Exercises 13 and 14, describe the ordered pairs in the relation determined by the Hasse diagram on the set A.

13. $A = \{1, 2, 3, 4\}$ 14. $A = \{1, 2, 3, 4\}$

In Exercises 15 and 16, determine the Hasse diagram of the partial order having the given digraph.

15.

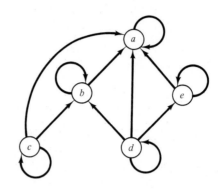

16.

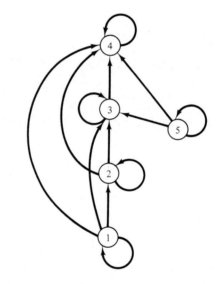

In Exercises 17 and 18, determine the Hasse diagram of the relations on $A = \{1, 2, 3, 4, 5\}$ whose matrix is shown.

17. $\begin{bmatrix} 1 & 1 & 1 & 1 & 1 \\ 0 & 1 & 1 & 1 & 1 \\ 0 & 0 & 1 & 1 & 1 \\ 0 & 0 & 0 & 1 & 1 \\ 0 & 0 & 0 & 0 & 1 \end{bmatrix}$ 18. $\begin{bmatrix} 1 & 0 & 1 & 1 & 1 \\ 0 & 1 & 1 & 1 & 1 \\ 0 & 0 & 1 & 1 & 1 \\ 0 & 0 & 0 & 1 & 0 \\ 0 & 0 & 0 & 0 & 1 \end{bmatrix}$

In Exercises 19 and 20, determine the matrix of the partial order whose Hasse diagram is given.

19.   **20.**

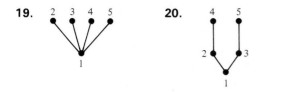

21. Let $A = Z^+ \times Z^+$ have lexicographic order. Answer each of the following as true or false.

   (a) $(2, 12) \prec (5, 3)$ (b) $(3, 6) \prec (3, 24)$

   (c) $(4, 8) \prec (4, 6)$ (d) $(15, 92) \prec (12, 3)$

   In Exercises 22–25, consider the partial order of divisibility on the set A. Draw the Hasse diagram of the poset and determine which posets are linearly ordered.

22. $A = \{1, 2, 3, 5, 6, 10, 15, 30\}$

23. $A = \{2, 4, 8, 16, 32\}$

24. $A = \{3, 6, 12, 36, 72\}$

25. $A = \{1, 2, 3, 4, 5, 6, 10, 12, 15, 30, 60\}$

26. Let $A = \{\Box, A, B, C, E, O, M, P, S\}$ have the usual alphabetical order, where $\Box$ represents a "blank" character and $\Box \leq x$ for all $x \in A$. Arrange the following in lexicographic order (as elements of $A \times A \times A \times A$).

   (a) MOP$\Box$ (e) BASE

   (b) MOPE   (f) ACE$\Box$

   (c) CAP$\Box$ (g) MACE

   (d) MAP$\Box$ (h) CAPE

   In Exercises 27 and 28, draw the Hasse diagram of a topological sorting of the given poset.

27.   **28.**

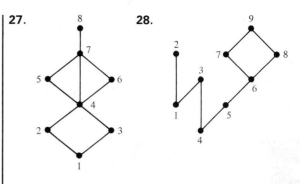

29. If $(A, \leq)$ is a poset and A' is a subset of A, show that $(A', \leq')$ is also a poset, where $\leq'$ is the restriction of $\leq$ to A'.

30. Show that if R is a linear order on the set A, then R^{-1} is also a linear order on A.

31. A relation R on a set A is called a **quasiorder** if it is transitive and irreflexive. Let $A = P(S)$ be the power set of a set S, and consider the following relation R on A: $U \, R \, T$ if and only if $U \subsetneq T$ (proper containment). Show that R is a quasiorder.

32. Let $A = \{x \mid x$ is a real number and $-5 \leq x \leq 20\}$. Show that the usual relation $<$ is a quasiorder (see Exercise 31) on A.

33. If R is a quasiorder on A (see Exercise 31), show that R^{-1} is also a quasiorder.

34. Let $B = \{2, 3, 6, 9, 12, 18, 24\}$ and let $A = B \times B$. Define the following relation on A: $(a, b) \prec (a', b')$ if and only if $a \mid a'$ and $b \leq b'$, where $\leq$ is the usual partial order. Show that $\prec$ is a partial order.

4.2 Extremal Elements of Partially Ordered Sets

Certain elements in a poset are of special importance for many of the properties and applications of posets. In this section we discuss these elements and in later sections we shall see the important role played by them. In this section we consider a poset $(A, \leq)$ with partial order $\leq$.

   An element $a \in A$ is called a **maximal element** of A if there is no element c in A such that $a < c$ (see Section 4.1). An element $b \in A$ is called a **minimal element** of A if there is no element c in A such that $c < b$.

It follows immediately that if $(A, \leq)$ is a poset and $(A, \geq)$ is its dual poset, an element $a \in A$ is a maximal element of $(A, \leq)$ if and only if a is a minimal element of $(A, \geq)$. Also, a is a minimal element of $(A, \leq)$ if and only if it is a maximal element of $(A, \geq)$.

Example 1   Consider the poset A whose Hasse diagram is shown in Fig. 1. The elements a_1, a_2, and a_3 are maximal elements of A and the elements b_1, b_2, and b_3 are minimal elements. Observe that since there is no line between b_2 and b_3, we can conclude neither that $b_3 \leq b_2$ nor that $b_2 \leq b_3$.

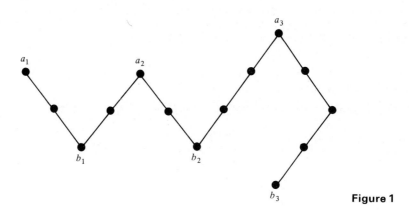

Figure 1

Example 2   Let A be the poset of all nonnegative real numbers with the usual partial order $\leq$. Then 0 is a minimal element of A. There are no maximal elements of A.

Example 3   The poset Z with the usual partial order $\leq$ has no maximal elements and has no minimal elements.

Theorem 1   Let A be a finite nonempty poset with partial order $\leq$. Then A has at least one maximal element and at least one minimal element.

   Proof. Let a be any element of A. If a is not maximal, we can find an element $a_1 \in A$ such that $a < a_1$. If a_1 is not maximal, we can find an element $a_2 \in A$ such that $a_1 < a_2$. This argument cannot be continued indefinitely, since A is a finite set. Thus we eventually obtain the finite chain

$$a < a_1 < a_2 < \cdots < a_{k-1} < a_k$$

which cannot be extended. Hence we cannot have $a_k < b$ for any $b \in A$, so a_k is a maximal element of $(A, \leq)$.

The same argument says that the dual poset $(A, \geq)$ has a maximal element, so $(A, \leq)$ has a minimal element.

By using the concept of a minimal element, we can give an algorithm for finding a topological sorting of a given finite poset $(A, \leq)$. We remark first that if $a \in A$, and $B = A - \{a\}$, then B is also a poset under the restriction of $\leq$ to $B \times B$ (see Section 2.2). We then have the following algorithm, which produces a linear array named SORT. We assume that SORT is ordered by increasing index, that is, $\text{SORT}[1] \prec \text{SORT}[2] \prec \cdots$. The relation $\prec$ on A defined in this way is a topological sorting of $(A, \leq)$.

ALGORITHM SORT
1. $I \leftarrow 1$
2. $S \leftarrow A$
3. WHILE $(S \neq \varnothing)$
   a. Choose a minimal element a of S
   b. $\text{SORT}[I] \leftarrow a$
   c. $I \leftarrow I + 1$
   d. $S \leftarrow S - \{a\}$
END OF ALGORITHM SORT

Example 4   Let $A = \{a, b, c, d, e\}$ and let the Hasse diagram of a partial order $\leq$ on A be as shown in Fig. 2(a). A minimal element of this poset is the vertex labeled d (we could also have chosen e). We put d in SORT[1] and in Fig. 2(b) we show the Hasse diagram of $A - \{d\} = S$. A minimal element of S is e, so that e becomes SORT[2], and $S - \{e\}$ is shown in Fig. 2(c). This process continues until we have exhausted A and filled SORT. Figure 2(f) shows the completed array SORT and the HASSE diagram of the poset corresponding to SORT. This is a topological sorting of $(A, \leq)$.

An element $a \in A$ is called a **greatest element** of A if $x \leq a$ for all $x \in A$. An element $a \in A$ is called a **least element** of A if $a \leq x$ for all $x \in A$.

As before, an element a of $(A, \leq)$ is a greatest (or least) element if and only if it is a least (or greatest) element of $(A, \geq)$.

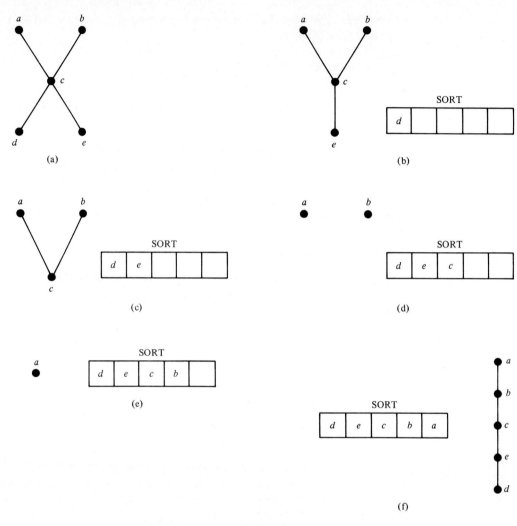

Figure 2

Example 5   Consider the poset defined in Example 2. Then 0 is a least element; there is no greatest element.

Example 6   Let $S = \{a, b, c\}$ and consider the poset $A = P(S)$ defined in Example 12 of Section 4.1. The empty set is a least element and the S is a greatest element.

Example 7   The poset Z with the usual partial order has neither a least nor a greatest element.

Theorem 2   A poset has at most one greatest element and at most one least element.

       Proof. Suppose that a and b are greatest elements of a poset A. Then since b is a greatest element, we have $a \leq b$. Similarly, since a is a greatest element, we have $b \leq a$. Hence $a = b$ by the antisymmetry property. Thus if the poset has a greatest element, it only has one such element. Since this fact is true for all posets, the dual poset $(A, \geq)$ has at most one greatest element, so $(A, \leq)$ also has at most one least element.

       The greatest element of a poset, if it exists, is denoted by I, and is often called the **unit element**. Similarly, the least element of a poset, if it exists, is denoted by 0, and is often called the **zero element**.

       Consider a poset A and a subset B of A. An element $a \in A$ is called an **upper bound** of B if $b \leq a$ for all $b \in B$. An element $a \in A$ is called a **lower bound** of B if $a \leq b$ for all b $\in B$.

Example 8   Consider the poset $A = \{a, b, c, d, e, f, g, h\}$, whose Hasse diagram is shown in Fig. 3. Find all upper and lower bounds of the following subsets of A: (a) $B_1 = \{a, b\}$; (b) $B_2 = \{c, d, e\}$.

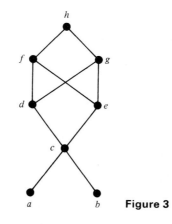

Figure 3

Solution.
(a) B_1 has no lower bounds; its upper bounds are $c, d, e, f, g,$ and h.
(b) The upper bounds of B_2 are $f, g,$ and h; its lower bounds are $c, a,$ and b.

       As Example 8 shows, a subset B of a poset may or may not have upper or lower bounds (in A). Moreover, an upper or lower bound of B may or may not belong to B itself.

       Let A be a poset and B a subset of A. An element $a \in A$ is called a **least upper bound** (LUB) of B if a is an upper bound of B and $a \leq a'$, whenever a' is an upper bound of B. Thus $a = $ LUB (B) if $b \leq a$ for all $b \in B$, and if whenever $a' \in A$ is also an upper bound of $B(b \leq a'$ for all $b \in B)$, then $a \leq a'$.

Similarly, an element $a \in A$ is called a **greatest lower bound** (GLB) of B if a is a lower bound of B and $a' \le a$, whenever a' is a lower bound of B. Thus $a = \text{GLB}(B)$ if $a \le b$ for all $b \in B$, and if whenever $a' \in A$ is also a lower bound of $B(a' \le b$ for all $b \in B)$, then $a' \le a$.

As usual, upper bounds in $(A, \le)$ correspond to lower bounds in $(A, \ge)$ (for the same set of elements), and lower bounds in $(A, \le)$ correspond to upper bounds in $(A, \ge)$. Similar statements hold for greatest lower bounds and least upper bounds.

Example 9 Let A be the poset considered in Example 8 with subsets B_1 and B_2 as defined in that example. Find all least upper bounds and all greatest lower bounds of (a) B_1; (b) B_2.

Solution.

(a) Since B_1 has no lower bounds, it has no greatest lower bounds. However,

$$\text{LUB}(B_1) = c$$

(b) Since the lower bounds of B_2 are c, a, and b, we find that

$$\text{GLB}(B_2) = c$$

The upper bounds of B_2 are f, g, and h. Since f and g are not comparable, we conclude that B_2 has no least upper bound.

Theorem 3 Let $(A, \le)$ be a poset. Then a subset B of A has at most one LUB and at most one GLB.

Proof. Similar to the proof of Theorem 2.

We conclude this section with some remarks about LUB and GLB in a poset A, as viewed from the Hasse diagram of A. Let $B = \{b_1, b_2, \ldots, b_r\}$. If $a = \text{LUB}(B)$, then a is the first vertex that can be reached from $b_1, b_2, \ldots,$ and b_r by an upward path. Similarly, if $a = \text{GLB}(B)$, then a is the first vertex that can be reached from $b_1, b_2, \ldots, b_r$ by a downward path.

Example 10 Let $A = \{1, 2, 3, 4, 5, \ldots, 11\}$ be the poset whose Hasse diagram is shown in Fig. 4. Find the LUB and the GLB of $B = \{6, 7, 10\}$, if they exist.

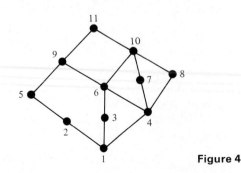

Figure 4

Solution. Exploring all upward paths from vertices 6, 7, and 10, we find that LUB $(B) = 10$. Similarly, by examining all downward paths from 6, 7, and 10, we find that GLB $(B) = 4$.

EXERCISE SET 4.2

In Exercises 1–8, determine all maximal and minimal elements of the poset.

8. $A = \{2, 3, 4, 6, 8, 24, 48\}$ with the partial order of divisibility.

In Exercises 9–16, determine the greatest and least elements, if they exist, of the poset.

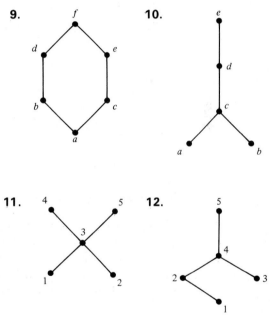

13. $A = \{x \mid x$ is a real number and $0 < x < 1\}$ with the usual partial order $\leq$.

14. $A = \{x \mid x$ is a real number and $0 \leq x \leq 1\}$ with the usual partial order $\leq$.

15. $A = \{2, 4, 6, 8, 12, 18, 24, 36, 72\}$ with the partial order of divisibility.

16. $A = \{2, 3, 4, 6, 12, 18, 24, 36\}$ with the partial order of divisibility.

5. $A = \mathbb{R}$ with the usual partial order $\leq$.

6. $A = \{x \mid x$ is a real number and $0 \leq x < 1\}$ with the usual partial order $\leq$.

7. $A = \{x \mid x$ is a real number and $0 < x \leq 1\}$ with the usual partial order $\leq$.

In Exercises 17–27, find, if they exist, (a) all upper bounds of B; (b) all lower bounds of B; (c) the least upper bound of B; (d) the greatest lower bound of B.

17.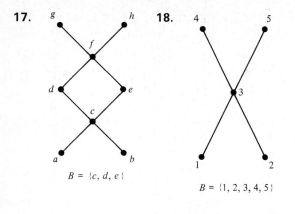

$B = \{c, d, e\}$

18.

$B = \{1, 2, 3, 4, 5\}$

19.

$B = \{b, c, d\}$

20.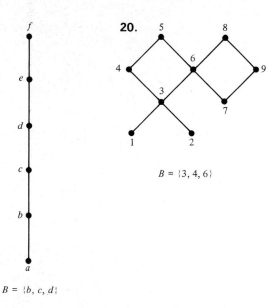

$B = \{3, 4, 6\}$

---

21. $(A, \leq)$ is the poset in Exercise 17; $B = \{b, g, h\}$.

22. $(A, \leq)$ is the poset in Exercise 20; $B = \{4, 6, 9\}$.

23. $(A, \leq)$ is the poset in Exercise 20; $B = \{3, 4, 8\}$.

24. $A = \mathbb{R}$ and $\leq$ denotes the usual partial order; $B = \{x \mid x \text{ is a real number and } 1 < x < 2\}$.

25. $A = \mathbb{R}$ and $\leq$ denotes the usual partial order; $B = \{x \mid x \text{ is a real number and } 1 \leq x < 2\}$.

26. $A = P(\{a, b, c\})$ and $\leq$ denotes the partial order of containment; $B = P(\{a, b\})$.

27. $A = \{2, 3, 4, 6, 8, 12, 24, 48\}$ and $\leq$ denotes the partial order of divisibility; $B = \{4, 6, 12\}$.

28. Construct the Hasse diagram of a topological sorting of the poset whose Hasse diagram is shown in Exercise 17. Use the algorithm SORT.

29. Construct the Hasse diagram of a topological sorting of the poset whose Hasse diagram is shown in Exercise 18. Use the algorithm SORT.

4.3 Lattices

A **lattice** is a poset $(L, \leq)$ in which every subset $\{a, b\}$ consisting of two elements has a least upper bound and a greatest lower bound. We denote LUB $(\{a, b\})$ by $a \vee b$ and call it the **join** of a and b. Similarly, we denote GLB $(\{a, b\})$ by $a \wedge b$ and call it the **meet** of a and b. Lattice structures often appear in computing and mathematical applications.

Example 1   Let S be a set and let $L = P(S)$. As we have seen, $\subseteq$, containment, is a partial order on L. If A and B are two elements of L, (that is, subsets of S), then the join of A and B is their union $A \cup B$, and the meet of A and B is their intersection $A \cap B$. Hence L is a lattice.

Example 2 Consider the poset $(Z^+, \leq)$, where for a and b in Z^+, $a \leq b$ if and only if $a \mid b$. Then L is a lattice in which the join and meet of a and b are their least common multiple and greatest common divisor, respectively (see Section 1.7). That is,

$$a \vee b = \text{LCM } (a, b) \quad \text{and} \quad a \wedge b = \text{GCD } (a, b)$$

Example 3 Let n be a positive integer and let D_n be the set of all positive divisors of n. Then D_n is a lattice under the relation of divisibility as considered in Example 2. Thus if $n = 20$, we have $D_{20} = \{1, 2, 4, 5, 10, 20\}$. The Hasse diagram of D_{20} is shown in Fig. 1(a). If $n = 30$, we have $D_{30} = \{1, 2, 3, 5, 6, 10, 15, 30\}$. The Hasse diagram of D_{30} is shown in Fig. 1(b).

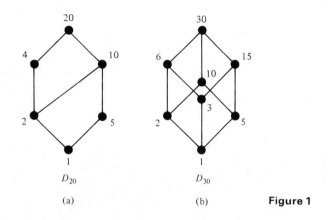

(a) (b) **Figure 1**

Example 4 Which of the Hasse diagrams in Fig. 2 represent lattices?

Solution. Hasse diagrams (a), (b), (d), and (e) represent lattices. Diagram (c) does not represent a lattice because $f \vee g$ does not exist. Diagram (f) does not represent a lattice because neither $d \wedge e$ nor $b \vee c$ exist. Diagram (g) does not represent a lattice because $c \wedge d$ does not exist.

Example 5 We have already observed in Example 4 of Section 4.1 that the set W of all equivalence relations on a set A is a poset under the partial order of set containment. We can now conclude that W is a lattice where the meet of the equivalence relations R and S is their intersection $R \cap S$ and their join is $(R \cup S)^\infty$, the transitive closure of their union (see Section 2.7).

Let $(L, \leq)$ be a poset and let $(L, \geq)$ be the dual poset. If $(L, \leq)$ is a lattice, we can show that $(L, \geq)$ is also a lattice. In fact, for any a and b in L, the least upper bound of a and b in $(L, \leq)$ is equal to the greatest lower bound of a and b in $(L, \geq)$. Similarly, the greatest lower bound of a and b in $(L, \leq)$ is equal to the least upper

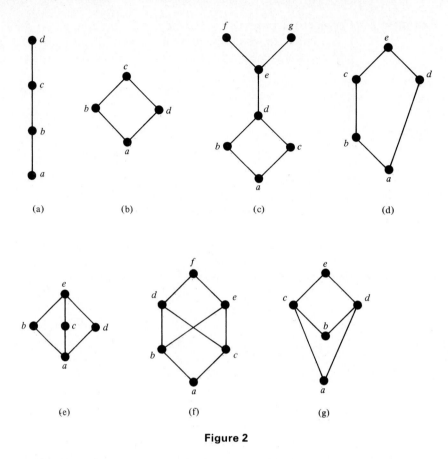

Figure 2

bound of a and b in $(L, \geq)$. If L is a finite set, this property can easily be seen by examining the Hasse diagrams of the poset and its dual.

Example 6 Let S be a set and $L = P(S)$, the set of all subsets of S. Then $(L, \subseteq)$ is a lattice, and its dual lattice is $(L, \supseteq)$, where $\subseteq$ is "contained in" and $\supseteq$ is "contains."

Let A and B belong to the poset $(L, \subseteq)$. Then $A \vee B$ is the set $A \cup B$. To see this, note that $A \subseteq A \cup B$, $B \subseteq A \cup B$, and if $A \subseteq C$ and $B \subseteq C$, then it follows that $A \cup B \subseteq C$. Similarly, we can show that the element $A \wedge B$ in $(L, \subseteq)$ is the set $A \cap B$. The discussion preceding this example then shows that in the poset $(L, \supseteq)$, the join $A \vee B$ is the set $A \cap B$, and the meet $A \wedge B$ is the set $A \cup B$.

Theorem 1 If $(L_1, \leq)$ and $(L_2, \leq)$ are lattices, then $(L, \leq)$ is a lattice, where $L = L_1 \times L_2$, and the partial order $\leq$ of L is the product partial order.

Proof. We denote the join and meet in L_1 by $\vee_1$ and $\wedge_1$, respectively, and the join and meet in L_2 by $\vee_2$ and $\wedge_2$, respectively. We already know from Theorem 1 of Section 4.1 that L is a poset. We now need to show that if (a_1, b_1) and $(a_2, b_2) \in L$, then $(a_1, b_1) \vee (a_2, b_2)$ and $(a_1, b_1) \wedge (a_2, b_2)$ exist in L. We leave it as an exercise to verify that

$$(a_1, b_1) \vee (a_2, b_2) = (a_1 \vee_1 a_2, b_1 \vee_2 b_2)$$

$$(a_1, b_1) \wedge (a_2, b_2) = (a_1 \wedge_1 a_2, b_1 \wedge_2 b_2)$$

Thus L is a lattice.

Example 7   Let L_1 and L_2 be the lattices shown in Fig. 3(a) and (b), respectively. Then $L = L_1 \times L_2$ is the lattice shown in Fig. 3(c).

Let $(L, \leq)$ be a lattice. A nonempty subset S of L is called a **sublattice** of L if $a \vee b \in S$ and $a \wedge b \in S$ whenever $a \in S$ and $b \in S$.

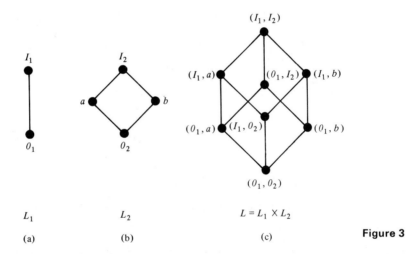

L_1 L_2 $L = L_1 \times L_2$

(a)   (b)   (c)   **Figure 3**

Example 8   The lattice D_n of all positive divisors of n (see Example 3) is a sublattice of the lattice Z^+ under the relation of divisibility (see Example 2).

Example 9   Consider the lattice L shown in Fig. 4(a). The partially ordered subset S_b shown in Fig. 4(b) is not a sublattice of L since $a \wedge b \notin S_b$ and $a \vee b \notin S_b$. The partially ordered subset S_c in Fig. 4(c) is not a sublattice of L since $a \vee b = c \notin S_c$. Observe, however, that S_c is a lattice when considered as a poset by itself. The partially ordered subset S_d in Fig. 4(d) is a sublattice of L.

Let $(L, \leq)$ be a lattice with join $\vee$ and meet $\wedge$ and let $(L', \leq')$ be a lattice with join $\vee'$ and meet $\wedge'$. A function $f: L \to L'$ that is one to one and onto is called an

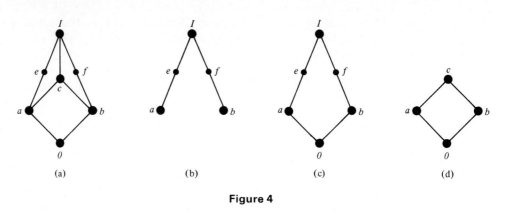

Figure 4

isomorphism from $(L, \leq)$ to $(L', \leq')$ if for any a and b in L, we have

$$f(a \vee b) = f(a) \vee' f(b) \quad \text{and} \quad f(a \wedge b) = f(a) \wedge' f(b)$$

If $f: L \to L'$ is an isomorphism, we say that L and L' are **isomorphic**.

If $(L, \leq)$ and $(L', \leq')$ are isomorphic lattices under the isomorphism $f: L \to L'$, it is not hard to show that

$$a \leq b \quad \text{if and only if} \quad f(a) \leq' f(b)$$

This means that two lattices L and L' are isomorphic if and only if either the Hasse diagrams of the two lattices are identical, or the vertices of the Hasse diagram of L can be relabeled in such a manner that the resulting Hasse diagram is identical to the Hasse diagram of L'.

Example 10 Let L be the lattice D_6, and let L' be the lattice $P(S)$, under the relation of containment, where $S = \{a, b\}$. The Hasse diagrams of these lattices are shown in Fig. 5. If we let $f: L \to L'$ be defined by

$$f(1) = \varnothing$$
$$f(6) = \{a, b\}$$
$$f(2) = \{a\}$$
$$f(3) = \{b\}$$

it is easy to verify that f is an isomorphism, so that L and L' are isomorphic.

In Exercise 13 we ask the student to determine all nonisomorphic lattices containing one, two, three, four, or five elements.

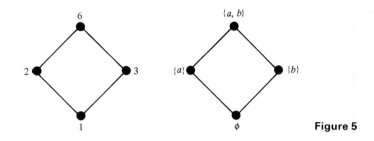

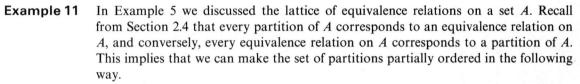

Figure 5

Example 11   In Example 5 we discussed the lattice of equivalence relations on a set A. Recall from Section 2.4 that every partition of A corresponds to an equivalence relation on A, and conversely, every equivalence relation on A corresponds to a partition of A. This implies that we can make the set of partitions partially ordered in the following way.

   If R_1 and R_2 are equivalence relations on A with corresponding partitions $P_1 = A/R_1$, $P_2 = A/R_2$, we define $P_1 \leq P_2$ if $R_1 \subseteq R_2$. The partitions then form a lattice, since the equivalence relations do. In Exercise 34 we describe the partial order of partitions directly in terms of the partitions themselves.

Properties of Lattices

Before proving a number of properties of lattices, we recall the meaning of $a \vee b$ and $a \wedge b$.

1. $a \leq a \vee b$ and $b \leq a \vee b$ ($a \vee b$ is an upper bound of a and b).
2. If $a \leq c$ and $b \leq c$, then $a \vee b \leq c$ ($a \vee b$ is the least upper bound of a and b).
1'. $a \wedge b \leq a$ and $a \wedge b \leq b$ ($a \wedge b$ is a lower bound of a and b).
2'. If $c \leq a$ and $c \leq b$, then $c \leq a \wedge b$ ($a \wedge b$ is the greatest lower bound of a and b).

Theorem 2   Let L be a lattice. Then for every a and b in L:

(a) $a \vee b = b$ if and only if $a \leq b$.
(b) $a \wedge b = a$ if and only if $a \leq b$.
(c) $a \wedge b = a$ if and only if $a \vee b = b$.

   Proof.
   (a) Suppose that $a \vee b = b$. Since $a \leq a \vee b = b$, we have $a \leq b$. Conversely, if $a \leq b$, then since $b \leq b$, b is an upper bound of a and b, so by definition of least upper bound we have $a \vee b \leq b$. Since $a \vee b$ is an upper bound, $b \leq a \vee b$, so $a \vee b = b$.
   (b) The proof is analogous to the proof of (a), and we leave it as an exercise for the student.
   (c) Follows from (a) and (b).

Example 12   Let L be a linearly ordered set. If a and $b \in L$, then either $a \leq b$ or $b \leq a$. It follows from Theorem 2 that L is a lattice, since every pair of elements has a least upper bound and a greatest lower bound.

Theorem 3   Let L be a lattice. Then

1. (a) $a \vee a = a$ ⎫
    (b) $a \wedge a = a$ ⎬   **(Idempotent Property)**

2. (a) $a \vee b = b \vee a$ ⎫
    (b) $a \wedge b = b \wedge a$ ⎬   **(Commutative Property)**

3. (a) $a \vee (b \vee c) = (a \vee b) \vee c$ ⎫
    (b) $a \wedge (b \wedge c) = (a \wedge b) \wedge c$ ⎬   **(Associative Property)**

4. (a) $a \vee (a \wedge b) = a$ ⎫
    (b) $a \wedge (a \vee b) = a$ ⎬   **(Absorption Property)**

Proof.

1. Follows from the definition of LUB and GLB.

2. The definition of LUB and GLB treat a and b symmetrically, so the results follow.

3. (a) From the definition of LUB we have $a \leq a \vee (b \vee c)$ and $b \vee c \leq a \vee (b \vee c)$. Moreover, $b \leq b \vee c$ and $c \leq b \vee c$, so by transitivity, $b \leq a \vee (b \vee c)$ and $c \leq a \vee (b \vee c)$. Thus $a \vee (b \vee c)$ is an upper bound of a and b, so by definition of least upper bound we have

$$a \vee b \leq a \vee (b \vee c)$$

Since $a \vee (b \vee c)$ is an upper bound of $a \vee b$ and c, we obtain

$$(a \vee b) \vee c \leq a \vee (b \vee c)$$

Similarly, $a \vee (b \vee c) \leq (a \vee b) \vee c$. By the antisymmetry of $\leq$, 3(a) follows.

(b) The proof is analogous to the proof of (a), and we omit it.

4. (a) Since $a \wedge b \leq a$ and $a \leq a$, we see that a is an upper bound of $a \wedge b$ and a, so $a \vee (a \wedge b) \leq a$. On the other hand, by the definition of LUB, we have $a \leq a \vee (a \wedge b)$, so $a \vee (a \wedge b) = a$.

(b) The proof is analogous to the proof of (a), and we omit it.

It follows from property 3 that we can write $a \vee (b \vee c)$ and $(a \vee b) \vee c$ merely as $a \vee b \vee c$, and similarly for $a \wedge b \wedge c$. Moreover, we can write

$$LUB \, (\{a_1, a_2, \ldots, a_n\}) \quad \text{as} \quad a_1 \vee a_2 \vee \cdots \vee a_n$$

$$GLB \, (\{a_1, a_2, \ldots, a_n\}) \quad \text{as} \quad a_1 \wedge a_2 \wedge \cdots \wedge a_n$$

since we can show by induction that these joins and meets are independent of the grouping of the terms.

Theorem 4   Let L be a lattice. Then for every a, b, and c in L:

1. If $a \leq b$, then
   (a) $a \vee c \leq b \vee c$
   (b) $a \wedge c \leq b \wedge c$
2. $a \leq c$ and $b \leq c$ if and only if $a \vee b \leq c$.
3. $c \leq a$ and $c \leq b$ if and only if $c \leq a \wedge b$.
4. If $a \leq b$ and $c \leq d$ then
   (a) $a \vee c \leq b \vee d$
   (b) $a \wedge c \leq b \wedge d$

Proof. The proof is left as an exercise.

Special Types of Lattices

A lattice L is said to be **bounded** if it has a greatest element I and a least element 0 (see Section 4.2).

Example 13   The lattice Z^+ under the partial order of divisibility, as defined in Example 2, is not a bounded lattice since it has a least element, the number 1, but no greatest element.

Example 14   The lattice Z under the partial order $\leq$ is not bounded since it has neither a greatest nor a least element.

Example 15   The lattice $P(S)$ of all subsets of a set S, as defined in Example 1, is bounded. Its greatest element is S and its least element is $\varnothing$.

If L is a bounded lattice, then for all $a \in A$

$$0 \leq a \leq I$$
$$a \vee 0 = a \qquad a \wedge 0 = 0$$
$$a \vee I = I \qquad a \wedge I = a$$

Theorem 5   Let $L = \{a_1, a_2, \ldots, a_n\}$ be a finite lattice. Then L is bounded.

Proof. The greatest element of L is $a_1 \vee a_2 \vee \cdots \vee a_n$ and its least element is $a_1 \wedge a_2 \wedge \cdots \wedge a_n$.

A lattice L is called **distributive** if for any elements a, b, and c in L we have the following **distributive laws**:

(a) $a \wedge (b \vee c) = (a \wedge b) \vee (a \wedge c)$
(b) $a \vee (b \wedge c) = (a \vee b) \wedge (a \vee c)$

If L is not distributive, we say that L is **nondistributive**.

We leave it as an exercise to show that the distributive law holds when any two of the elements a, b, or c are equal, or when any one of the elements is 0 or I. This observation reduces the number of cases that must be checked in verifying that the distributive law holds. However, verification of the distributive law is generally a tedious task.

Example 16   The lattice $P(S)$ is distributive, since union and intersection (the join and meet, respectively) satisfy the distributive law as shown in Section 1.3.

Example 17   The lattice shown in Fig. 6 is distributive, as can be seen by verifying the distributive law for all ordered triples chosen from the elements a, b, c, and d.

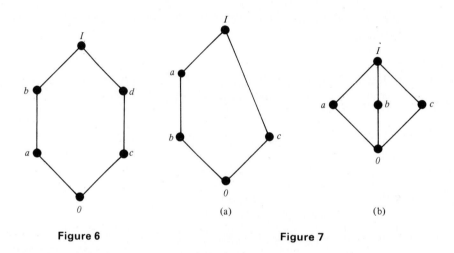

Figure 6   Figure 7

Example 18   Show that the lattices pictured in Fig. 7(a) and (b) are not distributive.

Solution.
(a) We have

$$a \wedge (b \vee c) = a \wedge I = a$$

while

$$(a \wedge b) \vee (a \wedge c) = b \vee 0 = b$$

(b) Observe that

$$a \wedge (b \vee c) = a \wedge I = a$$

while

$$(a \wedge b) \vee (a \wedge c) = 0 \vee 0 = 0$$

The nondistributive lattices discussed in Example 18 are useful for showing that a given lattice is nondistributive, as the following theorem, whose proof we omit, asserts.

Theorem 6   A lattice L is nondistributive if and only if it contains a sublattice that is isomorphic to one of the two lattices of Example 18.

Theorem 6 can be used quite efficiently by inspecting the Hasse diagram of L.

Let L be a bounded lattice with greatest element I and least element 0, and let $a \in L$. An element $a' \in L$ is called a **complement** of a if

$$a \vee a' = I \qquad \text{and} \qquad a \wedge a' = 0$$

Observe that

$$0' = I \qquad \text{and} \qquad I' = 0$$

Example 19   The lattice $L = P(S)$ is such that every element has a complement, since if $A \in L$, then its set complement $\bar{A}$ has the properties $A \vee \bar{A} = S$ and $A \wedge \bar{A} = \emptyset$.

Example 20   The lattices in Fig. 7 each have the property that every element has a complement. The element c in both cases has two complements, a and b.

Example 21   Consider the lattices D_{20} and D_{30} discussed in Example 3 and shown in Fig. 1(a) and (b). Observe that every element in D_{30} has a complement. For example, if $a = 5$, then $a' = 6$. However, the elements 2 and 10 in D_{20} have no complements.

Examples 20 and 21 show that an element a in a lattice need not have a complement and it may have more than one complement. However, for a bounded distributive lattice, the situation is more restrictive, as shown by the following theorem.

Theorem 7   Let L be a bounded distributive lattice. If a complement exists, it is unique.

Proof. Let a' and a'' be complements of the element $a \in L$. Then

$$a \vee a' = I \qquad a \vee a'' = I$$
$$a \wedge a' = 0 \qquad a \wedge a'' = 0$$

Using the distributive laws, we obtain

$$a' = a' \vee 0 = a' \vee (a \wedge a'') = (a' \vee a) \wedge (a' \vee a'')$$
$$= (a \vee a') \wedge (a' \vee a'')$$
$$= I \wedge (a' \vee a'') = a' \vee a''$$

Also,

$$a'' = a'' \vee 0 = a'' \vee (a \wedge a') = (a'' \vee a) \wedge (a'' \vee a')$$
$$= (a \vee a'') \wedge (a' \vee a'')$$
$$= I \wedge (a' \vee a'') = a' \vee a''$$

Hence

$$a' = a''$$

A lattice L is called **complemented** if it is bounded and if every element in L has a complement.

Example 22 $L = P(S)$ is complemented. Observe that in this case, each element of L has a unique complement, which can be seen directly or is implied by Theorem 7.

Example 23   The lattices discussed in Example 20 and shown in Fig. 7(a) and (b) are complemented. In this case, the complements are not unique.

EXERCISE SET 4.3

In Exercises 1–6, determine whether the Hasse diagram represents a lattice.

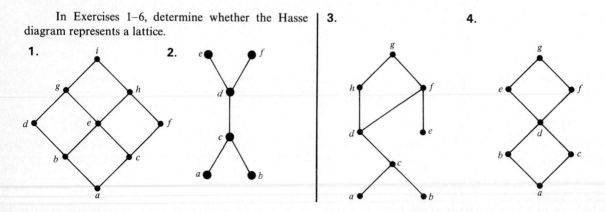

5.   **6.**

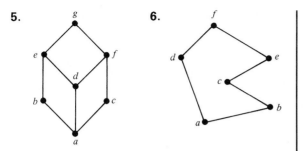

7. Is the poset $A = \{2, 3, 6, 12, 24, 36, 72\}$ under the relation of divisibility a lattice?

8. If L_1 and L_2 are the lattices shown below, draw the Hasse diagram of $L_1 \times L_2$, with the product partial order.

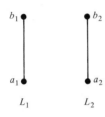

9. Let $L = P(S)$ be the lattice of all subsets of a set S under the relation of containment. Let T be a subset of S. Show that $P(T)$ is a sublattice of L.

10. Let L be a lattice and let a and b be elements of L such that $a \leq b$. The **interval** $[a, b]$ is defined as the set of all $x \in L$, such that $a \leq x \leq b$. Prove that $[a, b]$ is a sublattice of L.

11. Show that a subset of a linearly ordered poset is a sublattice.

12. Find all sublattices of D_{24} that contain at least five elements.

13. Give the Hasse diagrams of all nonisomorphic lattices that have one, two, three, four, or five elements.

14. Show that if a bounded lattice has two or more elements, then $0 \neq I$.

15. Prove part (b) of Theorem 2.

16. Show that the lattice Z^+ under the usual partial order $\leq$ is distributive.

17. Show that the lattice D_n is distributive.

18. Show that a linearly ordered poset is a distributive lattice.

19. Show that a sublattice of a distributive lattice is distributive.

20. Show that if L_1 and L_2 are distributive lattices, then $L = L_1 \times L_2$ is also distributive, where the order of L is the product of the orders in L_1 and L_2.

21. Is the dual of a distributive lattice also distributive?

22. Show that if $a \leq (b \wedge c)$ for some a, b, and c in a poset L, the distributive properties of a lattice are satisfied by a, b, and c.

23. Prove that if a and b are elements in a bounded, distributive lattice, and if a has a complement a', then

$$a \vee (a' \wedge b) = a \vee b$$
$$a \wedge (a' \vee b) = a \wedge b$$

24. Let L be a distributive lattice. Show that if $a \wedge x = a \wedge y$ or $a \vee x = a \vee y$ for all a, then $x = y$.

25. A lattice is said to be **modular** if for all a, b, c, $a \leq c$ implies that $a \vee (b \wedge c) = (a \vee b) \wedge c$.

(a) Show that a distributive lattice is modular.

(b) Show that

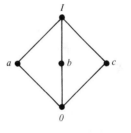

is a nondistributive lattice that is modular.

26. Find the complement of each element in D_{42}.

27. Find the complement of each element in D_{105}.

In Exercises 28-31, determine whether each lattice is distributive, complemented, or both.

28.   **29.**

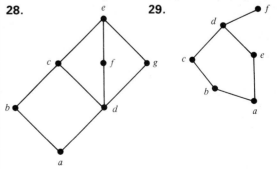

30.

31.

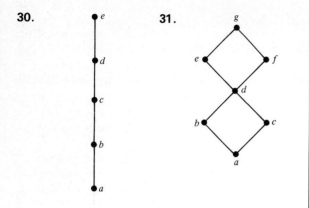

32. Let L be a bounded lattice with at least two elements. Show that no element of L is its own complement.

33. Consider the complemented lattice

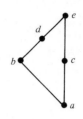

Give the complements of each element.

34. Let $P_1 = \{A_1, A_2, \ldots\}$, $P_2 = \{B_1, B_2, \ldots\}$ be two partitions of a set S. Show that $P_1 \leq P_2$ (see the definition in Example 11) if and only if each A_i is contained in some B_j.

4.4 Boolean Algebras

In this section we discuss a certain type of lattice that has a great many applications in computer science.

A **Boolean algebra** is a complemented distributive lattice with at least two elements. This excludes the trivial case of a single element poset.

Example 1 If S is a set, the poset $A = P(S)$ under the relation of containment is a Boolean algebra.

The Boolean algebra described in Example 1 is of crucial importance for the following reason. It can be shown, although we shall not do so in this book, that every finite Boolean algebra is isomorphic to the Boolean algebra $P(S)$, for a set S with n elements. Since $P(S)$ has 2^n elements if S has n elements, we obtain the following consequence of the result above: Every finite Boolean algebra has 2^n elements for some positive integer n.

Theorem 1 If $(L, \leq)$ is a Boolean algebra, then for all a and b in L

(a) $(a')' = a$ **Involution Property**

(b) $(a \vee b)' = a' \wedge b'$ $\left.\right\}$ **De Morgan's laws**
(c) $(a \wedge b)' = a' \vee b'$

Proof.
(a) We have

$$a \vee a' = I \quad \text{and} \quad a \wedge a' = 0$$

By the commutative property

$$a' \vee a = I \qquad \text{and} \qquad a' \wedge a = 0$$

so that a is a complement of a'. Since L is a bounded distributive lattice, it follows from Theorem 7 of Section 4.3 that a is the unique complement of a', so that $(a')' = a$.

(b) We have

$$
\begin{aligned}
(a \vee b) \vee (a' \wedge b') &= (b \vee a) \vee (a' \wedge b') && \text{(by the commutative property of $\vee$)} \\
&= ((b \vee a) \vee a') \wedge ((b \vee a) \vee b') && \text{(by the distributive property of $\vee$} \\
& && \text{with respect to $\wedge$)} \\
&= (b \vee (a \vee a')) \wedge (b \vee (a \vee b')) && \text{(by the associative property of $\vee$)} \\
&= b \vee ((a \vee a') \wedge (a \vee b')) && \text{(by distributivity of $\vee$ with respect to $\wedge$)} \\
&= b \vee (I \wedge (a \vee b')) && \text{(by definition of complement)} \\
&= b \vee (a \vee b') && \text{(by definition of I)} \\
&= b \vee (b' \vee a) = (b \vee b') \vee a = I \vee a = I && \text{(by commutativity and associativity of $\vee$).}
\end{aligned}
$$

Also (omitting reasons),

$$
\begin{aligned}
(a \vee b) \wedge (a' \wedge b') &= ((a \vee b) \wedge a') \wedge b' \\
&= ((a \wedge a') \vee (b \wedge a')) \wedge b' \\
&= (0 \vee (b \wedge a')) \wedge b' \\
&= (b \wedge a') \wedge b' = (a' \wedge b) \wedge b' \\
&= a' \wedge (b \wedge b') = a' \wedge 0 = 0
\end{aligned}
$$

Thus $a' \wedge b'$ is a complement of $a \vee b$. By uniqueness of complement, we conclude that

$$(a \vee b)' = a' \wedge b'$$

(c) The proof is left as an exercise.

Example 2 Consider the lattices shown in Fig. 7(a) and (b) of Section 4.3. These lattices are not Boolean algebras because complements are not unique. In each case, a and b are both complements of c. These lattices, discussed in Example 18 of Section 4.3, are also not distributive. To show that a given lattice is not a Boolean algebra, it is generally tedious to show, algebraically or graphically, that the distributive property fails to hold. A simpler approach is to show (if possible) that complements are not unique.

We now summarize the basic properties of a Boolean algebra $(L, \leq)$:

1. The partial order $\leq$ satisfies the reflexive, antisymmetric, and transitive properties.

2. The least upper bound $a \vee b$ and the greatest lower bound $a \wedge b$ exist for every pair of elements a and b in L.

3. $a \leq b$ if and only if $a \vee b = b$ if and only if $a \wedge b = a$.

4. (a) $a \vee a = a$
   (b) $a \wedge a = a$

5. (a) $a \vee b = b \vee a$
   (b) $a \wedge b = b \wedge a$

6. (a) $a \vee (b \vee c) = (a \vee b) \vee c$
   (b) $a \wedge (b \wedge c) = (a \wedge b) \wedge c$

7. (a) $a \vee (a \wedge b) = a$
   (b) $a \wedge (a \vee b) = a$

8. $0 \leq a \leq I$ for all $a \in L$

9. (a) $a \vee 0 = a$
   (b) $a \wedge 0 = 0$

10. (a) $a \vee I = I$
    (b) $a \wedge I = a$

11. (a) $a \wedge (b \vee c) = (a \wedge b) \vee (a \wedge c)$
    (b) $a \vee (b \wedge c) = (a \vee b) \wedge (a \vee c)$

12. Every element of L has a unique complement a'.

13. (a) $a \vee a' = I$
    (b) $a \wedge a' = 0$

14. (a) $0' = I$
    (b) $I' = 0$

15. $(a')' = a$

16. (a) $(a \vee b)' = a' \wedge b'$
    (b) $(a \wedge b)' = a' \vee b'$

Recall that most of the properties 1 to 16 listed above were seen to be satisfied by sets with the set operations union and intersection. We now see that the properties proved in Chapter 1 simply show that $P(S)$ is a Boolean algebra with LUB $(\{A, B\}) = A \cup B$ and GLB $(\{A, B\}) = A \cap B$.

Example 3   Consider the lattices D_{20} and D_{30} of all divisors of 20 and 30, respectively, under the relation of divisibility. These lattices have been discussed in Example 21 of Section 4.3, where it was shown that D_{20} is not a complemented lattice. Thus D_{20} is not a Boolean algebra. It is easy to verify that D_{30} is a Boolean algebra.

The following theorem gives necessary and sufficient conditions on n so that D_n is a Boolean algebra.

Theorem 2   Let

$$n = p_1^{r_1} p_2^{r_2} \cdots p_k^{r_k}$$

where $p_1, p_2, \ldots, p_k$ are k distinct primes and each r_i is in Z^+. The lattice D_n of all positive divisors of n under the relation of divisibility is a Boolean algebra if and only if $r_1 = r_2 = \cdots = r_k = 1$.

Proof. First, observe that D_n is a bounded lattice with greatest element n and least element 1. If $a \in D_n$, and a has a complement a', then by definition of complement $a \vee a' = n$ and $a \wedge a' = 1$. Thus by definition of the operations in D_n, we have LCM $(a, a') = n$ and GCD $(a, a') = 1$. Recall from Section 1.7 that for any two integers x and y, GCD $(x, y) \cdot$ LCM $(x, y) = xy$. Thus if a has a complement a', we must have $aa' = n \cdot 1 = n$, or $a' = n/a$. Since $a \wedge a' = 1$, we have GCD $(a, n/a) = 1$. If this condition does not hold, then a cannot have a complement.

Now suppose that some $r_j > 1$, so that $n = p_j^{r_j} q$. Then $n/p_j = p_j^{r_j - 1} q$, so GCD $(p_j, n/p_j) \neq 1$. Thus p_j cannot have a complement in L_n, and L_n is not a Boolean algebra.

Conversely, if all r_j equal 1, then n is the product of the distinct primes $p_1, p_2, \ldots, p_k$. Let $S = \{p_1, p_2, \ldots, p_k\}$. If $T \subseteq S$ and a_T is the product of the primes in T, then $a_T \mid n$. Any divisor of n must be of the form a_T for some subset T of S (where we let $a_\varnothing = 1$). The reader may verify that if V and T are subsets of S, $V \subseteq T$ if and only if $a_V \mid a_T$. Also, it follows from the proof of Theorem 6 of Section 1.7 that $a_{V \cap T} = a_V \wedge a_T$ [that is, GCD (a_V, a_T)] and $a_{V \cup T} = a_V \vee a_T$ [that is, LCM (a_V, a_T)]. Thus the function $f: P(S) \rightarrow D_n$ given by $f(T) = a_T$ is an isomorphism from $P(S)$ to D_n. Since $P(S)$ is a Boolean algebra, so is D_n. Incidentally, this shows that the complement of every element a_T exists and is just $a_{\bar{T}}$, where $\bar{T}$ is the set complement of T in S.

Example 4   We illustrate the last part of the proof above with the lattice D_{30}. Since $30 = 2 \cdot 3 \cdot 5$, we know we will obtain a Boolean algebra. Here $S = \{2, 3, 5\}$, and we now list the subsets of S together with the corresponding divisors of 30.

T (subset of S)	a_T (product of elements of T)
$\varnothing$	1
$\{2\}$	2
$\{3\}$	3
$\{5\}$	5
$\{2, 3\}$	6
$\{2, 5\}$	10
$\{3, 5\}$	15
$\{2, 3, 5\}$	30

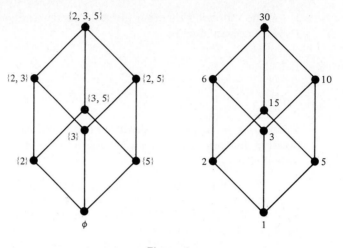

Figure 1

The corresponding lattices are shown in Fig. 1. We see that D_{30} and $P(S)$ can be put in one-to-one correspondence in such a way that they are isomorphic. Then, since $P(S)$ is a Boolean algebra, D_{30} is one as well.

Example 5   We shall now define an important Boolean algebra. Let $B = \{0, 1\}$ be a poset with the partial order $\leq$, where $0 \leq 0$, $0 \leq 1$, $1 \leq 1$. We display the LUB and GLB of every pair of elements of B as follows:

$\vee$	0	1
0	0	1
1	1	1

$\wedge$	0	1
0	0	0
1	0	1

Moreover, we see that

$$0' = 1 \quad \text{and} \quad 1' = 0$$

It is then easy to show that the poset B is a Boolean algebra, where $0_B = 0$ and $I_B = 1$, by checking that it is a distributive complemented lattice. The reader may also recall that this set and the least upper bound and greatest lower bound have been considered in Section 1.8 from a different point of view.

There are two other ways of viewing Example 5. The first way ties the example to logic (see the Appendix). Let 0 denote the collection of all false statements, and let 1 denote the collection of all true statements. A statement is merely a declarative sentence which is either true or false (see the Appendix). Now suppose that the symbol ∨ stands for the conjunction *or*, the symbol ∧ stands for the conjunction *and*, and the complement ' stands for *not*. Then the tables for ∨ and ∧ given above express facts about the truth or falsity of compound statements, based on the truth or falsity of their components. Frequently, we use F in place of 0 and T in place of 1. Then the tables above can be written as follows:

or	F	T		and	F	T
F	F	T		F	F	F
T	T	T		T	F	T

The first expresses the fact that a compound statement p or q is true if *either* (or both) of the propositions p or q is true. The second expresses the fact that the compound statement p and q is true only if *both* p and q are true.

Because of this analogy, tables like those given above for ∨ and ∧ are often called **truth tables**. This is true even when they arise in abstract Boolean algebras or circuit design, and they have no actual connection with logic. Similarly, properties involving the complement ' are often described using the logical phrase "not."

The second way of viewing Example 5 is schematically. Figure 2 shows the schematic form of the tables for ∨ and ∧ given above, and a schematic representation of '. The idea is that the inputs x and y (or just x in the case of complement) can each be 0 or 1, and the corresponding output is determined by ∨, ∧, or ' [In Fig. 2(c), 0 input gives 1 output, and 1 input gives 0 output.] The schematic form for ∨ is called an **or gate**, that for ∧ is called an **and gate**, and the form for ' is called an **inverter**. Again, the logical names are simply due to the analogy discussed above.

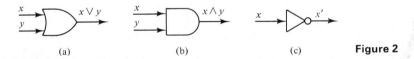

(a)   (b)   (c)   **Figure 2**

These schematic forms actually convey no more information than the tables originally given. They are used to emphasize the fact that if inputs and outputs represent voltages in a computer, then the functions ∨, ∧, and ' can be implemented by electronic circuitry. It is hard to overestimate the importance of this result for computer science. Once again because of analogy with logic, diagrams using the symbols of Fig. 2 are often called **circuit logic diagrams**.

Theorem 3   If L_1 and L_2 are Boolean algebras, then $L_1 \times L_2$, with the product partial order, is a Boolean algebra.

Proof. By Theorem 1 of Section 4.3, we already know that $L_1 \times L_2$ is a lattice. The rest of the proof is left as an exercise.

Theorem 3 can be generalized to show that if $L_1, L_2, \ldots,$ and L_n are Boolean algebras, then $L_1 \times L_2 \times \cdots \times L_n$ is a Boolean algebra.

Example 6   Let B be the Boolean algebra considered in Example 5. Then by the generalized Theorem 3, the poset

$$B^n = \{(b_1, b_2, \ldots, b_n) \mid b_i \in B \text{ for } i = 1, 2, \ldots, n\}$$

is also a Boolean algebra. If

$$a = (a_1, a_2, \ldots, a_n) \in B^n \qquad \text{and} \qquad b = (b_1, b_2, \ldots, b_n) \in B^n$$

then

$$a \vee b = (a_1 \vee b_1, a_2 \vee b_2, \ldots, a_n \vee b_n)$$
$$a \wedge b = (a_1 \wedge b_1, a_2 \wedge b_2, \ldots, a_n \wedge b_n)$$
$$a' = (a'_1, a'_2, \ldots, a'_n)$$

Moreover,

$$0_{B^n} = (0, 0, \ldots, 0) \qquad \text{and} \qquad I_{B^n} = (1, 1, \ldots, 1)$$

Functions on the Boolean Algebra B^n

Let $B = \{0, 1\}$ and let B^n be as defined above. Then we may discuss functions f from B^n to B. For such functions f, and for every n-tuple $(x_1, x_2, \ldots, x_n) \in B^n$, $f(x_1, \ldots, x_n)$ is either 0 or 1. Such functions may be viewed as functions of n "variables," each of which takes on only the values 0 and 1. It is customary to list such functions in tables, giving each possible n-tuple $(x_1, x_2, \ldots, x_n)$ and the corresponding value of the function f. Figure 3(a) shows a display for a particular function f of three variables; that is, $f: B^3 \to B$. Any possible distribution of 2^m 0's and 1's can be the set of values of some function $f: B^m \to B$.

The reason that such functions are important is that, as shown schematically in Fig. 3(b), they may be used to represent the output requirements of a circuit for any possible input values. Thus each x_i represents an input circuit capable of carrying two indicator voltages (one voltage for 0 and a different voltage for 1). The function f represents the desired output response in all cases. Such requirements occur at the design stage of all combinational and sequential computer circuitry. As before, tables such as shown in Fig. 3(a) are often called **truth tables** for f, because of the analogy with propositional logic.

Note carefully that the specification of a function $f: B^n \to B$ simply lists circuit requirements. It gives no indication of how these requirements can be met. One

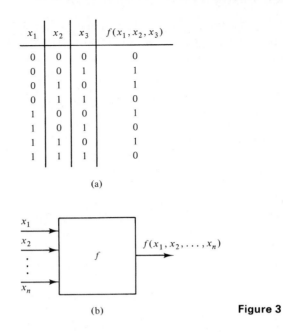

x_1	x_2	x_3	$f(x_1, x_2, x_3)$
0	0	0	0
0	0	1	1
0	1	0	1
0	1	1	0
1	0	0	1
1	0	1	0
1	1	0	1
1	1	1	0

(a)

(b)   **Figure 3**

important way of generating functions is with Boolean expressions, which we consider next.

Boolean Expressions

Let $x_1, x_2, \ldots, x_n$ be a set of n symbols or variables. A **Boolean expression** $E(x_1, x_2, \ldots, x_n)$ or **Boolean polynomial** in $x_1, x_2, \ldots, x_n$ is defined recursively as follows:

1. $x_1, x_2, \ldots, x_n$ are Boolean expressions.
2. The symbols 0 and I are Boolean expressions.
3. If $E_1(x_1, x_2, \ldots, x_n)$ and $E_2(x_1, x_2, \ldots, x_n)$ are Boolean expressions in $x_1, x_2, \ldots, x_n$, then so are $E_1(x_1, x_2, \ldots, x_n) \vee E_2(x_1, x_2, \ldots, x_n)$, $E_1(x_1, x_2, \ldots, x_n) \wedge E_2(x_1, x_2, \ldots, x_n)$, and $(E_1(x_1, x_2, \ldots, x_n))'$.
4. There are no Boolean expressions other than those that can be obtained by using rules 1, 2, and 3.

Example 7   The following are Boolean expressions in x, y, and z.

$$E_1(x, y, z) = (x \vee y) \wedge z$$

$$E_2(x, y, z) = (x \vee y') \vee (y \wedge I)$$

$$E_3(x, y, z) = (x \vee (y' \wedge z)) \vee (x \wedge y \wedge I)$$

$$E_4(x, y, z) = (x \vee y \vee z') \wedge (x' \wedge z) \wedge (y' \vee 0)$$

Ordinary polynomials in several variables, such as $x^2y + z^4$, $xy + yz + x^2y^2$, $x^3y^3 + xz^4$, and so on, are generally interpreted as expressions representing algebraic computations with unspecified numbers. As such, they are subject to the usual rules of arithmetic. Thus the polynomials $x^2 + 2x + 1$ and $(x + 1)(x + 1)$ are considered equivalent, and so are $x(xy + yz)(x + z)$ and $x^3y + 2x^2yz + xyz^2$, since in each case we can turn one into the other with algebraic manipulation.

Similarly, Boolean polynomials or expressions may be interpreted as representing Boolean computations with unspecified elements of some Boolean algebra. As such, these expressions are subject to the rules of Boolean arithmetic, and two expressions are considered equivalent if we can turn one into the other by Boolean manipulations. Thus, $x \wedge (y \vee z)$ and $(x \wedge y) \vee (x \wedge z)$ are equivalent (since distributivity holds in a Boolean algebra).

Suppose now that L is a fixed Boolean algebra and consider $x_1, x_2, \ldots, x_n$ as variables that can only be replaced by elements of L. The result of such a replacement is an element of L, so we have evaluated the Boolean expression for the particular assignment of the variables $x_1, x_2, \ldots, x_n$.

Example 8   Let $L = \{0, a, b, I\}$ be the Boolean algebra shown in Fig. 4 and consider the Boolean expressions defined in Example 7. If x is replaced by a, y by b, and z by 0, then

$$E_1(a, b, 0) = (a \vee b) \wedge 0$$

$$= I \wedge 0 = 0$$

$$E_3(a, b, 0) = a \vee (b' \wedge 0) \vee (a \wedge b \wedge I)$$

$$= a \vee 0 \vee 0 = a$$

Figure 4

Let $E(x_1, x_2, \ldots, x_n)$ be a Boolean expression in the n variables $x_1, x_2, \ldots, x_n$. If $(a_1, a_2, \ldots, a_n)$ is a particular ordered n-tuple of elements from a Boolean algebra L, then $E(a_1, a_2, \ldots, a_n)$ is an element of L. Thus $E(x_1, x_2, \ldots, x_n)$ may be interpreted as a formula for a function f from L^n to L, where $f(a_1, \ldots, a_n) = E(a_1, \ldots, a_n)$ for all $(a_1, a_2, \ldots, a_n)$ in L^n.

This is similar to the usual algebraic situation in which, for example, a polynomial $x^2 + y$ could be considered a function from $\mathbb{R} \times \mathbb{R}$ to $\mathbb{R}$ [we would write it as $z = f(x, y) = x^2 + y$].

Thus an expression $E(x_1, x_2, \ldots, x_n)$ yields, by substitution of values, a function from L^n to L. Functions from L^n to L are called **Boolean functions of n variables**. Although not every real function of n real variables arises from a polynomial expression, we will show in the next section that in the Boolean algebra $B = \{0, 1\}$, every Boolean function does arise from a Boolean expression. This fact is fundamental for the construction of circuits to implement a desired Boolean function.

Two Boolean expressions $E_1(x_1, x_2, \ldots, x_n)$ and $E_2(x_1, x_2, \ldots, x_n)$ are said to be **equivalent** if we can obtain one expression from the other by using a finite number of properties of a Boolean algebra. Since a Boolean expression determines a unique Boolean function, it follows that $E_1(x_1, x_2, \ldots, x_n)$ and $E_2(x_1, x_2, \ldots, x_n)$ are equivalent if and only if their corresponding Boolean functions are equal.

Example 9   Show that the Boolean expressions

$$E_1(x, y, z) = (x \wedge y \wedge z') \vee (x \wedge y' \wedge z')$$

and

$$E_2(x, y, z) = x \wedge z'$$

are equivalent.

Solution.

$$
\begin{aligned}
E_1(x, y, z) &= (x \wedge y \wedge z') \vee (x \wedge y' \wedge z') \\
&= (x \wedge z' \wedge y) \vee (x \wedge z' \wedge y') \\
&= (x \wedge z') \wedge (y \vee y') \quad \text{(distributive property)} \\
&= (x \wedge z') \wedge I \\
&= x \wedge z' = E_2(x, y, z)
\end{aligned}
$$

We shall now restrict our attention to the Boolean algebra $B = \{0, 1\}$ defined in Example 5. Then, as above, a Boolean function from B^n to B is obtained from each Boolean expression in n variables.

Example 10   Consider the Boolean expression

$$E(x_1, x_2, x_3) = (x_1 \wedge x_2) \vee (x_1 \vee (x_2' \wedge x_3))$$

Describe the Boolean function $f: B^3 \to B$ determined by the Boolean expression $E(x_1, x_2, x_3)$.

x_1	x_2	x_3	$f(x_1, x_2, x_3) = (x_1 \wedge x_2) \vee (x_1 \vee (x_2' \wedge x_3))$
0	0	0	0
0	0	1	1
0	1	0	0
0	1	1	0
1	0	0	1
1	0	1	1
1	1	0	1
1	1	1	1

Figure 5

Solution. The Boolean function $f: B^3 \to B$ is described by substituting all the 2^3 ordered triples of values from B for x_1, x_2, and x_3. The truth table for the resulting function is shown in Fig. 5.

We have now discussed general functions $f: B^n \to B$, and we have shown how certain of these functions arise from Boolean expressions. The latter are very important since they can be expressed entirely in terms of the functions $\vee$, $\wedge$, and $'$. Thus we may express such functions schematically in a logic diagram, that is, by using the symbols of Fig. 2. Those symbols correspond to basic "building-block" circuits, so a function f given by a Boolean expression contains a description of circuitry that will implement it.

Example 11 Let $f(x, y, z) = (x \wedge y) \vee (y \wedge z')$. Figure 6(a) shows the truth table for the corresponding function $f: B^3 \to B$. Figure 6(b) shows the logic diagram for f; it describes some circuitry to implement f. Many different expressions may produce the same function; therefore, there may be many equivalent logic diagrams for f.

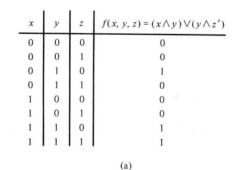

x	y	z	$f(x, y, z) = (x \wedge y) \vee (y \wedge z')$
0	0	0	0
0	0	1	0
0	1	0	1
0	1	1	0
1	0	0	0
1	0	1	0
1	1	0	1
1	1	1	1

(a)

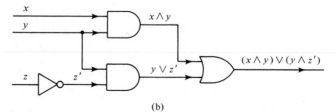

(b)

Figure 6

In the next section we show that every function $f: B^n \to B$ can be implemented by some Boolean expression. Thus, whatever requirements are set forth by f, there is circuitry that will implement these requirements.

EXERCISE SET 4.4

In Exercises 1–10, determine whether the poset is a Boolean algebra.

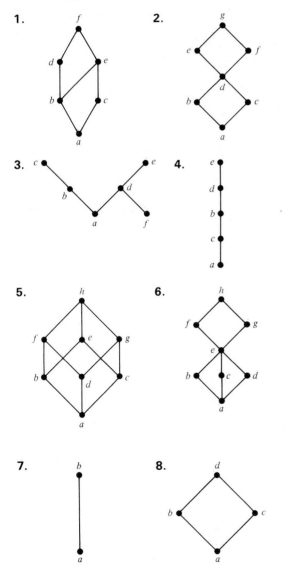

1.

2.

3.

4.

5.

6.

7.

8.

9. D_{210}

10. D_{60}

11.   Show directly that there are no Boolean algebras having three elements.

12.   Show that in a Boolean algebra, for any a and b, $a \leq b$ if and only if $b' \leq a'$.

13.   Show that in a Boolean algebra, for any a and b, $a = b$ if and only if $(a \wedge b') \vee (a' \wedge b) = 0$.

14.   Show that in a Boolean algebra, for any a, b, and c:

   (a)   If $a \leq b$, then $a \vee c \leq b \vee c$.
   (b)   If $a \leq b$, then $a \wedge c \leq b \wedge c$.

15.   Show that in a Boolean algebra the following statements are equivalent for any a and b.

   (a) $a \vee b = b$
   (b) $a \wedge b = a$
   (c) $a' \vee b = I$
   (d) $a \wedge b' = 0$
   (e) $a \leq b$

16.   Show that in a Boolean algebra, for any a and b:

$$(a \wedge b) \vee (a \wedge b') = a$$

17.   Show that in a Boolean algebra, for any a and b,

$$b \wedge (a \vee (a' \wedge (b \vee b'))) = b$$

18.   Show that in a Boolean algebra, for any a, b, and c,

$$(a \wedge b \wedge c) \vee (b \wedge c) = b \wedge c$$

19.   Show that in a Boolean algebra, for any a, b, and c,

$$((a \vee c) \wedge (b' \vee c))' = (a' \vee b) \wedge c'$$

20.   Show that in a Boolean algebra, for any a, b, and c, if $a \leq b$ then

$$a \vee (b \wedge c) = b \wedge (a \vee c)$$

21. Show that in a Boolean algebra, for any a, b, and c, if $a \leq c \leq b$ then

$$(a \wedge b) \vee (a \wedge b \wedge c) \vee (b \wedge c) \vee (a \wedge c) = c$$

22. Show that in a Boolean algebra, for any a and b,

$$(a \wedge b)' = a' \vee b'$$

23. Consider the Boolean expression

$$E(x, y, z) = x \wedge (y \vee z')$$

If $B = \{0, 1\}$, compute the truth table of the function $f: B^3 \to B$ defined by E.

24. Consider the Boolean expression

$$E(x, y, z) = (x \vee y) \wedge (z \vee x')$$

If $B = \{0, 1\}$, compute the truth table of the function $f: B^3 \to B$ defined by E.

25. Consider the Boolean expression

$$E(x, y, z) = (x \wedge y') \vee (y \wedge (x' \vee y))$$

If $B = \{0, 1\}$, compute the truth table of the function $f: B^3 \to B$ defined by E.

26. Construct a logic diagram implementing the function f of Exercise 23.

27. Construct a logic diagram implementing the function f of Exercise 24.

28. Construct a logic diagram implementing the function f of Exercise 25.

4.5 Implementation of Boolean Functions

In the preceding section we considered functions from B^n to B, where B is the Boolean algebra $\{0, 1\}$. We noted that such functions can represent input–output requirements for models of many practical computer circuits. We also pointed out that if the function is given by some Boolean expression, then we can construct a logic diagram for it, and thus model the implementation of the function. In this section we show that all functions from B^n to B are given by Boolean expressions, and thus logic diagrams can be constructed for any such function. Our discussion illustrates a method for finding a Boolean expression that produces a given function.

If $f: B^n \to B$, we will let $S(f) = \{b \in B^n \mid f(b) = 1\}$. We then have the following result.

Theorem 1   Let f, f_1, and f_2 be three functions from B^n to B.

(a) If $S(f) = S(f_1) \cup S(f_2)$, then $f(b) = f_1(b) \vee f_2(b)$ for all b in B.
(b) If $S(f) = S(f_1) \cap S(f_2)$, then $f(b) = f_1(b) \wedge f_2(b)$ for all b in B.

($\vee$ and $\wedge$ are LUB and GLB, respectively, in B.)

Proof.
(a) Let $b \in B^n$. If $b \in S(f)$, then, by definition of $S(f)$, $f(b) = 1$. Since $S(f) = S(f_1) \cup S(f_2)$, either $b \in S(f_1)$ or $b \in S(f_2)$, or both. In any case, $f_1(b) \vee f_2(b) = 1$. Now if $b \notin S(f)$, then $f(b) = 0$. Also, we must have $b \notin S(f_1)$ and $b \notin S(f_2)$, so $f_1(b) = 0$ and $f_2(b) = 0$. This means that $f_1(b) \vee f_2(b) = 0$. Thus, for all $b \in B^n$, $f(b) = f_1(b) \vee f_2(b)$.
(b) This part is proved in a manner completely analogous to that used in part (a).

Recall that a function $f: B^n \to B$ can be viewed as a function $f(x_1, x_2, \ldots, x_n)$ of n variables, each of which may assume the values 0 or 1. If $E(x_1, x_2, \ldots, x_n)$ is a Boolean expression, then the function that it produces is generated by substituting all combinations of 0's and 1's for the x_i's in the expression.

Example 1 Let $f_1: B^2 \to B$ be produced by the expression $E(x, y) = x'$, and let $f_2: B^2 \to B$ be produced by the expression $E(x, y) = y'$. Then the truth tables of f_1 and f_2 are shown in Fig. 1(a) and (b), respectively. Let $f: B^2 \to B$ be the function whose truth table is shown in Fig. 1(c). Clearly, $S(f) = S(f_1) \cup S(f_2)$, since f_1 is 1 at the elements $(0, 0)$, and $(0, 1)$ of B^2, f_2 is 1 at the elements $(0, 0)$ and $(1, 0)$ of B^2, and f is 1 at the elements $(0, 0)$, $(0, 1)$, and $(1, 0)$ of B^2. By Theorem 1, $f = f_1 \vee f_2$, so a Boolean expression that produces f is $x' \vee y'$. This is easily verified.

It is not hard to show that any function $f: B^n \to B$, for which $S(f)$ has exactly one element, is produced by a Boolean expression.

x	y	$f_1(x, y)$
0	0	1
0	1	1
1	0	0
1	1	0

(a)

x	y	$f_2(x, y)$
0	0	1
0	1	0
1	0	1
1	1	0

(b)

x	y	$f(x, y)$
0	0	1
0	1	1
1	0	1
1	1	0

(c)

Figure 1

Example 2 Let $f: B^2 \to B$ be the function whose truth table is shown in Fig. 2(a). This function is equal to 1 only at the element $(0, 1)$ of B^2, that is, $S(f) = \{(0, 1)\}$. Thus $f(x, y) = 1$ only when $x = 0$ and $y = 1$. This is also true for the expression $E(x, y) = x' \wedge y$, so f is produced by this expression. The following table shows the correspondence between functions that are 1 at just one element, and the Boolean expressions that produce these functions.

$S(f)$	Expression producing f
$\{(0, 0)\}$	$x' \wedge y'$
$\{(0, 1)\}$	$x' \wedge y$
$\{(1, 0)\}$	$x \wedge y'$
$\{(1, 1)\}$	$x \wedge y$

x	y	z	$f(x, y, z)$
0	0	0	0
0	0	1	0
0	1	0	0
0	1	1	1
1	0	0	0
1	0	1	0
1	1	0	0
1	1	1	0

x	y	$f(x, y)$
0	0	0
0	1	1
1	0	0
1	1	0

(a)   (b)   **Figure 2**

The function $f: B^3 \to B$ whose truth table is shown in Fig. 2(b) has $S(f) = \{(0, 1, 1)\}$, that is, f equals 1 only when $x = 0$, $y = 1$, and $z = 1$. This is also true for the Boolean expression $x' \wedge y \wedge z$, which must therefore produce f.

If $b \in B^n$, then b is a sequence $(c_1, c_2, \ldots, c_n)$ of length n, where each c_k is 0 or 1. Let E_b be the Boolean expression $\bar{x}_1 \wedge \bar{x}_2 \wedge \cdots \wedge \bar{x}_n$, where $\bar{x}_k = x_k$ when $c_k = 1$ and $\bar{x}_k = \bar{x}'_k$ when $c_k = 0$. Such an expression is called a **minterm**. Example 2 illustrates the fact that any function $f: B^n \to B$, for which $S(f)$ is a single element of B^n, is produced by a minterm expression. In fact, if $S(f) = \{b\}$, it is easily seen that the minterm expression E_b produces f. We then have the following result.

Theorem 2   Any function $f: B^n \to B$ is produced by a Boolean expression.

Proof. Let $S(f) = \{b_1, b_2, \ldots, b_k\}$, and for each i, let $f_i : B^n \to B$ be the function defined by

$$f_i(b_i) = 1$$
$$f_i(b) = 0 \qquad \text{if } b \neq b_i$$

Then $S(f_i) = \{b_i\}$, so $S(f) = S(f_1) \cup \cdots \cup S(f_n)$ and by Theorem 1,

$$f = f_1 \vee f_2 \vee \cdots \vee f_n$$

By the discussion given above, each f_i is produced by the minterm E_{b_i}. Thus f is produced by the Boolean expression

$$E_{b_1} \vee E_{b_2} \vee \cdots \vee E_{b_n}$$

and this completes the proof.

Example 3   Consider the function $f: B^3 \to B$ whose truth table is shown in Fig. 3. Since $S(f) = \{(0, 1, 1), (1, 1, 1)\}$, Theorem 2 shows that f is produced by the Boolean

x	y	z	$f(x, y, z)$
0	0	0	0
0	0	1	0
0	1	0	0
0	1	1	1
1	0	0	0
1	0	1	0
1	1	0	0
1	1	1	1

Figure 3

expression $E(x, y, z) = E_{(0, 1, 1)} \vee E_{(1, 1, 1)} = (x' \wedge y \wedge z) \vee (x \wedge y \wedge z)$. This expression, however, is not the simplest Boolean expression that produces f. Using properties of Boolean algebras, we have

$$(x' \wedge y \wedge z) \vee (x \wedge y \wedge z) = (x' \vee x) \wedge (y \wedge z)$$

$$= 1 \wedge (y \wedge z) = y \wedge z$$

Thus f is also produced by the simple expression $y \wedge z$.

The process of writing a function as an "or" combination of minterms, and simplifying the resulting expression, can be systematized in various ways. We will demonstrate a graphical procedure utilizing what is known as a **Karnaugh map.** This procedure is easy for human beings to use with functions $f: B^n \to B$, if n is not too large. We will illustrate the method for $n = 2$, 3, and 4. If n is large, or if a programmable algorithm is desired, other techniques may be preferable.

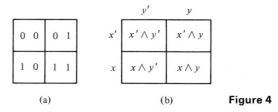

(a)   (b)   **Figure 4**

We consider first the case where $n = 2$, so that f is a function of two variables, say x and y. In Fig. 4(a) we show a 2×2 matrix of "squares," with each square containing one possible input b from B^2. In Fig. 4(b) we have replaced each input b with the corresponding minterm E_b. The labeling of the squares in Fig. 4 is for reference only. In the future we will not exhibit these labels, but we will assume that the reader remembers their locations. In Fig. 4(b) we note that the x variable appears everywhere in the first row as x' and everywhere in the second row as x. We label these rows accordingly, and we perform a similar labeling of the columns.

Example 4   Let $f: B^2 \to B$ be the function whose truth table is shown in Fig. 5(a). In Fig. 5(b) we have arranged the values of f in the appropriate squares, and we have kept the row and column labels. The resulting 2×2 array of 0's and 1's is called the **Karnaugh map of** f. Since $S(f) = \{(0, 0), (0, 1)\}$, the corresponding expression for f is $(x' \wedge y') \vee (x' \wedge y) = x' \wedge (y' \vee y) = x'$.

The outcome of Example 4 is typical. When the 1-values of a function $f: B^2 \to B$ exactly fill one row, or one column, the label of that row or column gives the Boolean expression for f. Of course, we already know that if the 1-values of f fill just one square, then f is produced by the corresponding minterm. It can be shown that the larger the rectangle of 1-values of f, the smaller the expression for f will be. Finally, if the 1-values of f do not lie in a rectangle, we can decompose these values into the union of (possibly overlapping) rectangles. Then by Theorem 1 the Boolean expression for f can be found by computing the expressions corresponding to each rectangle, and combining them with "$\vee$" symbols.

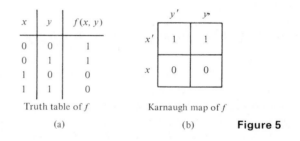

x	y	$f(x, y)$
0	0	1
0	1	1
1	0	0
1	1	0

Truth table of f

(a)

Karnaugh map of f

(b)   **Figure 5**

Example 5   Consider the function $f: B^2 \to B$ whose truth table is shown in Fig. 6(a). In Fig. 6(b) we show the Karnaugh map of f and decompose the 1-values into the two indicated rectangles. The expression for the function having 1's in the horizontal rectangle is x' (verify). The function having all its 1's in the vertical rectangle corresponds to the expression y' (verify). Thus f corresponds to the expression $x' \vee y'$. In Fig. 6(c) we show a different decomposition of the 1-values of f into rectangles. This decomposition is also correct, but it leads to the more complex expression $y' \vee (x' \wedge y)$. We see that the decomposition into rectangles is not unique, and that one should try to use the largest possible rectangles.

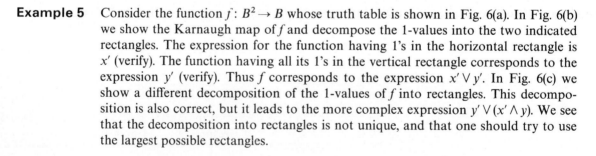

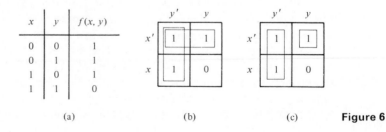

x	y	$f(x, y)$
0	0	1
0	1	1
1	0	1
1	1	0

(a)   (b)   (c)   **Figure 6**

We now turn to the case of a function $f: B^3 \to B$, which we consider to be a function of x, y, and z. We could proceed as in the case of two variables and construct a "cube" of side 2 to contain the values of f. This would work, but three-dimensional figures are awkward to draw and use, and the idea would not generalize. Instead, we use a rectangle of size 2×4. In Figs. 7(a) and (b), respectively, we show the inputs (from B^3) and corresponding minterms for each square of such a rectangle.

Consider the rectangular areas shown in Fig. 8. If the 1-values for a function $f: B^3 \to B$ exactly fill one of the rectangles shown, then the Boolean expression for this function is one of the six expressions x, y, z, x', y', or z', as indicated in Fig. 8.

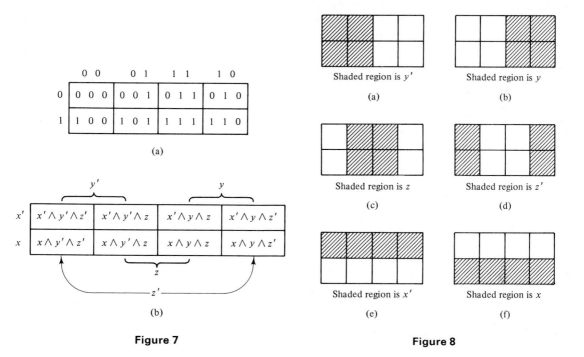

Figure 7

Figure 8

Consider the situation shown in Fig. 8(a). Theorem 1(a) shows that f can be computed by joining all the minterms corresponding to squares of the region with the symbol $\vee$. Thus f is produced by

$$(x' \wedge y' \wedge z') \vee (x' \wedge y' \wedge z) \vee (x \wedge y' \wedge z') \vee (x \wedge y' \wedge z)$$
$$= ((x' \vee x) \wedge (y' \wedge z')) \vee ((x' \vee x) \wedge (y' \wedge z))$$
$$= (1 \wedge (y' \wedge z')) \vee (1 \wedge (y' \wedge z))$$
$$= (y' \wedge z') \vee (y' \wedge z)$$
$$= y' \wedge (z' \vee z) = y' \wedge 1 = y'$$

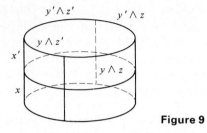

Figure 9

A similar computation shows that the other five regions are correctly labeled.

If we think of the left and right edges of our basic rectangle as "glued together" to make a cylinder, as we show in Fig. 9, we can say that the six large regions shown in Fig. 8 consist of any two adjacent columns of the cylinder, or of the top or bottom half-cylinder.

The six basic regions shown in Fig. 8 are the only ones whose corresponding Boolean expressions need be considered. That is why we used them to label Fig. 7(b), and we keep them as labels for all Karnaugh maps of functions from B^3 to B. Theorem 1(b) tells us that if the 1-values of a function $f: B^3 \to B$ form exactly the intersection of two or three of the basic six regions, then a Boolean expression for f can be computed by combining the expressions for these basic regions with $\wedge$ symbols.

Thus if the 1-values of the function f are as shown in Fig. 10(a), then we get

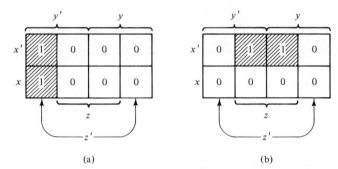

(a)   (b)

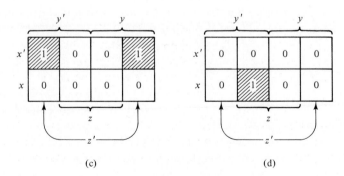

(c)   (d)   **Figure 10**

them by intersecting the regions shown in Fig. 8(a) and (d). The Boolean expression for f is therefore $y' \wedge z'$. Similar derivations can be given for the other three columns. If the 1-values of f are as shown in Fig. 10(b), we get them by intersecting the regions of Fig. 8(c) and (e), so a Boolean expression for f is $z \wedge x'$. In a similar fashion we can compute the expression for any function whose 1-values fill two horizontally adjacent squares. There are eight such functions if we again consider the rectangle to be formed into a cylinder. Thus we include the case where the 1-values of f are as shown in Fig. 10(c). The resulting Boolean expression is $z' \wedge x'$.

If we intersect three of the basic regions, and the intersection is not empty, the intersection must be a single square, and the resulting Boolean expression is a minterm. In Fig. 10(d), the 1-values of f form the intersection of the three regions shown in Fig. 8(a), (c), and (f). The corresponding minterm is $y' \wedge z \wedge x$. Thus one need not remember the placement of minterms in Fig. 7(b), but instead may reconstruct it.

We have seen how to compute a Boolean expression for any function $f: B^3 \to B$ whose 1-values form a rectangle of adjacent squares (in the cylinder) of length $2^n \times 2^m$, $n = 1, 2$; $m = 1, 2$. In general, if the set of 1-values of f do not form such a rectangle, we may write this set as the union of such rectangles. Then a Boolean expression for f is computed by combining the expressions associated with each rectangle with $\vee$ symbols. This is true by Theorem 1(a). The discussion above shows that the larger the rectangles that are chosen, the simpler will be the resulting Boolean expression.

x	y	z	$f(x, y, z)$
0	0	0	1
0	0	1	1
0	1	0	0
0	1	1	1
1	0	0	1
1	0	1	0
1	1	0	0
1	1	1	1

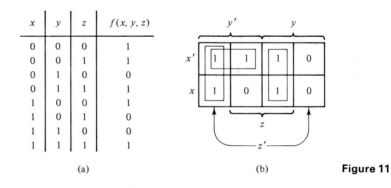

(a)   (b)   **Figure 11**

Example 6   Consider the function f whose truth table and corresponding Karnaugh map are shown in Fig. 11. The placement of the 1's can be derived by locating the corresponding inputs in Fig. 7(a). One decomposition of the 1-values of f is shown in Fig. 11(b). From this we see that a Boolean expression for f is $(y' \wedge z') \vee (x' \wedge y') \vee (y \wedge z)$.

Example 7   Figure 12 shows the truth table and corresponding Karnaugh map for a function f. The decomposition into rectangles shown in Fig. 12(b) uses the idea that the first and last columns are considered adjacent (by wrapping around the cylinder). Thus

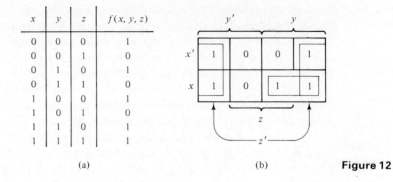

x	y	z	$f(x, y, z)$
0	0	0	1
0	0	1	0
0	1	0	1
0	1	1	0
1	0	0	1
1	0	1	0
1	1	0	1
1	1	1	1

(a)   (b)   **Figure 12**

the symbols are left "open ended" to signify that they join in one 2×2 rectangle, corresponding to z'. The resulting Boolean expression is $z' \vee (x \wedge y)$ (verify).

Finally, without additional comment, we present in Fig. 13 the distribution of inputs and corresponding labeling of rectangles for the case of a function $f : B^4 \to B$, considered as a function of x, y, z, and w. Here again, we consider the first and last columns to be adjacent, and the first and last rows to be adjacent, both by "wrap around," and we look for rectangles with sides of length some power of 2, so the length is 1, 2, or 4. The expression corresponding to such rectangles is given by "intersecting" the large labeled rectangles of Fig. 14.

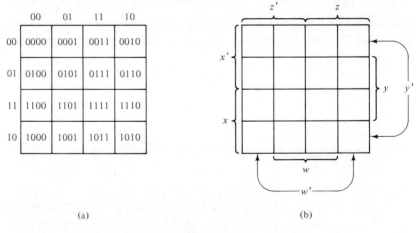

	00	01	11	10
00	0000	0001	0011	0010
01	0100	0101	0111	0110
11	1100	1101	1111	1110
10	1000	1001	1011	1010

(a)   (b)

Figure 13

Example 8   Figure 15 shows the Karnaugh map of a function $f : B^4 \to B$. The 1-values are placed by considering the location of inputs in Fig. 13(a). Thus $f(0101) = 1$, $f(0001) = 0$, and so on.

     The center 2×2 square represents the Boolean expression $w \wedge y$ (verify).

     The four corners also form a square of side 2 since the right and left edges and

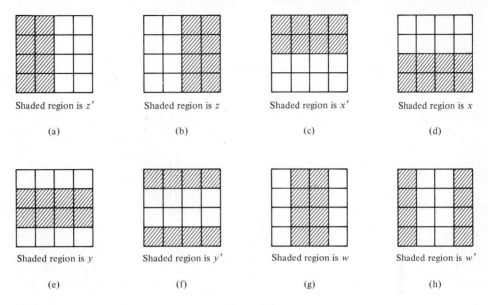

Shaded region is z'	Shaded region is z	Shaded region is x'	Shaded region is x
(a)	(b)	(c)	(d)

Shaded region is y	Shaded region is y'	Shaded region is w	Shaded region is w'
(e)	(f)	(g)	(h)

Figure 14

the top and bottom edges are considered adjacent. From a geometric point of view, we can see that if we wrap the rectangle around horizontally, getting a cylinder, then when we further wrap around vertically, we will get a torus or "inner tube." On this inner tube, the four corners form a square of side 2 which represents the Boolean expression $w' \wedge y'$ (verify).

It then follows that the decomposition above leads to the Boolean expression

$$(w \wedge y) \vee (w' \wedge y')$$

for f.

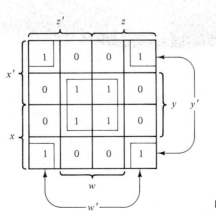

Figure 15

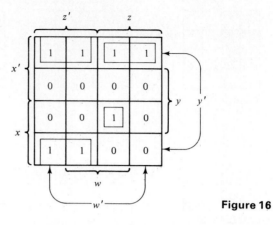

Figure 16

Example 9   We show, in Fig. 16, the Karnaugh map of a function $f: B^4 \to B$. The decomposition of 1-values into rectangles of sides 2^n, shown in this figure, again uses the "wrap around" property of top and bottom rows. The resulting expression for f is (verify)

$$(z' \wedge y') \vee (x' \wedge y' \wedge z) \vee (x \wedge y \wedge z \wedge w)$$

The first term comes from the 2×2 square formed by joining the 1×2 rectangle in the upper left-hand corner and the 1×2 rectangle in the lower left-hand corner. The second comes from the rectangle of size 1×2 in the upper right-hand corner, and the last is a minterm corresponding to the isolated square.

EXERCISE SET 4.5

In Exercises 1–6, construct Karnaugh maps for the functions whose truth tables are given.

1.

x	y	$f(x, y)$
0	0	1
0	1	0
1	0	0
1	1	1

2.

x	y	$f(x, y)$
0	0	1
0	1	0
1	0	1
1	1	0

3.

x	y	z	$f(x, y, z)$
0	0	0	1
0	0	1	1
0	1	0	0
0	1	1	0
1	0	0	1
1	0	1	0
1	1	0	1
1	1	1	0

4.

x	y	z	$f(x, y, z)$
0	0	0	0
0	0	1	1
0	1	0	1
0	1	1	1
1	0	0	0
1	0	1	0
1	1	0	0
1	1	1	1

5.

x	y	z	w	$f(x, y, z, w)$
0	0	0	0	0
0	0	0	1	0
0	0	1	0	1
0	0	1	1	0
0	1	0	0	0
0	1	0	1	0
0	1	1	0	1
0	1	1	1	0
1	0	0	0	0
1	0	0	1	0
1	0	1	0	0
1	0	1	1	1
1	1	0	0	0
1	1	0	1	0
1	1	1	0	1
1	1	1	1	1

6.

x	y	z	w	$f(x, y, z, w)$
0	0	0	0	1
0	0	0	1	0
0	0	1	0	1
0	0	1	1	0
0	1	0	0	0
0	1	0	1	1
0	1	1	0	1
0	1	1	1	0
1	0	0	0	0
1	0	0	1	0
1	0	1	0	0
1	0	1	1	0
1	1	0	0	1
1	1	0	1	0
1	1	1	0	1
1	1	1	1	0

In Exercises 7–11, Karnaugh maps of functions are given, and a decomposition of 1-values into rectangles is shown. Write the Boolean expression for these functions which arise from the maps and rectangular decompositions.

9.

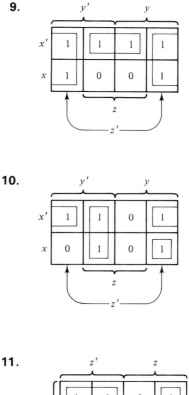

10.

11.

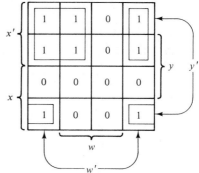

In Exercises 12–17, use the Karnaugh map method to find a Boolean expression for the function f.

12. Let f be the function of Exercise 1.
13. Let f be the function of Exercise 2.
14. Let f be the function of Exercise 3.
15. Let f be the function of Exercise 4.
16. Let f be the function of Exercise 5.
17. Let f be the function of Exercise 6.

7. **8.**

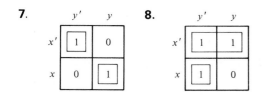

KEY IDEAS FOR REVIEW

☐ Partial order on a set: relation that is reflexive, antisymmetric, and transitive.

☐ Partially ordered set or poset: set together with a partial order.

☐ Linearly ordered set: partially ordered set in which every pair of elements are comparable.

☐ Theorem: If A and B are posets, then $A \times B$ is a poset with the product partial order.

☐ Dual of a poset $(A, \leq)$: The poset $(A, \geq)$, where $\geq$ denotes the inverse of $\leq$.

☐ Hasse diagram: see page 170.

☐ Topological sorting: see page 172.

☐ Maximal (minimal) element of a poset: see page 175.

☐ Theorem: A finite nonempty poset has at least one maximal element and at least one minimal element.

☐ Greatest (least) element of a poset A: see page 177.

☐ Theorem: A poset has at most one greatest element and at most one least element.

☐ Upper (lower) bound of subset B of poset A: Element $a \in A$ such that $b \leq a$ $(a \leq b)$ for all $b \in B$.

☐ Least upper bound (greatest lower bound) of subset B of poset A: Element $a \in A$ such that a is an upper (lower) bound of B and $a \leq a'$ $(a' \leq a)$, where a' is any upper (lower) bound of B.

☐ Lattice: a poset in which every subset consisting of two elements has a LUB and a GLB.

☐ Theorem: If L_1 and L_2 are lattices, then $L = L_1 \times L_2$ is a lattice.

☐ Theorem: Let L be a lattice, and $a \in L, b \in L, c \in L$. Then

    (a) $a \vee b = b$ if and only if $a \leq b$.

    (b) $a \wedge b = a$ if and only if $a \leq b$.

    (c) $a \wedge b = a$ if and only if $a \vee b = b$.

☐ Theorem: Let L be a lattice. Then

    1. (a) $a \vee a = a$
       (b) $a \wedge a = a$ (Idempotent Property)

    2. (a) $a \vee b = b \vee a$
       (b) $a \wedge b = b \wedge a$ (Commutative Property)

    3. (a) $a \vee (b \vee c) = (a \vee b) \vee c$
       (b) $a \wedge (b \wedge c) = (a \wedge b) \wedge c$ (Associative Property)

    4. (a) $a \vee (a \wedge b) = a$
       (b) $a \wedge (a \vee b) = a$ (Absorption Property)

☐ Theorem: Let L be a lattice, and $a \in L, b \in L, c \in L$.

    1. If $a \leq b$, then

(a) $a \vee c \leq b \vee c$

(b) $a \wedge c \leq b \wedge c$

2. $a \leq c$ and $b \leq c$ if and only if $a \vee b \leq c$.

3. $c \leq a$ and $c \leq b$ if and only if $c \leq a \wedge b$.

4. If $a \leq b$ and $c \leq d$, then

(a) $a \vee c \leq b \vee d$

(b) $a \wedge c \leq b \wedge d$

☐ Bounded lattice: has a greatest element I and a least element 0.

☐ Theorem: A finite lattice is bounded.

☐ Distributive lattice: satisfies the Distributive Laws:

$$a \wedge (b \vee c) = (a \wedge b) \vee (a \wedge c); \qquad a \vee (b \wedge c) = (a \vee b) \wedge (a \vee c)$$

☐ Complement of a: Element $a' \in L$ (bounded lattice) such that

$$a \vee a' = I \qquad \text{and} \qquad a \wedge a' = 0$$

☐ Theorem: Let L be a bounded distributive lattice. If a complement exists, it is unique.

☐ Complemented lattice: bounded lattice in which every element has a complement.

☐ Boolean algebra: complemented, distributive lattice.

☐ Properties of a Boolean algebra: see page 196.

☐ Truth tables: see page 199.

☐ Theorem: If B_1 and B_2 are Boolean algebras, then $B_1 \times B_2$ is a Boolean algebra.

☐ Boolean expression: see page 201.

☐ Minterm: a Boolean expression of the form $\bar{x}_1 \wedge \bar{x}_2 \wedge \cdots \wedge \bar{x}_n$, where each $\bar{x}_k$ is either x_k or x'_k.

☐ Theorem: Any function $f \colon B^n \to B$ is produced by a Boolean expression.

☐ Karnaugh map: see page 209.

FURTHER READING

Abbott, J. C., *Sets, Lattices, and Boolean Algebras*, Allyn and Bacon, Boston, 1969.

Arnold, B. H., *Logic and Boolean Algebra*, Prentice-Hall, Englewood Cliffs, N.J., 1962.

Harrison, Michael A., *Introduction to Switching Theory and Automata Theory*, McGraw-Hill, New York, 1965.

Hohn, F. E., *Applied Boolean Algebra*, 2nd ed., Macmillan, New York, 1966.

Whitesitt, J. E., *Boolean Algebra and Its Applications*, Addison-Wesley, Reading, Mass., 1961.

Trees and Languages

Prerequisites: Chapters 1, 2, and 3.

5.1 Trees

In this chapter we study a special type of relation which is exceptionally useful in a variety of computer science applications, and which is usually represented by its digraph. These relations are essential for the construction of data bases and language compilers, to name just two important areas. They are called trees or sometimes rooted trees, because of the appearance of their digraphs.

Let A be a finite set, and let T be a relation on A. We say that T is a **tree** if there is a vertex v_0 in A with the property that there exists a unique path in T from v_0 to every other vertex in A, but no path from v_0 to v_0.

We show below that the vertex v_0, described in the definition above, is unique. It is often called the **root** of the tree T, and T is then referred to as a **rooted tree**. We write (T, v_0) to denote a rooted tree T with root v_0.

If (T, v_0) is a rooted tree on the set A, an element v of A will often be referred to as a **vertex** in T. This terminology simplifies discussion, since it often happens that the underlying set A of T is of no importance.

To help us see the nature of trees, we will prove some simple properties satisfied by trees.

Theorem 1   Let (T, v_0) be a rooted tree. Then

(a) There are no cycles in T.
(b) v_0 is the only root of T.
(c) Each vertex in T, other than v_0, has in-degree one, and v_0 has in-degree zero.

Proof.

(a) Suppose that there is a cycle q in T, beginning and ending at vertex v. By the definition of a tree, we know that $v \neq v_0$, and there must be a path p from v_0 to v. Then $p \circ q$ (see Section 2.3) is a path from v_0 to v which is different from p, and this contradicts the definition of a tree.

(b) If v_0' is another root of T, there is a path p from v_0 to v_0' and a path q from v_0' to v_0 (since v_0' is a root). Then $p \circ q$ is a cycle from v_0 to v_0, and this is impossible by definition. Hence the vertex v_0 is the unique root.

(c) Let w_1 be a vertex in T other than v_0. Then there is a unique path $v_0, \cdots, v_k, w_1$ from v_0 to w_1 in T. This means that $(v_k, w_1) \in T$, so w_1 has in-degree at least one. If the in-degree of w_1 is more than one, there must be distinct vertices w_2 and w_3, such that (w_2, w_1) and (w_3, w_1) are both in T. If $w_2 \neq v_0$ and $w_3 \neq v_0$, there are paths p_2 from v_0 to w_2 and p_3 from v_0 to w_3, by definition. Then $p_2 \circ (w_2, w_1)$ and $p_3 \circ (w_3, w_1)$ are two different paths from v_0 to w_1, and this contradicts the definition of a tree with root v_0. Hence, the in-degree of w_1 is one. We leave it as an exercise to complete the proof if $w_2 = v_0$ or $w_3 = v_0$, and to show that v_0 has in-degree zero.

Theorem 1 summarizes the geometric properties of a tree. With these properties in mind, we can see how the digraph of a typical tree must look.

Let us first draw the root v_0. No edges enter v_0, but several may leave, and we draw these edges downward. The terminal vertices of the edges beginning at v_0 will be called the **level 1** vertices, while v_0 will be said to be at **level 0**. Also, v_0 is sometimes called the **parent** of these level 1 vertices, and the level 1 vertices are called the **offspring** of v_0. This is shown in Fig. 1(a). Each vertex at level 1 has no other edges entering it, by part (c) of Theorem 1, but each of these vertices may have edges leaving the vertex. The edges leaving a vertex of level 1 are drawn downward and terminate at various vertices which are said to be at **level 2**. Figure 1(b) shows the situation at this point. A parent–offspring relationship holds also for these levels (and at every consecutive pair of levels). For example, v_3 would be called the parent of the three offspring v_7, v_8, and v_9. The offspring of any one vertex are sometimes called **siblings**.

The process above continues for as many levels as are required to complete the digraph. If we view the digraph upside down, we will see why these relations are called trees. The number of levels of a tree is called the **height** of the tree.

We should note that a tree may have infinitely many levels, and that any level other than level 0 may contain an infinite number of vertices. In fact, any vertex could have infinitely many offspring. However, in all our future discussions, trees will be assumed to have a finite number of vertices. Thus the trees will always have a bottom (highest-numbered) level. The vertices of the tree that have no offspring are, for geometric reasons, called the **leaves** of the tree.

The vertices of a tree that lie at any one level simply form a set of vertices in A. Often, however, it is useful to suppose that the offspring of each vertex of the tree are linearly ordered. Thus if a vertex v has four offspring, we may assume that they are ordered, so that we may refer to them as the first, second, third, or fourth

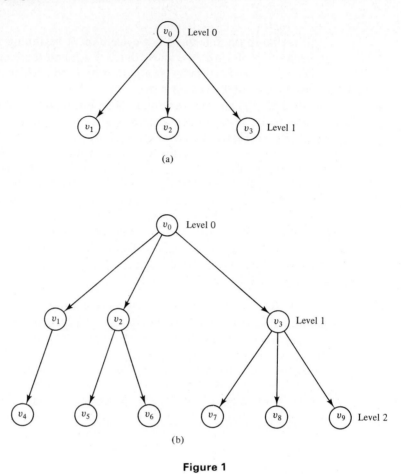

Figure 1

offspring of v. Whenever we draw the digraph of a tree, we automatically assume some ordering at each level, by arranging offspring from left to right. Such a tree will be called an **ordered tree**. Generally, ordering of offspring in a tree is not explicitly mentioned. If ordering is needed, it is usually introduced at the time when the need arises, and it often is specified by the way the digraph of the tree is drawn. The following relational properties of trees are easily verified.

Theorem 2   Let (T, v_0) be a rooted tree on a set A. Then:

(a) T is irreflexive.

(b) T is asymmetric.

(c) If $(a, b) \in T$ and $(b, c) \in T$, then $(a, c) \notin T$, for all a, b, and c in A.

Proof. The proof is left as an exercise.

Example 1   Let A be the set of all female descendants of a given human female v_0. We now define the following relation T on A: If v_1 and v_2 are elements of A, then $v_1 \; T \; v_2$ if and only if v_1 is the mother of v_2. The relation T on A is a rooted tree with root v_0.

Example 2   Let $A = \{v_0, v_1, v_2, v_3, v_4, v_5, v_6, v_7, v_8, v_9, v_{10}\}$ and let $T = \{(v_2, v_3), (v_2, v_1), (v_4, v_5), (v_4, v_6), (v_5, v_8), (v_6, v_7), (v_4, v_2), (v_7, v_9), (v_7, v_{10})\}$. Show that T is a rooted tree, and identify the root.

Solution. Since no paths begin at vertices v_1, v_3, v_8, v_9, and v_{10}, these vertices cannot be roots of a tree. There are no paths from vertices v_6, v_7, v_2, and v_5 to vertex v_4, so we must eliminate these vertices as possible roots. Thus if T is a rooted tree, its root must be vertex v_4. It is easy to show that there is a path from v_4 to every other vertex. For example, the path v_4, v_6, v_7, v_9 leads from v_4 to v_9, since $(v_4, v_6), (v_6, v_7)$, and (v_7, v_9) are all in T. We draw the digraph of T, beginning with vertex v_4, and with edges shown downward. The result is shown in Fig. 2. A quick inspection of this digraph shows that paths from vertex v_4 to every other vertex are unique, and there are no paths from v_4 to v_4. Thus T is a tree with root v_4.

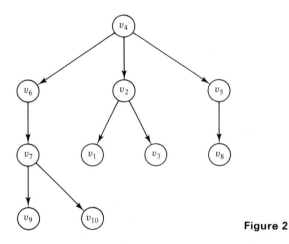

Figure 2

If n is a positive integer, we will say that a tree T is an **n-tree** if every vertex has at most n offspring. If all vertices of T, other than the leaves, have exactly n offspring, we will say that T is a **complete n-tree**. In particular, a 2-tree will often be called a **binary tree**, and a complete 2-tree will often be called a **complete binary tree**.

Binary trees are extremely important since there are quite efficient methods of implementing them and searching through them on computers. We will see some of these methods in Section 5.5, and we will also see that any tree can be reorganized as a binary tree.

Let (T, v_0) be a rooted tree on the set A, and let v be a vertex of T. Let B be the set consisting of v and all of its **descendants**, that is, all vertices of T that can be reached by a path beginning at v. Observe that $B \subseteq A$. Let $T(v)$ be the restriction of the relation T to B, that is, $T \cap (B \times B)$ (see Section 2.2). Then we have the following result.

Theorem 3 If (T, v_0) is a rooted tree and $v \in T$, then $T(v)$ is also a rooted tree with root v. We will say that $T(v)$ is the **subtree** of T beginning at v.

Proof. By definition of $T(v)$, we see that there is a path from v to every other vertex in $T(v)$. If there is a vertex w in $T(v)$, such that there are two distinct paths q and q' from v to w, and if p is the path in T from v_0 to v, then $p \circ q$ and $p \circ q'$ would be two distinct paths in T from v_0 to w. This is impossible, since T is a tree with root v_0. Thus each path from v to another vertex w in $T(v)$ must be unique. Also if q is a cycle at v in $T(v)$, then q is also a cycle in T. This contradicts Theorem 1(a), therefore q cannot exist. It follows that $T(v)$ is a tree with root v.

Example 3 Consider the tree T of Example 2. This tree has root v_4, and is shown in Fig. 2. In Fig. 3 we have drawn the subtrees $T(v_5)$, $T(v_2)$, and $T(v_6)$ of T.

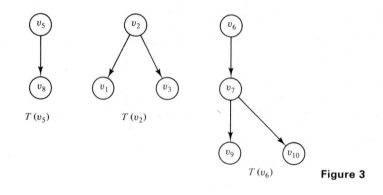

$T(v_5)$ $T(v_2)$

$T(v_6)$ **Figure 3**

EXERCISE SET 5.1

In Exercises 1–8, each relation R is defined on the set A. In each case determine if R is a tree, and if it is, find the root.

1. $A = \{a, b, c, d, e, f\}$
   $R = \{(a, d), (b, c), (c, a), (d, e)\}$
2. $A = \{a, b, c, d, e\}$
   $R = \{(a, b), (b, e), (c, d), (d, b), (c, a)\}$
3. $A = \{a, b, c, d, e, f\}$
   $R = \{(a, b), (c, e), (f, a), (f, c), (f, d)\}$
4. $A = \{1, 2, 3, 4, 5, 6\}$
   $R = \{(2, 1), (3, 4), (5, 2), (6, 5), (6, 3)\}$

5. $A = \{1, 2, 3, 4, 5, 6\}$
   $R = \{(1, 1), (2, 1), (2, 3), (3, 4), (4, 5), (4, 6)\}$
6. $A = \{1, 2, 3, 4, 5, 6\}$
   $R = \{(1, 2), (1, 3), (4, 5), (4, 6)\}$
7. $A = \{t, u, v, w, x, y, z\}$
   $R = \{(t, u), (u, w), (u, x), (u, v), (v, z), (v, y)\}$
8. $A = \{u, v, w, x, y, z\}$
   $R = \{(u, x), (u, v), (w, v), (x, z), (x, y)\}$

In Exercises 9–12, consider the rooted tree (T, v_0) shown in Fig. 4.

9. Compute the tree $T(v_2)$.

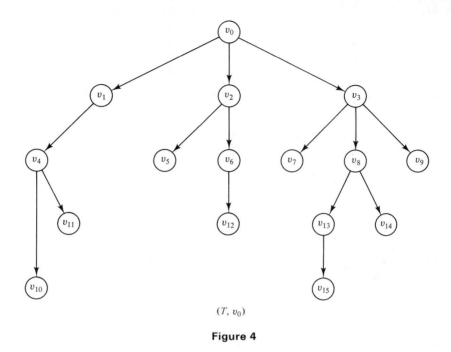

(T, v_0)

Figure 4

10. Compute the tree $T(v_3)$.
11. Compute the tree $T(v_8)$.
12. Compute the tree $T(v_1)$.

13. Prove Theorem 2.

14. Let T be a tree. Suppose that T has r vertices and s edges. Find a formula relating r to s.

15. Show that if (T, v_0) is a rooted tree, then v_0 has in-degree zero.

5.2 Labeled Trees

It is sometimes useful to label the vertices or edges of a digraph to indicate that the digraph is being used for a particular purpose. This is especially true for many uses of trees in computer science. We will now give a series of examples in which the sets of vertices of the trees are not important, but rather the utility of the tree is best emphasized by the labels on these vertices. Thus we will represent the vertices simply as dots, and show the label of each vertex next to the dot representing that vertex.

Example 1   We have expressed our algorithms in this text as pseudocode statements, and have avoided the use of flowcharts. The idea of a flowchart is simply to give a pictorial representation of the order of execution of steps in a program, and we will assume that the reader has at least a rudimentary familiarity with this concept.

Consider the algorithm expressed in pseudocode in Fig. 1. Assume that the algorithm has access to integers N and M, a positive integer MAX, and a print subroutine PRINT. The algorithm is to return M^N (in variable S), when defined (0^0 is not defined), and print an error statement otherwise. Also, an error statement is to be printed when $M^N > $ MAX. In Fig. 2 we show the usual flowchart for the algorithm in Fig. 1. We have suppressed the detail of the loops, in order to emphasize the results of selection. In Fig. 3 we represent this flowchart as a labeled tree. Vertices are labeled with sections of pseudocode which are to be processed at the corresponding places in the algorithm. Edges represent transitions from one set of statements to another.

1. $S \leftarrow 1$
2. IF $(N > 0)$ THEN
   a. FOR $I = 1$ THRU N
     1. $S \leftarrow S \times M$
   b. IF $(S > $ MAX$)$ THEN
     1. CALL PRINT ('ANSWER IS TOO LARGE')
3. ELSE
   a. IF $(M = 0)$ THEN
     1. CALL PRINT ('ANS. UNDEFINED')
   b. ELSE
     1. IF $(N = 0)$ THEN
       a. $S \leftarrow 1$
     2. ELSE
       a. MRECIP $\leftarrow 1 \div M$
       b. $N \leftarrow (-1) \times N$
       c. FOR $I = 1$ THRU N
         1. $S \leftarrow S \times$ MRECIP

Figure 1

The FOR loop is treated as a self-contained section. Vertices have also been added to represent explanations of the processing. The underlying tree structure of this program is apparent from the figure.

We should note that a more detailed analysis would expand the loops graphically, and the resulting digraph would not be a tree. In any case, the graphical display and analysis of programs is a matter that has been highly developed in recent years and has proved to be quite useful.

Example 2   Consider the fully parenthesized, algebraic expression

$$(3 - (2 \times x)) + ((x - 2) - (3 + x))$$

We assume, in such an expression, that no operation such as $-, +, \times, \div$, can be performed until both of its arguments have been evaluated, that is, until all

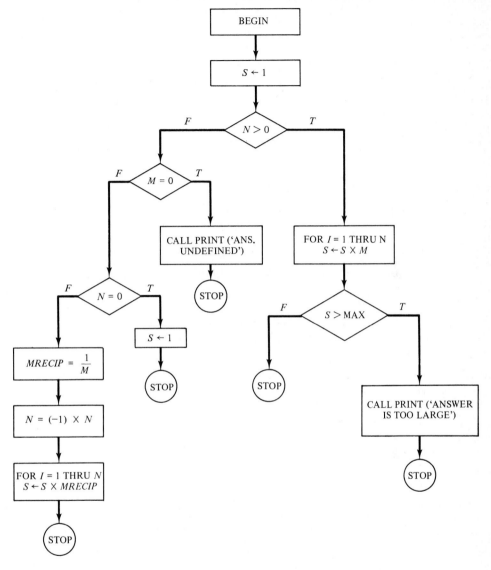

Figure 2

computations inside both the left and right arguments have been performed. Thus we cannot perform the central addition until we have evaluated $(3 - (2 \times x))$ and $((x - 2) - (3 + x))$. We cannot perform the central subtraction in $((x - 2) - (3 + x))$ until we evaluate $(x - 2)$ and $(3 + x)$, and so on. It is easy to see that each such expression has a **central operator**, corresponding to the last computation which can be performed. Thus " + " is central to the main expression above, " − " is central to $(3 - (2 \times x))$, and so on. An important graphical representation of such an ex-

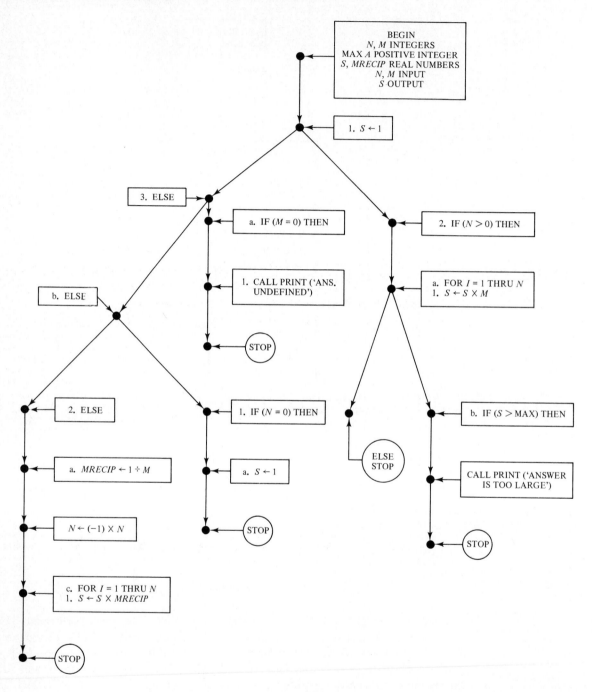

Figure 3

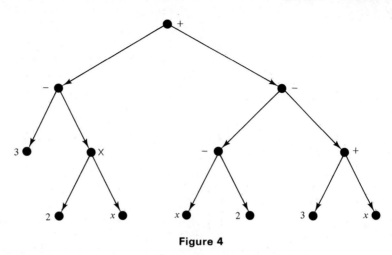

Figure 4

pression is as a labeled binary tree. In this tree the root is labeled with the central operator of the main expression. The two offspring of the root are labeled with the central operator of the expressions for the left and right arguments, respectively. If either argument is a constant or variable, instead of an expression, this constant or

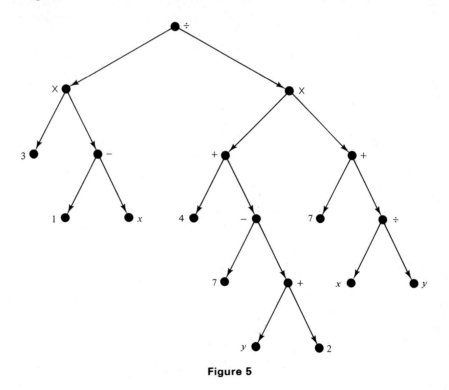

Figure 5

variable is used to label the corresponding offspring vertex. This process continues until the expression is exhausted. Fig. 4 shows the tree for the original expression of this example. To illustrate the technique further, we have shown in Fig. 5 the tree corresponding to the full parenthesized expression

$$(3 \times (1 - x)) \div ((4 + (7 - (y + 2))) \times (7 + (x \div y)))$$

Our final example of a labeled tree is important for the computer implementation of a tree data structure. We start with an n-tree (T, v_0). Each vertex in T has at most n offspring. We imagine that each vertex *potentially* has exactly n offspring, which would be ordered from 1 to n, but that some of the offspring in the sequence may be missing. The remaining offspring are labeled with the position they occupy in the hypothetical sequence. Thus the offspring of any vertex are labeled with distinct numbers from the set $\{1, 2, \ldots, n\}$.

Such a labeled digraph is sometimes called **positional**, and we will also use this term. Note that positional trees are also ordered trees. When drawing the digraphs of a positional tree, we will imagine that the n offspring positions for each vertex are arranged symmetrically below that vertex, and we place in its appropriate position each offspring that actually occurs.

Figure 6 shows the digraph of a positional 3-tree, with all actually occurring positions labeled. If offspring 1 of any vertex v actually exists, the edge from v to that offspring is drawn sloping to the left. Offspring 2 of any vertex v is drawn vertically downward from v, whenever it occurs. Similarly, offspring labeled 3 will be drawn to the right. Naturally, the root is not labeled, since it is not an offspring.

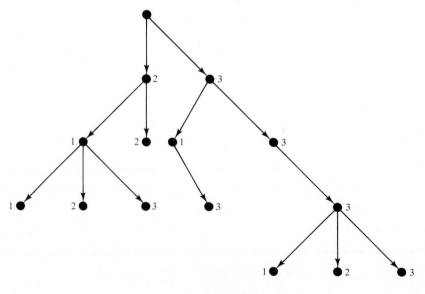

Figure 6

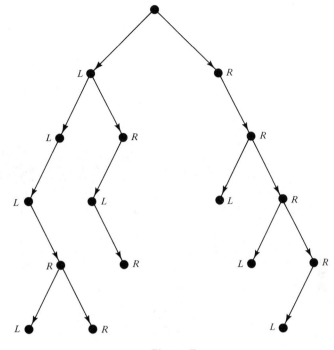

Figure 7

The positional binary tree is of special importance. In this case, for obvious reasons, the positions for potential offspring often are labeled *left* and *right*, instead of 1 and 2. Figure 7 shows the digraph of a positional binary tree, with offspring labeled "*L*" for left and "*R*" for right. Labeled trees may have several sets of labels, all in force simultaneously. We will usually omit the left–right labels on a positional binary tree, in order to emphasize other useful labels. The positions of the offspring will then be indicated by the direction of the edges, as we have drawn them in Fig. 7.

EXERCISE SET 5.2

In Exercises 1–10, construct the tree of the algebraic expression.

1. $(7 + (6 - 2)) - (x - (y - 4))$

2. $(x + (y - (x + y))) \times ((3 \div (2 \times 7)) \times 4)$

3. $3 - (x + (6 \times (4 \div (2 - 3))))$

4. $(((2 \times 7) + x) \div y) \div (3 - 11)$

5. $((2 + x) - (2 \times x)) - (x - 2)$

6. $(11 - (11 \times (11 + 11))) + (11 + (11 \times 11))$

7. $(3 - (2 - (11 - (9 - 4)))) \div (2 + (3 + (4 + 7)))$

8. $(x \div y) \div ((x \times 3) - (z \div 4))$

9. $((2 \times x) + (3 - (4 \times x))) + (x - (3 \times 11))$

10. $((1 + 1) + (1 - 2)) \div ((2 - x) + 1)$

11. Construct the digraphs of all distinct binary positional trees having three or fewer edges, and exactly two levels.

12. How many distinct binary positional trees are there having exactly two levels?

13. How many distinct positional 3-trees are there having exactly two levels?

14. Construct the digraphs of all distinct positional 3-trees having two or fewer edges.

5.3 Languages

If S is a set, then in Section 1.2 we considered the set S^* consisting of all finite strings of elements from the set S. There are many possible interpretations of the elements of S^*, depending on the nature of S. If we think of S as a set of "words," then S^* may be regarded as the collection of all possible "sentences" formed from words in S. Of course, such "sentences" do not have to be meaningful or even sensibly constructed. We may think of a language as a complete specification, at least in principle, of three things. First, there must be a set S consisting of all "words" which are to be regarded as being part of the language. Second, a subset of S^* must be designated as the set of "properly constructed sentences" in the language. The meaning of this term will depend very much on the language being constructed. Finally, it must be determined which of the properly constructed sentences have meaning, and what the meaning is.

Suppose, for example, that S consists of all English words. The specification of a properly constructed sentence involves the complete rules of English grammar; the meaning of a sentence is determined by this construction and by the meaning of the words. The sentence

"Going to the store John George to sing"

is a string in S^*, but is not a properly constructed sentence. The arrangement of nouns and verb phrases is illegal. On the other hand, the sentence

"Noiseless blue sounds sit cross-legged under the mountaintop"

is properly constructed, but completely meaningless.

For another example, S may consist of the integers, the symbols $+$, $-$, $\times$, and $\div$, and left and right parentheses. We will obtain a language if we designate as proper those strings in S^* which represent unambiguously parenthesized algebraic expressions. Thus

$$((3 - 2) + (4 \times 7)) \div 9 \qquad \text{and} \qquad (7 - (8 - (9 - 10)))$$

are properly constructed "sentences" in this language. On the other hand, $(2 - 3)) + 4$, $4 - 3 - 2$, and $)2 + (3 -) \times 4$ are not properly constructed. The first has too many parentheses, the second has too few (we do not know which subtraction to perform first), and the third has parentheses and numbers completely out of place. All properly constructed expressions have meaning except those involving division by zero. The meaning of an expression is the rational number it represents. Thus the meaning of $((2 - 1) \div 3) + (4 \times 6)$ is $73/3$, while $2 + (3 \div 0)$ and $(4 + 2) - (0 \div 0)$ are not meaningful.

The specification of the proper construction of sentences is called the **syntax** of a language. The specification of the meaning of sentences is called the **semantics** of a language. Among those languages that are of fundamental importance in computer science are the programming languages. These include BASIC, FORTRAN, ALGOL, PASCAL, APL, LISP, SNOBOL, ADA, FORTH, and many other general- and special-purpose languages. When students are taught to program in some programming language, they are actually taught the syntax of the language. In a compiled language such as FORTRAN, most mistakes in syntax are detected by the compiler, and appropriate error messages are generated. The semantics of a programming language forms a much more difficult and advanced topic of study. The meaning of a line of programming is taken to be the entire sequence of events that take place inside the computer as a result of executing or interpreting that line.

We will not deal with semantics at all. We will study the syntax of a class of languages called phrase structure grammars. Although these are not nearly complex enough to include "real" languages such as English, they are general enough to encompass many languages of importance in computer science. This includes most aspects of programming languages, although the complete specification of some higher-level programming languages exceeds the scope of these grammars. On the other hand, phrase structure grammars are simple enough to be studied precisely, since the syntax is determined by substitution rules. Those grammars that will occupy most of our attention lead to interesting examples of labeled trees.

A **phrase structure grammar** G is defined to be a 4-tuple $(V, S, v_0, \mapsto)$, where V is a finite set, S is a subset of V, $v_0 \in V - S$, and $\mapsto$ is a finite relation on V^*. The idea here is that S is, as discussed above, the set of all allowed "words" in the language, and V consists of S together with some other symbols. The element v_0 of V is a starting point for the substitutions which will shortly be discussed. Finally, the relation $\mapsto$ on V^* is a replacement relation, in the sense that if $w \mapsto w'$, we may replace w by w' whenever the string w occurs, either alone or as a substring of some other string. Traditionally, the statement $w \mapsto w'$ is called a **production** of G. Then w and w' are termed the **left** and **right** sides of the production, respectively. We assume that no production of G has the empty string Λ as its left side. We will call $\mapsto$ the **production relation** of G.

With these ingredients, we can introduce a substitution relation, denoted by $\Rightarrow$, on V^*. We let $x \Rightarrow y$ mean that $x = l \cdot w \cdot r$, $y = l \cdot w' \cdot r$, and $w \mapsto w'$, where l and r are completely arbitrary strings in V^*. In other words, $x \Rightarrow y$ means that y results from x by using one of the allowed productions to replace part or all of x. The relation $\Rightarrow$ is usually called **direct derivability**. Finally, we let $\Rightarrow^*$ be the reachability relation of $\Rightarrow$ (see Section 2.3), and we decree that a string w in S^* is a syntactically correct sentence if and only if $v_0 \Rightarrow^* w$. In more detail, this says that a string w is a properly constructed sentence if w is in S^*, not just in V^*, and if we can get from v_0 to w by making a finite number of substitutions. This may seem complicated, but it is really quite a simple idea, as the following examples will show.

If $G = (V, S, v_0, \mapsto)$ is a phrase structure grammar, we will call S the set of **terminal symbols**, and $N = V - S$ the set of **nonterminal symbols**. Note that $V = S \cup N$.

The reader should be warned that other texts have slight variations of the definitions and notations which we have used for phrase structure grammars.

Example 1　Let S = {John, Jill, drives, jogs, carelessly, rapidly, frequently}, N = {sentence, noun, verbphrase, verb, adverb}, and let $V = S \cup N$. Let v_0 = "sentence," and suppose that the relation $\mapsto$ on V^* is described by listing all productions as follows.

$$\text{sentence} \mapsto \text{noun verbphrase}$$

$$\text{noun} \mapsto \text{John}$$

$$\text{noun} \mapsto \text{Jill}$$

$$\text{verbphrase} \mapsto \text{verb adverb}$$

$$\text{verb} \mapsto \text{drives}$$

$$\text{verb} \mapsto \text{jogs}$$

$$\text{adverb} \mapsto \text{carelessly}$$

$$\text{adverb} \mapsto \text{rapidly}$$

$$\text{adverb} \mapsto \text{frequently}$$

The set S contains all the allowed words in the language; N consists of words that describe parts of sentences, but which are not actually contained in the language.

We claim that the sentence "Jill drives frequently", which we will denote by w, is an allowable or syntactically correct sentence in this language. To prove this, we consider the following sequence of strings in V^*.

sentence		
noun	verbphrase	
Jill	verbphrase	
Jill	verb	adverb
Jill	drives	adverb
Jill	drives	frequently

Now each of these strings follows from the preceding one, by using a production to make a partial or complete substitution. In other words, each string is related to the following string by the relation $\Rightarrow$, so sentence $\Rightarrow^* w$. By definition then, w is syntactically correct since, for this example, v_0 is "sentence." In phrase structure grammars, correct syntax simply refers to the process by which a sentence is formed, nothing else.

It should be noted that the sequence of substitutions which produces a valid sentence, a sequence that will be called the **derivation** of the sentence, is not unique.

The following derivation produces the sentence w of Example 1, but is not identical with the derivation given there.

sentence		
noun	verbphrase	
noun	verb	adverb
noun	verb	frequently
noun	drives	frequently
Jill	drives	frequently

The set of all properly constructed sentences that can be produced using a grammar G, is called the **language** of G, and is denoted by $L(G)$. The language of the grammar given in Example 1 is a somewhat simpleminded sublanguage of English, and it contains exactly 12 sentences. The reader can verify that "John jogs carelessly" is in the language $L(G)$ of this grammar, while "Jill frequently jogs" is not in $L(G)$.

It is also true that many different phrase structure grammars may produce the same language; that is, they may have exactly the same set of syntactically correct sentences. Thus a grammar cannot be reconstructed from its language. In Section 5.4 we will give examples in which different grammars are used to construct the same language.

Example 1 illustrates the process of derivation of a sentence in a phrase structure grammar. Another method that may sometimes be used to show the derivation process is the construction of a **derivation tree** for the sentence. The starting symbol, v_0, is taken as the label for the root of this tree. The level 1 vertices correspond to, and are labeled in order by the various words involved in the first substitution for v_0. Then the offspring of each vertex, at every succeeding level, are labeled by the various words (if any) that are substituted for that vertex the next time it is subjected to substitution. Consider, for example, the first derivation of sentence w in Example 1. Its derivation tree begins with "sentence," and the next-level vertices correspond to "noun" and "verbphrase" since the first substitution replaces the word "sentence" with the string "noun verbphrase." This part of the tree is shown in Fig. 1(a). Next, we substitute "Jill" for "noun," and the tree becomes as shown in Fig. 1(b). The next two substitutions "verb adverb" for "verbphrase," and "drives" for "verb," extend the tree as shown in Fig. 1(c) and (d). Finally, the tree is completed with the substitution of "frequently" for "adverb." The finished derivation tree is shown in Fig. 1(e).

The second derivation sequence, given in Example 1 for the sentence w, yields a derivation tree in the stages shown in Fig. 2. Notice that the same tree results in both figures. Thus these two derivations yield the same tree, and the differing orders of substitution simply create the tree in different ways. The sentence being derived labels the leaves of the resulting tree.

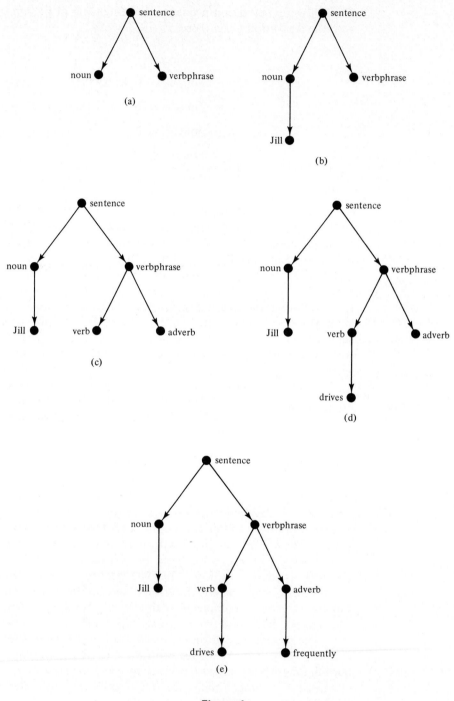

Figure 1

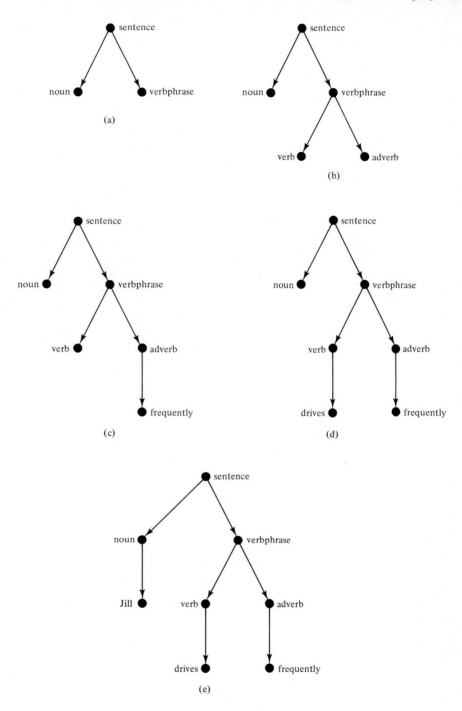

Figure 2

Example 2   Let $V = \{v_0, w, a, b, c\}$, $S = \{a, b, c\}$, and let $\mapsto$ be the relation on V^* given by:

1. $v_0 \mapsto aw$
2. $w \mapsto bbw$
3. $w \mapsto c$

Consider the phrase structure grammar $G = (V, S, v_0, \mapsto)$. To derive a sentence of $L(G)$, it is necessary to perform successive substitutions, using (1), (2), and (3) above, until all symbols are eliminated other than the terminal symbols a, b, and c. Since we begin with the symbol v_0, we must first use production (1), or we could never eliminate v_0. This first substitution results in the string aw. We may now use (2) or (3) to substitute for w. If we use production (2), the result will contain a "w." Thus one application of (2) to aw produces the string ab^2w (a symbol b^n means n consecutive b's). If we then use (2) again, we will have the string ab^4w. We may use production (2) any number of times, but we will finally have to use production (3) to eliminate the symbol w. Once we use (3), only terminal symbols remain, so the process ends. We may summarize this analysis by saying that $L(G)$ is the subset of S^* corresponding to the regular expression $a(bb)^*c$ (see Section 1.6). Thus the word ab^6c is in the language of G, and its derivation tree is shown in Fig. 3. Note that, unlike the tree of Example 2, the derivation tree for ab^6c is not a binary tree.

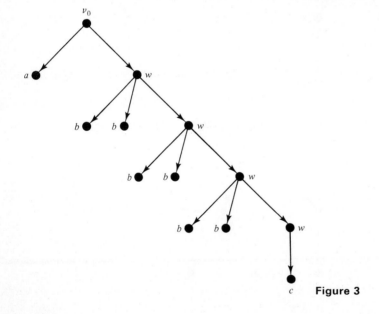

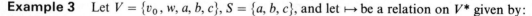

c **Figure 3**

Example 3   Let $V = \{v_0, w, a, b, c\}$, $S = \{a, b, c\}$, and let $\mapsto$ be a relation on V^* given by:

1. $v_0 \mapsto av_0 b$
2. $v_0 b \mapsto bw$
3. $abw \mapsto c$

Let $G = (V, S, v_0, \mapsto)$ be the corresponding phrase structure grammar. As we did in Example 2, we determine the form of allowable sentences in $L(G)$.

Since we must begin with the symbol v_0 alone, we must use production (1) first. We may continue to use (1) any number of times, but we must eventually use production (2) to eliminate v_0. Repeated use of (1) will result in a string of the form $a^n v_0 b^n$; that is, there are equal numbers of a's and b's. When (2) is used, the result is a string of the form $a^m (abw) b^m$ with $m \geq 0$. At this point the only production that can be used is (3), and we must use it to remove the nonterminal symbol w. The use of (3) finishes the substitution process and produces a string in S^*. Thus the allowable sentences $L(G)$ of the grammar G all have the form $w = a^n cb^n$, where $n \geq 0$. In this case it can be shown that $L(G)$ does not correspond to a regular expression over S.

Another interesting feature of the grammar in Example 3 is that the derivations of the sentences cannot be expressed as trees. Our construction of derivation trees works only when the left-hand sides of all productions used consist of single, nonterminal symbols. The left-hand sides of the productions in Example 3 do not have this simple form. Although it is possible to construct a graphical representation of these derivations, the resulting digraph would not be a tree. Many other problems can arise if no restrictions are placed on the productions. For this reason, a classification of phrase structure grammars has been devised.

Let $G = (V, S, v_0, \mapsto)$ be a phrase structure grammar. Then we say that G is:

Type 0 if no restrictions are placed on the productions of G

Type 1 if for any production $w_1 \mapsto w_2$, the length of w_1 is less than or equal to the length of w_2 (where the **length** of a string is the number of words in that string)

Type 2 if the left-hand side of each production is a single, nonterminal symbol and the right-hand side consists of one or more symbols

Type 3 if the left-hand side of each production is a single, nonterminal symbol, and the right-hand side has one or more symbols including at most one nonterminal symbol, which must be at the extreme right of the string

In each of the preceding types, we permit the inclusion of the trivial production $v_0 \mapsto \Lambda$, where Λ represents the empty string. This is an exception to the defining rule for types 1, 2, and 3, but it is included so that the empty string can be made part of the language. This avoids constant consideration of unimportant special cases.

It follows from the definition that each type of grammar is a special case of the type preceding it. Example 1 is a type 2 grammar, Example 2 is type 3, and Example 3 is type 0. Grammars of types 0 or 1 are quite difficult to study, and little is known about them. They include many pathological examples which are of no known practical use. We will restrict further consideration of grammars to types 2 and 3. These types have derivation trees for the sentences of their languages, and they are sufficiently complex to describe many aspects of actual programming languages.

Type 2 grammars are sometimes called **context-free grammars**, since the symbols on the left of the productions are substituted for wherever they occur. On the other hand, a production of the type $l \cdot w \cdot r \mapsto l \cdot w' \cdot r$ (which could not occur in a type 2 grammar), is called **context sensitive**, since w' is substituted for w only in the context where it is surrounded by the strings l and r. Type 3 grammars have a very close relationship with finite-state machines, which will be studied in Chapter 7. Type 3 grammars are also called **regular grammars.**

A language will be called **type 2** or **type 3** if there is a grammar of type 2 or type 3 which produces it. This concept can cause problems. Even if a language is produced by a nontype 2 grammar, it is possible that some type 2 grammar also produces this same language. In this case, the language is type 2. The same situation may arise in the case of type 3 grammars.

The process we have considered in this section, namely deriving a sentence within a grammar, has a converse process. The converse process involves taking a sentence, and verifying that it is syntactically correct in some grammar G, by constructing a derivation tree that will produce it. This process is called **parsing** the sentence, and the resulting derivation tree is often called the **parse tree** of the sentence. Parsing is of fundamental importance for compilers, and other forms of language translation. A sentence in one language is parsed to show its structure, and a tree is constructed. The tree is then searched and, at each step, corresponding sentences are generated in another language. In this way a FORTRAN program, for example, is compiled into a machine language program. The contents of this section and the next two sections are essential to the compiling process, but the complete details must be left to a more advanced course.

Exercise Set 5.3

In Exercises 1–7, a grammar G will be specified. In each case describe precisely the language, $L(G)$, produced by this grammar; that is, describe all syntactically correct "sentences."

1. $G = (V, S, v_0, \mapsto)$
   $V = \{v_0, v_1, x, y, z\}, S = \{x, y, z\}$
   $\mapsto$: $v_0 \mapsto x\, v_0$
   $\quad\quad v_0 \mapsto y\, v_1$
   $\quad\quad v_1 \mapsto y\, v_1$
   $\quad\quad v_1 \mapsto z$

2. $G = (V, S, v_0, \mapsto)$
   $V = \{v_0, a\}, S = \{a\}$
   $\mapsto$: $v_0 \mapsto a\, a\, v_0$
   $\quad\quad v_0 \mapsto a\, a$

3. $G = (V, S, v_0, \mapsto)$
   $V = \{v_0, a, b\}, S = \{a, b\}$
   $\mapsto$: $v_0 \mapsto a\, a\, v_0$
   $\quad\quad v_0 \mapsto a$
   $\quad\quad v_0 \mapsto b$

4. $G = (V, S, v_0, \mapsto)$
   $V = \{v_0, x, y, z\}, S = \{x, y, z,\}$
   $\mapsto$: $v_0 \mapsto x\, v_0\, v_0$
   $\quad\quad v_0 \mapsto y\, v_0$
   $\quad\quad v_0 \mapsto z$

5. $G = (V, S, v_0, \mapsto)$
   $V = \{v_0, v_1, v_2, a, +, (,)\}$
   $S = \{(,), a, +\}$
   $\mapsto$: $v_0 \mapsto (v_0)$ (where left and right parentheses
   $\quad\quad\quad\quad\quad$ are symbols from S)
   $\quad\quad v_0 \mapsto a + v_1$
   $\quad\quad v_1 \mapsto a + v_2$
   $\quad\quad v_2 \mapsto a + v_2$
   $\quad\quad v_2 \mapsto a$

6. $G = (V, S, v_0, \mapsto)$
   $V = \{v_0, v_1, a, b\}, S = \{a, b\}$
   $\mapsto$: $v_0 \mapsto a\, v_1$
   $\quad\quad v_1 \mapsto b\, v_0$
   $\quad\quad v_1 \mapsto a$

7. $G = (V, S, v_0, \mapsto)$
$V = \{v_0, v_1, v_2, x, y, z\}$, $S = \{x, y, z\}$
$\mapsto : v_0 \mapsto v_0 v_1$
$\quad\quad v_0 v_1 \mapsto v_2 v_0$
$\quad\quad v_2 v_0 \mapsto x\ y$
$\quad\quad v_2 \mapsto x$
$\quad\quad v_1 \mapsto z$

8. For each grammar in Exercises 1–7, state whether the grammar is type 1, 2, or 3.

9. Let $G = (V, S, I, \mapsto)$, where

$V = \{I, L, D, W, a, b, c, 0, 1, 2, 3, 4, 5, 6, 7, 8, 9\}$

$S = \{a, b, c, 0, 1, 2, 3, 4, 5, 6, 7, 8, 9\}$

$\mapsto$ is given by:

1. $\quad I \mapsto L$
2. $\quad I \mapsto LW$
3. $\quad W \mapsto LW$
4. $\quad W \mapsto DW$
5. $\quad W \mapsto L$
6. $\quad W \mapsto D$
7. $\quad L \mapsto a$
8. $\quad L \mapsto b$
9. $\quad L \mapsto c$
10. $\quad D \mapsto 0$
11. $\quad D \mapsto 1$

$\quad\quad .$
$\quad\quad .$
$\quad\quad .$

19. $\quad D \mapsto 9$

Which of the following statements are true for this grammar?

(a) $ab092 \in L(G)$
(b) $2a3b \in L(G)$
(c) $aaaa \in L(G)$
(d) $I \Rightarrow a$
(e) $I \Rightarrow^* ab$
(f) $DW \Rightarrow 2$
(g) $DW \Rightarrow^* 2$
(h) $W \Rightarrow^* 2abc$
(i) $W \Rightarrow^* ba2c$

10. If G is the grammar of Exercise 9, describe $L(G)$.

11. Draw a derivation tree for $ab3$ in the grammar of Exercise 9.

12. Draw a derivation tree for the string $x^2 y^2 z$ in the grammar of Exercise 1.

13. Draw a derivation tree for the string aba^2 in the grammar of Exercise 6.

14. Draw a derivation tree for the string a^8 in the grammar of Exercise 2.

15. Give two distinct derivations (sequences of substitutions starting at v_0) for the string $xyzyz \in L(G)$, where G is the grammar of Exercise 4.

16. Let G be the grammar of Exercise 5. Can you give two distinct derivations (see Exercise 15) for the string $((a + a + a))$?

17. Let G be the grammar of Exercise 9. Give two distinct derivations (see Exercise 15) of the string $a100$.

In Exercises 18–24, construct a phrase structure grammar G such that the language, $L(G)$, of G is equal to the language L.

18. $L = \{a^n b^n \mid n \geq 1\}$

19. $L = \{$strings of 0's and 1's with an equal number $n \geq 0$ of 0's and 1's$\}$

20. $L = \{a^n b^m \mid n \geq 1, m \geq 1\}$

21. $L = \{a^n b^n \mid n \geq 1\}$

22. $L = \{a^n b^m \mid n \geq 1, m \geq 3\}$

23. $L = \{x^n y^m \mid n \geq 2, m$ nonnegative and even$\}$

24. $L = \{x^n y^m \mid n$ even, m positive and odd$\}$

25. Let $G = (V, S, v_0, \mapsto)$ where
$V = \{v_0, v_1, v_2, a, b, c\}$, $S = \{a, b, c\}$
$\mapsto : v_0 \mapsto aav_0$
$\quad\quad v_0 \mapsto bv_1$
$\quad\quad v_1 \mapsto cv_2 b$
$\quad\quad v_1 \mapsto cb$
$\quad\quad v_2 \mapsto bbv_2$
$\quad\quad v_2 \mapsto bb$

State which of the following are in $L(G)$.

(a) $aabcb$
(b) $abbcb$
(c) $aaaabcbb$
(d) $aaaabcbbb$
(e) $abcbbbbb$

5.4 Representations of Special Grammars and Languages

BNF Notation

For type 2 grammars (which include type 3 grammars) there are some useful, alternative methods of displaying the productions. A commonly encountered alternative is called the **BNF notation** (for Backus–Naur Form). We know that the left-hand sides of all productions in a type 2 grammar are single, nonterminal symbols. For any such symbol w, we combine all productions having w as left-hand side. The symbol w remains on the left, and all right-hand sides associated with w are listed together, separated by the symbol "|". The relational symbol "$\mapsto$" is replaced by the symbol "$::=$". Finally, the nonterminal symbols, wherever they occur, are enclosed in pointed brackets "$\langle \;\; \rangle$". This has the additional advantage that nonterminal symbols may be permitted to have embedded spaces. Thus $\langle$word1 word2$\rangle$ shows that the string between the brackets is to be treated as one "word", not as two words. That is, we may use the space as a convenient and legitimate "letter" in a word, as long as we use pointed brackets to delimit the words.

Example 1   In BNF notation, the productions of Example 1 of Section 5.3 appear as follows.

$$\langle \text{sentence} \rangle \quad ::= \langle \text{noun} \rangle \, \langle \text{verbphrase} \rangle$$

$$\langle \text{noun} \rangle \qquad ::= \text{John} \,|\, \text{Jill}$$

$$\langle \text{verbphrase} \rangle ::= \langle \text{verb} \rangle \, \langle \text{adverb} \rangle$$

$$\langle \text{verb} \rangle \qquad ::= \text{drives} \,|\, \text{jogs}$$

$$\langle \text{adverb} \rangle \qquad ::= \text{carelessly} \,|\, \text{rapidly} \,|\, \text{frequently}$$

Example 2   In BNF notation, the productions of Example 2 of Section 5.3 appear as follows.

$$\langle v_0 \rangle ::= a \langle w \rangle$$

$$\langle w \rangle ::= bb \langle w \rangle \,|\, c$$

Note that the left-hand side of a production may also appear in one of the strings on the right-hand side. Thus in the second line of Example 2, $\langle w \rangle$ appears on the left, and it appears in the string $bb\langle w \rangle$ on the right. When this happens, we say that the corresponding production $w \mapsto bbw$ is **recursive**. If a recursive production has w as left-hand side, we will say that the production is **normal** if w appears only once on the right-hand side, and is the rightmost symbol. Other nonterminal symbols may also appear on the right-side. The recursive production $w \mapsto bbw$ given in

Example 2 is normal. Note that any recursive production that appears in a type 3 (regular) grammar is normal, by the definition of type 3.

Example 3   BNF notation is often used to specify actual programming languages. Both ALGOL and PASCAL had their grammars given in BNF initially. In this example, we consider a small subset common to these two grammars. This subset describes the syntax of decimal numbers, and can be viewed as a minigrammar whose corresponding language consists precisely of all properly formed decimal numbers.

Let $S = \{0, 1, 2, 3, 4, 5, 6, 7, 8, 9, \bullet \}$. Let V be the union of S with set

$$N = \{\text{decimal-number, decimal-fraction, unsigned-integer, digit}\}$$

Then let G be the grammar with symbol sets V and S, with starting symbol "decimal-number," and with productions given in BNF form as follows:

1. $\langle$decimal-number$\rangle ::= \langle$unsigned-integer$\rangle \mid \langle$decimal-fraction$\rangle \mid$
   $\qquad\qquad\qquad\qquad \langle$unsigned-integer$\rangle \langle$decimal-fraction$\rangle$
2. $\langle$decimal-fraction$\rangle ::= \bullet \langle$unsigned-integer$\rangle$
3. $\langle$unsigned-integer$\rangle ::= \langle$digit$\rangle \mid \langle$digit$\rangle\langle$unsigned-integer$\rangle$
4. $\langle$digit$\rangle ::= 0 \mid 1 \mid 2 \mid 3 \mid 4 \mid 5 \mid 6 \mid 7 \mid 8 \mid 9$

Figure 1 shows the derivation tree, in this grammar, for the decimal number

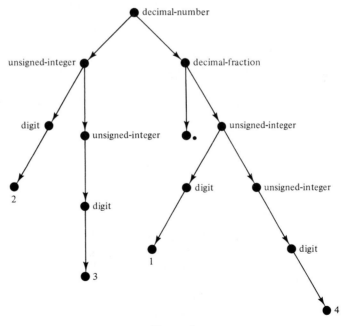

Figure 1

23.14. Notice that the BNF statement numbered 3 is recursive in the second part of its right-hand side. That is, the production "unsigned-integer $\mapsto$ digit unsigned-integer" is recursive, and it is also normal. In general, we know that many different grammars may produce the same language. If the line numbered 3 above were replaced by the line

3′. $\langle$unsigned-integer$\rangle::= \langle$digit$\rangle\,|\,\langle$unsigned-integer$\rangle\langle$digit$\rangle$

we would have a different grammar which produced exactly the same language, that is, the correctly formed decimal numbers. However, this grammar contains productions that are recursive but not normal.

Example 4   As in Example 3, we give a grammar that specifies a piece of several actual programming languages. In these languages, an identifier (a name for a variable, function, subroutine, and so on) must be composed of letters and digits, and must begin with a letter. The following grammar, with productions given in BNF, has precisely these identifiers as its language.

$$G = (V, S, \text{identifier}, \mapsto)$$

$$N = \{\text{identifier, remaining, digit, letter}\}$$

$$S = \{a, b, c, \ldots, z, 0, 1, 2, 3, \ldots, 9\}, \qquad V = N \cup S$$

1. $\langle$identifier$\rangle::= \langle$letter$\rangle\,|\,\langle$letter$\rangle\langle$remaining$\rangle$
2. $\langle$remaining$\rangle::= \langle$letter$\rangle\,|\,\langle$digit$\rangle\,|\,\langle$letter$\rangle\langle$remaining$\rangle\,|\,\langle$digit$\rangle\langle$remaining$\rangle$
3. $\langle$letter$\rangle::= a\,|\,b\,|\,c\,|\cdots|\,z$
4. $\langle$digit$\rangle::= 0\,|\,1\,|\,2\,|\,3\,|\,4\,|\,5\,|\,6\,|\,7\,|\,8\,|\,9$

Again we see that the productions "remaining $\mapsto$ letter remaining" and "remaining $\mapsto$ digit remaining", occuring in BNF statement 2, are recursive and normal.

Syntax diagrams

A second alternative method for displaying the productions in some type 2 grammars is the **syntax diagram**. This is a pictorial display of the productions that allows the user to view the substitutions dynamically, that is, to view them as movement through the diagram. We will illustrate, in Fig. 2, the diagrams that result from translating typical sets of productions, usually all of the productions appearing on the right-hand side of some BNF statement.

A BNF statement that involves just a single production, such as $\langle w\rangle::= \langle w_1\rangle\langle w_2\rangle\langle w_3\rangle$, will result in the diagram shown in Fig. 2(a). The symbols (words) that make up the right-hand side of the production are drawn in sequence from left

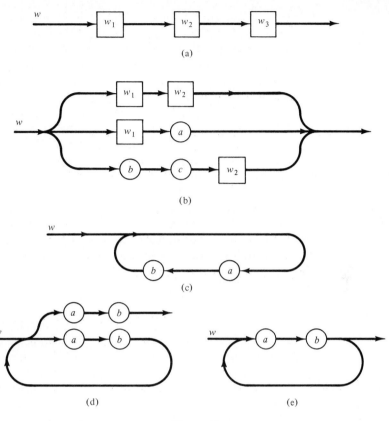

Figure 2

to right. The arrows indicate the direction in which to move to accomplish a substitution, while the label "w" indicates that we are substituting for the symbol w. Finally, the rectangles enclosing w_1, w_2, and w_3 denote the fact that these are nonterminal symbols. If terminal symbols were present, they would instead be enclosed in circles or ellipses. Figure 2(b) shows the situation when there are several productions with the same left-hand side. This figure is a syntax diagram translation of the following BNF specification:

$$\langle w \rangle ::= \langle w_1 \rangle \langle w_2 \rangle \,|\, \langle w_1 \rangle a \,|\, bc \langle w_2 \rangle$$

(where, by convention, a, b, and c must be terminal symbols). Here the diagram shows that when we substitute for w, by moving through the figure in the direction of the arrows, we may take any one of three paths. This corresponds to the three alternative substitutions for the symbol w. Now consider the following normal, recursive production, in BNF form:

$$\langle w \rangle ::= ab \langle w \rangle$$

The syntax diagram for this production is shown in Fig. 2(c). If we go through the loop once, we encounter a, then b, and we then return to the starting point designated by w. This represents the recursive substitution of abw for w. Several trips around the diagram represent several successive substitutions. Thus, if we traverse the diagram three times and return to the starting point, we see that w will be replaced by $abababw$ in three successive substitutions. This is typical of the way in which movement through a syntax diagram represents the substitution process.

The remarks above show how to construct a syntax diagram for a normal recursive production. Nonnormal recursive productions do not lead to the simple diagrams discussed above, but we may sometimes replace nonnormal, recursive productions by normal recursive productions, and obtain a grammar that produces the same language. Since recursive productions in regular grammars must be normal, syntax diagrams can always be used to represent regular grammars.

We also note that syntax diagrams for a language are by no means unique. They will not only change when different, equivalent productions are used, but they may be combined and simplified in a variety of ways. Consider the following BNF specification:

$$\langle w \rangle ::= ab \mid ab\langle w \rangle$$

If we construct the syntax diagram for w, using exactly the rules discussed above, we will obtain the diagram of Fig. 2(d). This shows that we can "escape" from w, that is, eliminate w entirely, only by passing through the upper path. On the other hand, we may first traverse the lower loop any number of times. Thus any movement through the diagram which eventually results in the complete elimination of w by successive substitutions will produce a string of terminal symbols of the form $(ab)^n$, $n \geq 1$.

It is easily seen that the simpler diagram of Fig. 2(e), produced by combining the paths of Fig. 2(d) in an obvious way, is an entirely equivalent syntax diagram. These types of simplifications are performed whenever possible.

Example 5   The syntax diagrams of Fig. 3(a) represent the BNF statements of Example 2, constructed with our original rules for drawing syntax diagrams. A slightly more aesthetic version is shown in Fig. 3(b).

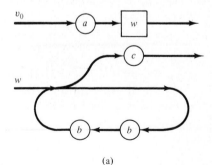

(a)

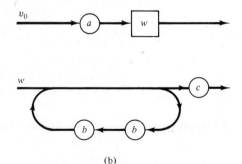

(b)

Figure 3

Example 6   Consider the BNF statements 1, 2, 3, and 4 of Example 4. The direct translation into syntax diagrams is shown in Fig. 4. In Fig. 5 we combine the first two diagrams of Fig. 4, and simplify the result. We thus eliminate the symbol "remaining", and we arrive at the customary syntax diagrams for identifiers.

Example 7   The productions of Example 3, for well-formed decimal numbers, are shown in syntax diagram form in Fig. 6. We show in Fig. 7 the result of substituting the diagram for "unsigned-integer" into that for "decimal-number" and "decimal-fraction." In Fig. 8 the process of substitution is carried one step further. Although this is not usually done, it does illustrate the fact that one can be quite flexible in designing syntax diagrams.

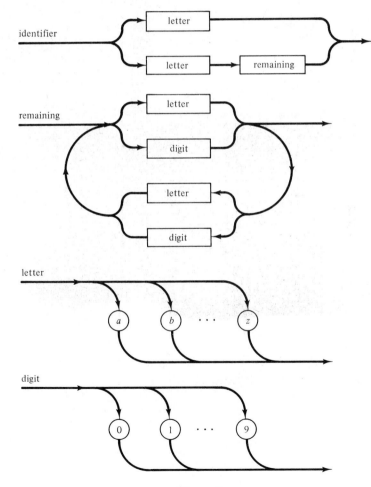

Figure 4

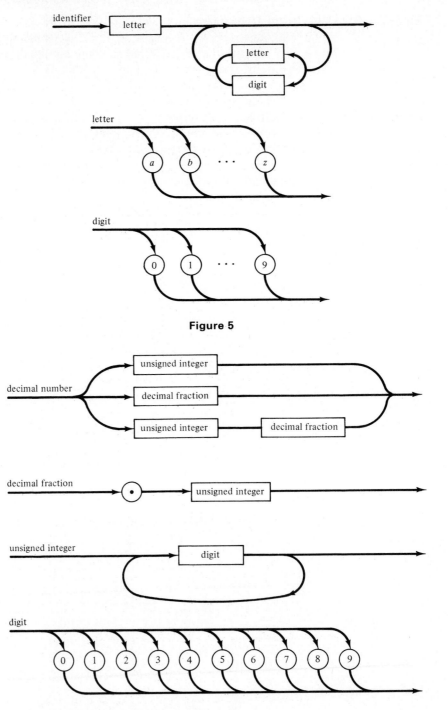

Figure 5

Figure 6

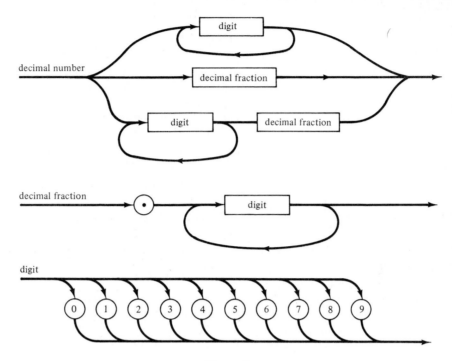

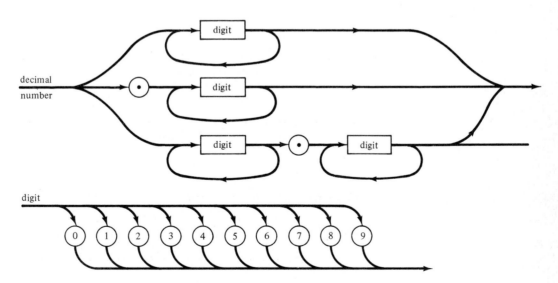

Figure 7

Figure 8

If we were to take the extreme case, and combine the diagrams of Fig. 8 into one huge diagram, that diagram would contain only terminal symbols. In that case a valid "decimal-number" would be any string that resulted from moving through the diagram, recording each symbol encountered in the order in which it was encountered, and eventually exiting to the right.

Regular grammars and regular expressions

There is a close connection between the language of a regular grammar, and a regular expression (see Section 1.6). We state the following theorem without proof.

Theorem 1   Let S be a finite set, and $L \subseteq S^*$. Then L is a regular set if and only if $L = L(G)$ for some regular grammar $G = (V, S, v_0, \mapsto)$.

Theorem 1 tells us that the language $L(G)$ of a regular grammar G must be the set corresponding to some regular expression over S, but it does not tell us how to find such a regular expression. If the relation $\mapsto$ of G is specified in BNF or syntax diagram form, we may compute the regular expression desired in a reasonably straightforward way. Suppose, for example, that $G = (V, S, v_0, \mapsto)$ and that $\mapsto$ is specified by a set of syntax diagrams. As we previously mentioned, it is possible to combine all of the syntax diagrams into one large diagram that represents v_0, and involves only terminal symbols. We will call the result the **master diagram** of G. Consider the following rules of correspondence between regular expressions and parts, or segments, of the master diagram of G.

1. Terminal symbols of the diagram correspond to themselves, as regular expressions.
2. If a segment D of the diagram is composed of two segments D_1 and D_2 in sequence, as shown in Fig. 9(a), and if D_1 and D_2 correspond to regular expressions α_1 and α_2 respectively, then D corresponds to $\alpha_1\alpha_2$.

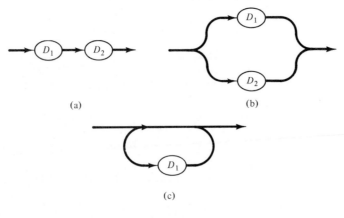

(a)

(b)

(c)

Figure 9

3. If a segment D of the diagram is composed of alternative segments D_1 and D_2, as shown in Fig. 9(b), and if D_1 and D_2 correspond to regular expressions α_1 and α_2 respectively, then D corresponds to $\alpha_1 \vee \alpha_2$.

4. If a segment D of the diagram is a loop through a segment D_1, as shown in Fig. 9(c), and if D_1 corresponds to the regular expression α, then D corresponds to α^*.

Rules 2 and 3 extend to any finite number of segments D_i of the diagram. Using the foregoing rules, we may construct the single expression that corresponds to the master diagram as a whole. This expression is the regular expression that corresponds to $L(G)$.

Example 8   Consider the syntax diagram shown in Fig. 10(a). It is composed of three alternative segments; the first corresponding to the expression "a", the second to the expression "b" and the third, a loop, corresponding to the expression "c^*". Thus the entire diagram corresponds to the regular expression "$a \vee b \vee c^*$".

The diagram shown in Fig. 10(b) is composed of three sequential segments. The first segment is itself composed of two alternative subsegments, and it corresponds to the regular expression "$a \vee b$". The second component segment of the diagram corresponds to the regular expression "c", and the third component, a loop, corresponds to the regular expression "d^*". Thus the overall diagram corresponds to the regular expression "$(a \vee b)cd^*$".

Finally, consider the syntax diagram shown in Fig. 10(c). This is one large loop through a segment that corresponds to the regular expression "$a \vee bc$". Thus the entire diagram corresponds to the regular expression "$(a \vee bc)^*$".

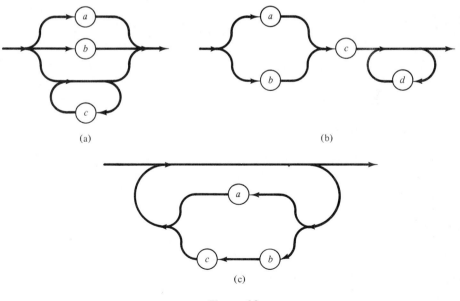

(a)

(b)

(c)

Figure 10

Example 9   Consider the grammar G given in BNF in Example 2. Syntax diagrams for this grammar were discussed in Example 5, and shown in Fig. 3(b). If we substitute the diagram representing w into the diagram that represents v_0, we get the master diagram for this grammar. This is easily visualized, and it shows that $L(G)$ corresponds to the regular expression "$a(bb)^*c$", as we stated in Example 5, section 5.3.

Example 10   Consider the grammar G of Examples 4 and 6. Then $L(G)$ is the set of legal identifiers, whose syntax diagrams are shown in Fig. 5. In Fig. 11 we show the master diagram that results from combining the diagrams of Fig. 4. It follows that a regular expression corresponding to $L(G)$ is "$(a \vee b \vee \cdots \vee z)(a \vee b \vee \cdots \vee z \vee 0 \vee 1 \vee \cdots \vee 9)^*$".

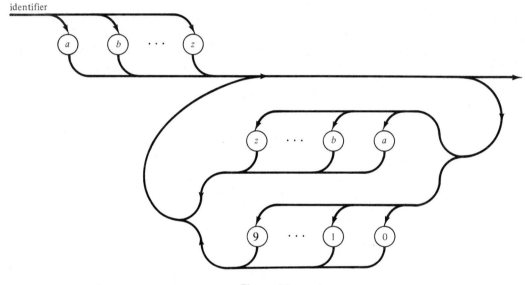

Figure 11

The type of diagram segments illustrated in Fig. 9 can be combined to produce syntax diagrams for any regular grammar. Thus we may always proceed as illustrated above to find the corresponding regular expression. With practice one can learn to compute this expression directly from multiple syntax diagrams or BNF, thus avoiding the need to make a master diagram. In any event, complex cases may prove too cumbersome for hand analysis.

EXERCISE SET 5.4

In each of Exercises 1–5, we have referenced a grammar described in the exercises of a previous section. In each case give the BNF and corresponding syntax diagrams for the productions of the grammar.

1. Section 5.3, Exercise 1.

2. Section 5.3, Exercise 2.

3. Section 5.3, Exercise 6.

4. Section 5.3, Exercise 9.

5. Section 5.3, Exercise 25.

6. Give the BNF for the productions of Exercise 3, Section 5.3.

7. Give the BNF for the productions of Exercise 4, Section 5.3.

8. Give the BNF for the productions of Exercise 5, Section 5.3.

9. Give the BNF for the productions of Exercise 9, Section 5.3.

In Exercises 10 and 11, give a BNF representation for the syntax diagram shown. The symbols a, b, c, and d are supposed to be terminal symbols of some grammar. You may provide nonterminal symbols as needed (in addition to v_0), to use in the BNF productions. You may use several BNF statements if needed.

In each of Exercises 12-16, we have referenced a grammar G, described in the exercises of a previous section. In each case find a regular expression that corresponds to the language $L(G)$.

12. Section 5.3, Exercise 2.

13. Section 5.3, Exercise 3.

14. Section 5.3, Exercise 5.

15. Section 5.3, Exercise 6.

16. Section 5.3, Exercise 9.

17. Find the regular expression that corresponds to the syntax diagram of Exercise 10.

18. Find the regular expression that corresponds to the syntax diagram of Exercise 11.

10.

11.

5.5 Tree Searching

There are many occasions when it is useful to consider each vertex of a tree T exactly once in some specified order. As each successive vertex is encountered, we may wish to take some action or perform some computation, appropriate to the application being represented by the tree. For example, if the tree T is labeled, the label on each vertex may be displayed. If T is the tree of an algebraic expression, then at each vertex we may want to perform the computation indicated by the operator which labels that vertex. For another example, we may take T to be the parse tree of a sentence in some grammar. Then, as we reach each vertex, we may wish to generate corresponding statements in some other language, to which we are translating. Performing appropriate tasks at a vertex will be called **visiting** the vertex. This is a convenient, nonspecific term which allows us to write algorithms without giving the details of what constitutes a "visit" in each particular case.

The process of visiting each vertex of a tree, in some specified order, will be called **searching** the tree, or performing a **tree search**. In some texts, this process is called **walking** or **traversing** the tree.

Let us consider tree searches on binary positional trees. Recall that in a binary positional tree, each vertex has two "potential" offspring. We denote these potential offspring by v_L (the left offspring) and v_R (the right offspring), and either or both may be missing. If a binary tree T is not positional, it may always be labeled so that it becomes positional.

Let T be a binary positional tree with root v. Then if v_L exists, the subtree $T(v_L)$ (see Section 5.1) will be called the **left subtree** of T, and if v_R exists, the subtree $T(v_R)$ will be called the **right subtree** of T.

Note that $T(v_L)$, if it exists, is a positional binary tree with root v_L, and similarly $T(v_R)$ is a positional binary tree with root v_R. This notation allows us to specify searching algorithms in a natural and powerful recursive form. Recall that recursive algorithms are those which refer to themselves. We first describe a method of searching called a **preorder search**. Suppose that the details of "visiting" any vertex of a tree are described by a subroutine VISIT, which, for the moment, we leave unspecified. Then consider the following pseudocode subroutine for searching a positional binary tree T with root v.

> SUBROUTINE PREORDER(T,v)
> 1. CALL VISIT(v)
> 2. IF (v_L exists) THEN
>   a. CALL PREORDER($T(v_L),v_L$)
> 3. IF (v_R exists) THEN
>   a. CALL PREORDER($T(v_R),v_R$)
> END OF SUBROUTINE PREORDER

Informally, we see that a preorder search of a tree consists of the following three steps:

1. Visit the root.
2. Search the left subtree if it exists.
3. Search the right subtree if it exists.

Example 1 Let T be the labeled, positional, binary tree whose digraph is shown in Fig. 1(a). The root of this tree is the vertex labeled "A." Suppose that for any vertex v of T, VISIT(v) simply prints out the label of v. Let us now apply the preorder search algorithm to this tree. Note first that if a tree consists only of one vertex, its root, then a search of this tree simply prints out the label of the root. In Fig. 1(b) we have placed boxes around the subtrees of T, and numbered these subtrees (in the corner of the boxes) for convenient reference.

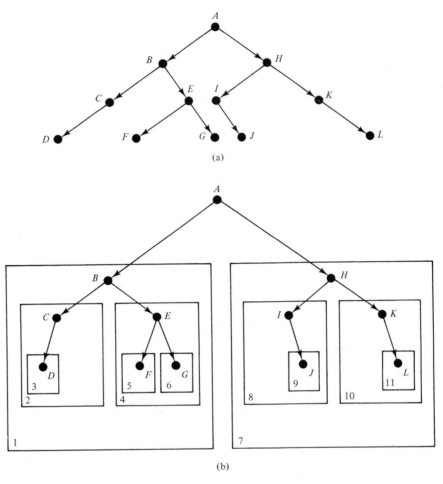

(a)

(b)

Figure 1

According to PREORDER, applied to T, we will visit the root and print "A," then search subtree 1, then subtree 7. Applying PREORDER to subtree 1 results in visiting the root of subtree 1 and printing "B," then searching subtree 2, and finally searching subtree 4. The search of subtree 2 first prints the symbol "C," then searches subtree 3. Subtree 3 has just one vertex, and so as previously mentioned, a search of this tree yields just the symbol "D." Up to this point, the search has yielded the string "$ABCD$." Note that we have had to interrupt the search of each tree (except subtree 3, which is a leaf of T) in order to apply the search procedure to a subtree. Thus we cannot finish the search of T, by searching subtree 7, until we apply the search procedure to subtrees 2 and 4. We could not complete the search of subtree 2 until we search subtree 3, and so on. The bookkeeping brought about by these interruptions produces the labels in the desired order, and recursion is a simple way to specify this bookkeeping.

Returning to the search, we have completed searching subtree 2, and we now must search subtree 4, since this is the right subtree of tree 1. Thus we print "*E*," and search subtrees 5 and 6 in order. These searches produce "*F*" and "*G*." The search of subtree 1 is now complete, and we go to subtree 7. Applying the same procedure, we can see that the search of subtree 7 will ultimately produce the string "*HIJKL*." The result, then, of the complete search of *T*, is to print the string "*ABCDEFGHIJKL*."

Example 2   Consider the completely parenthesized expression $(a - b) \times (c + (d \div e))$. Figure 2(a) shows the digraph of the labeled, positional binary tree representation of this expression. We apply the search procedure PREORDER to this tree, as we did to the tree in Example 1. Figure 2(b) shows the various subtrees encountered in the search. Proceeding as in Example 1, and supposing again that VISIT(*v*) simply prints out

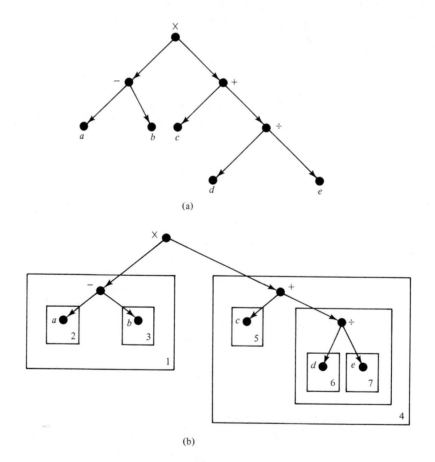

(b)

Figure 2

the label of v, we see that the string $\times - ab + c \div de$ is the result of the search. This is the **prefix** or **Polish form** of the given algebraic expression. Once again, the numbering of the boxes in Fig. 2(b) shows the order in which the subroutine PRE-ORDER is applied to subtrees.

The Polish form of an algebraic expression is interesting because it represents the expression unambiguously, without the need for parentheses. To evaluate an expression in Polish form, proceed as follows. Move from left to right until we find a string of the form Fxy, where F is the symbol for a binary operation (say $+$, $-$, $\times$, and so on) and x and y are numbers. Evaluate xFy, and substitute the answer for the string Fxy. Continue this procedure until only one number remains.

For example, in the expression above, suppose that $a = 6$, $b = 4$, $c = 5$, $d = 2$, and $e = 2$. Then we are to evaluate $\times - 6 4 + 5 \div 2 2$. This is done in the following sequence of steps.

1. $\times - 6 4 + 5 \div 2 2$
2. $\times 2 + 5 \div 2 2$ (since the first string of the correct type is $- 6 4$, and $6 - 4 = 2$)
3. $\times 2 + 5 1$ (replacing $\div 2 2$ by $2 \div 2 = 1$)
4. $\times 2 6$ (replacing $+ 5 1$ by $5 + 1 = 6$)
5. 12 (replacing $\times 2 6$ by 2×6)

This example is one of the primary reasons for calling this type of search the **preorder search**.

Consider now the following two pseudocodes for searching a positional binary tree T with root v.

SUBROUTINE INORDER(T,v)
    1. IF (v_L exists) THEN
       a. CALL INORDER($T(v_L)$, v_L)
    2. CALL VISIT(v)
    3. IF (v_R exists) THEN
       a. CALL INORDER($T(v_R),v_R$)
END OF SUBROUTINE INORDER

SUBROUTINE POSTORDER(T,v)
    1. IF (v_L exists) THEN
       a. CALL POSTORDER($T(v_L),v_L$)
    2. IF (v_R exists) THEN
       a. CALL POSTORDER($T(v_R),v_R$)
    3. CALL VISIT(v)
END OF SUBROUTINE POSTORDER

As indicated by the naming of the subroutines, these searches are called respectively the **inorder** and **postorder** searches.

Example 3   Consider the tree of Fig. 1(a), and apply the subroutine INORDER to search it. First we must search subtree 1. This requires us to first search subtree 2, and this in turn requires us to search subtree 3. As before, a search of a tree with only one vertex simply prints the label of the vertex. Thus "D" is the first symbol printed. The search of subtree 2 continues by printing "C," then stops since there is no right subtree at C. We then visit the root of subtree 1 and print "B," and proceed to the search of subtree 4, which yields "F," "E," and "G," in that order. We then visit the root of T and print "A," and proceed to search subtree 7. The reader may complete the analysis of the search of subtree 7 to show that it yields the string "$IJHKL$." Thus the complete search yields the string "$DCBFEGAIJHKL$."

Suppose now that we apply subroutine POSTORDER to search the same tree, with the same subroutine VISIT. Again, the search of a tree with just one vertex will yield the label of that vertex. In general, we must search both the left and the right subtrees of a tree with root v, before we print out the label at v.

Referring again to Fig. 1(b), we see that both subtree 1 and subtree 7 must be searched before "A" is printed. Subtrees 2 and 4 must be searched before "B" is printed, and so on.

The search of subtree 2 requires us to search subtree 3, and "D" is the first symbol printed. The search of subtree 2 continues by printing "C". We now search subtree 4 yielding "F," "G," and "E." We next visit the root of subtree 1 and print "B." Proceeding with the search of subtree 7 we print the symbols "J," "I," "L," "K," and "H." Finally, we visit the root of T and print "A." Thus we print out the string "$DCFGEBJILKHA$."

Example 4   Let us now apply the inorder and postorder searches to the algebraic expression tree of Example 2 (see Fig. 2). The use of INORDER produces the string $a - b \times c + d \div e$. Notice that this is exactly the expression that we began with in Example 2, with all parentheses removed. Since the algebraic symbols lie between their arguments, this is often called the **infix notation**, and this explains the name INORDER. The expression above is ambiguous without parentheses. It could have come from the expression $a - (b \times ((c + d) \div e))$, which would have produced a different tree. Thus the tree cannot be recovered from the output of search procedure INORDER, while it can be shown that the tree is recoverable from the Polish form produced by PREORDER. For this reason, Polish notation is often better for computer applications, although infix form is more familiar to human beings.

The use of search procedure POSTORDER on this tree produces the string $ab - cde \div + \times$. This is the **postfix** or **reverse Polish** form of the expression. It is evaluated in a manner similar to that used for Polish form, except that the arithmetic symbol is *after* its arguments rather than *before* them. If $a = 2$, $b = 1$, $c = 3$, $d = 4$, $e = 2$, the expression above is evaluated in the following sequence of steps.

1. $2\ 1 - 3\ 4\ 2 \div\ +\ \times$
2. $1\ 3\ 4\ 2 \div\ +\ \times$ (replacing $2\ 1 -$ by $2 - 1 = 1$)

3. $1\ 3\ 2 + \times$ (replacing $4\ 2 \div$ by $4 \div 2 = 2$)

4. $1\ 5 \times$ (replacing $3\ 2 +$ by $3 + 2 = 5$)

5. 5 (replacing $1\ 5 \times$ by $1 \times 5 = 5$)

Reverse Polish form is also parenthesis-free, and from it one can recover the tree of the expression. It is used even more frequently than the Polish form.

Computer Representation of Binary Positional Trees

In Section 2.5 we discussed an idealized information storage unit called a cell. A cell contains two items. One is data of some sort and the other is a pointer to the next cell; that is, an address where the next cell is located. A collection of such cells, linked together by their pointers, is called a linked list. The discussion in Section 2.5 included both a graphical representation of linked lists, and an implementation of them that used arrays.

We will need an extended version of this concept, called a **doubly linked list**, in which each cell contains two pointers and a data item. We use the pictorial symbol ←[• | |•]→ to represent these new cells. The center space represents data storage and the two pointers, called the **left pointer** and the **right pointer**, are represented as before by dots and arrows. Once again we use the symbol •— for a pointer signifying no additional data. Sometimes a doubly linked list is arranged so that each cell points to both the next cell and the previous cell. This is useful if we want to search through a set of data items in either direction. Our use of doubly linked lists is quite different. We will use them to represent binary positional labeled trees. Each cell will correspond to a vertex, and the data part can contain a label for the vertex, or a pointer to such a label. The left and right pointers will direct us to the left and right offspring vertices, if they exist. If either offspring fails to exist, the corresponding pointer will be •—.

We implement this representation by using three arrays; LEFT holds the pointers to the left offspring, RIGHT holds the pointers to the right offspring, and DATA holds information or labels related to each vertex, or pointers to such information. The value 0, used as a pointer, will signify that the corresponding offspring does not exist. To the linked list and the arrays we add a starting entry that points to the root of the tree.

Example 5   We consider again the positional binary tree shown in Fig. 1(a). In Fig. 3(a) we represent this tree as a doubly linked list, in symbolic form. In Fig. 3(b) we show the implementation of this list as a sequence of three arrays (see also Section 2.5). The first row of these arrays is just a starting point whose left pointer points to the root of the tree. As an example of how to interpret the three arrays, consider the sixth entry in the array DATA, which is E. The sixth entry in LEFT is 7, which means that the left offspring of E is the seventh entry in DATA, or F. Similarly, the sixth entry in RIGHT is 8, so the right offspring of E is the eighth entry in DATA, or G.

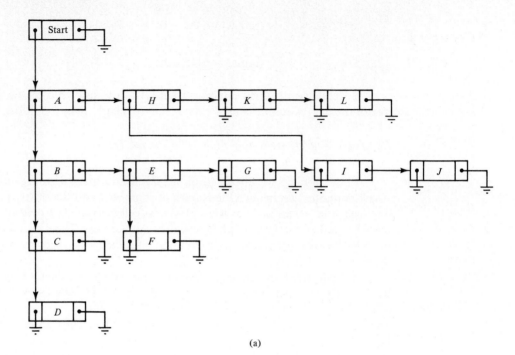

(a)

Index	Left	Data	Right
1	2	⊠	0
2	3	A	9
3	4	B	6
4	5	C	0
5	0	D	0
6	7	E	8
7	0	F	0
8	0	G	0
9	10	H	12
10	0	I	11
11	0	J	0
12	0	K	13
13	0	L	0

(b)

Figure 3

Example 6   Now consider the tree of Fig. 2(a). We represent this tree in Fig. 4(a) as a doubly linked list. As before, Fig. 4(b) shows the implementation of this linked list in three arrays. Again, the first entry is a starting point whose left pointer points to the root of the tree. We have listed the vertices in a somewhat unnatural order, to show that, if the pointers are correctly determined, any ordering of vertices can be used.

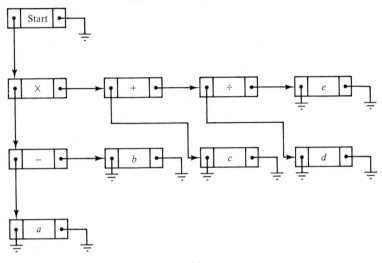

(a)

Index	Left	Data	Right
1	9	✕	0
2	0	a	0
3	0	b	0
4	0	c	0
5	0	d	0
6	0	e	0
7	4	+	10
8	2	−	3
9	8	✕	7
10	5	÷	6

(b)

Figure 4

Searching General Trees

Until now, we have only shown how to search binary positional trees. We now show that any ordered tree T (see Section 5.1) may be represented as a binary positional tree which, although different from T, captures all the structure of T and can be used to recreate T. With the binary positional description of the tree, we may apply the computer representation and search methods previously developed. Since any tree may be ordered, we can use this technique on any (finite) tree.

Let T be any ordered tree and let A be the set of vertices of T. Define a binary positional tree $B(T)$, on the set of vertices A, as follows. If $v \in A$, then the left offspring v_L of v in $B(T)$ is the first offspring of v in T, if it exists. The right offspring v_R of v in $B(T)$ is the next sibling of v in T (in the given order of siblings in T), if it exists.

Example 7   Figure 5(a) shows the digraph of a labeled tree T. We assume that each set of siblings is ordered from left to right, as they are drawn. Thus the offspring of vertex 1, namely vertices 2, 3, and 4, are ordered with vertex 2 first, 3 second, and 4 third. Similarly, the first offspring of vertex 5 is vertex 11, the second is vertex 12, and the third is vertex 13.

In Fig. 5(b) we show the digraph of the corresponding binary positional tree, $B(T)$. To obtain Fig. 5(b), we simply draw a left edge from each vertex v to its first offspring (if v has offspring). Then we draw a right edge from each vertex v to its next sibling (in the order given), if v has a next sibling. Thus the left edge from vertex 2, in Fig. 5(b), goes to vertex 5, because vertex 5 is the first offspring of vertex 2 in the tree T. Also, the right edge from vertex 2, in Fig. 5(b), goes to vertex 3, since

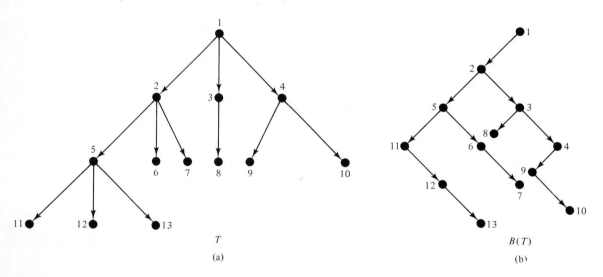

T

(a)

$B(T)$

(b)

Figure 5

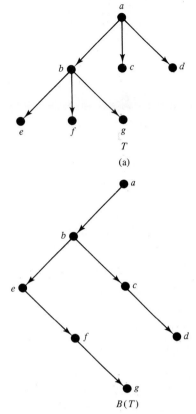

T

(a)

B(T)

(b)

Index	Left	Data	Right
1	2	⊠	0
2	3	a	0
3	6	b	4
4	0	c	5
5	0	d	0
6	0	e	7
7	0	f	8
8	0	g	0

(c)

Figure 6

vertex 3 is the next sibling in line (among all offspring of vertex 1). A doubly-linked-list representation of $B(T)$ is sometimes simply referred to as a **linked-list representation of T**.

Example 8 Figure 6(a) shows the digraph of another labeled tree, with siblings ordered from left to right, as indicated. Figure 6(b) shows the digraph of the corresponding tree $B(T)$, and Fig. 6(c) gives an array representation of $B(T)$. As mentioned above, the data in Fig. 6(c) would be called a linked-list representation of T.

EXERCISE SET 5.5

In Exercises 1–4, the digraphs of labeled, binary positional trees are shown. In each case we suppose that visiting a node results in printing out the label of that node. For each exercise, show the result of performing a preorder search of the tree whose digraph is shown.

1.

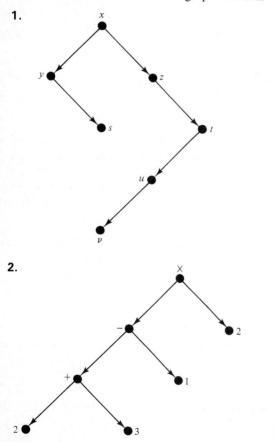

2.

3.

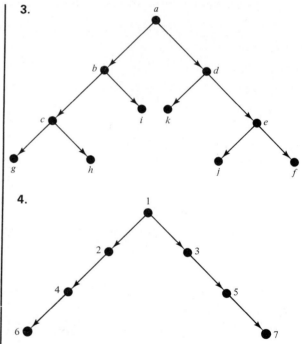

4.

In Exercises 5–12, visiting a node means printing out the label of the node.

5. Show the result of performing an inorder search of the tree in Exercise 1.

6. Show the result of performing an inorder search of the tree in Exercise 2.

7. Show the result of performing an inorder search of the tree in Exercise 3.

8. Show the result of performing an inorder search of the tree in Exercise 4.

9. Show the result of performing a postorder search of the tree in Exercise 1.

10. Show the result of performing a postorder search of the tree in Exercise 2.

11. Show the result of performing a postorder search of the tree in Exercise 3.

12. Show the result of performing a postorder search of the tree in Exercise 4.

13. Consider the following tree digraph and list of words. Suppose that visiting a node of this tree means printing out the word corresponding to the number which labels that node. Print out the sentence that results from doing a postorder search of the tree.

1.	ONE	7.	I
2.	COW	8.	A
3.	SEE	9.	I
4.	NEVER	10.	I
5.	PURPLE	11.	SAW
6.	NEVER	12.	HOPE

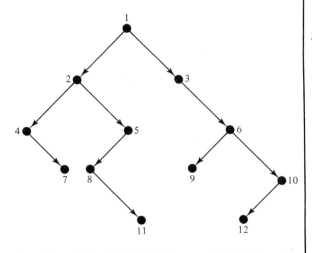

14. Give arrays LEFT, DATA, and RIGHT describing the tree of Exercise 1 as a doubly linked list.

15. Give arrays LEFT, DATA, and RIGHT describing the tree of Exercise 2 as a doubly linked list.

16. Give arrays LEFT, DATA, and RIGHT describing the tree of Exercise 3 as a doubly linked list.

17. Give arrays LEFT, DATA, and RIGHT describing the tree of Exercise 4 as a doubly linked list.

18. Give arrays LEFT, DATA, and RIGHT describing the tree of Exercise 13 as a doubly linked list. In DATA place the numbers labeling the nodes. These are pointers to the actual information used in visiting the vertices.

19. Consider the labeled tree whose digraph is shown below. Draw the digraph of the corresponding binary positional tree $B(T)$. Label the vertices of $B(T)$ to show their correspondence to vertices of T.

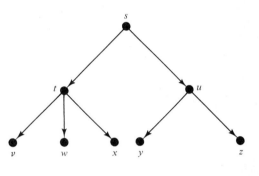

20. We give below, in array form, the doubly-linked-list representation of a labeled tree T (not binary). Draw the digraph of both the labeled binary tree $B(T)$ actually stored in the arrays, and the labeled tree T, of which $B(T)$ is the binary representation.

INDEX	LEFT	DATA	RIGHT
1	2		0
2	3	a	0
3	4	b	5
4	6	c	7
5	8	d	0
6	0	e	10
7	0	f	0
8	0	g	11
9	0	h	0
10	0	i	9
11	0	j	12
12	0	k	0

21. Consider the digraph of a labeled binary positional tree given below. If this tree is the binary form $B(T)$ of some tree T, draw the digraph of the labeled tree T.

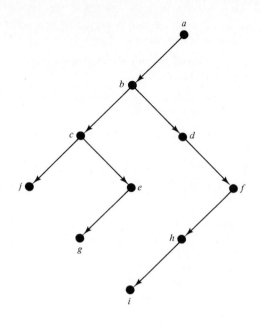

5.6 Undirected Trees

An **undirected tree** is simply the symmetric closure of a tree (see Section 2.6); that is, it is a tree with all edges made bidirectional. As is the custom with symmetric relations, we represent an undirected tree by its graph rather than by its digraph. The graph of an undirected tree T will have a single line without arrows connecting vertices a and b whenever (a, b) and (b, a) belong to T. The set $\{a, b\}$, where (a, b) and (b, a) are in T, is called an undirected edge of T (see Section 2.4). Thus each undirected edge $\{a, b\}$ corresponds to two ordinary edges, (a, b) and (b, a). The lines in the graph of an undirected tree T correspond to the undirected edges in T.

Example 1 Figure 1(a) shows the graph of an undirected tree T. In Fig. 1(b) and (c), we show digraphs of ordinary trees T_1 and T_2, respectively, which have T as symmetric closure. This merely shows that an undirected tree will, in general, correspond to many directed trees. Labels are included to show the correspondence of underlying vertices in the three relations. Note that the graph of T in Fig. 1(a) has six lines (undirected edges), although the relation T contains 12 pairs.

We want to present some useful alternative definitions of an undirected tree, and to do so we must make a few remarks about symmetric relations.

Let R be a symmetric relation and let $p = v_1, v_2, \ldots, v_n$ be a path in R. We will say that p is **simple** if no two edges of p correspond to the same undirected edge. If, in addition, v_1 equals v_n (so that p is a cycle), we will call p a **simple cycle.**

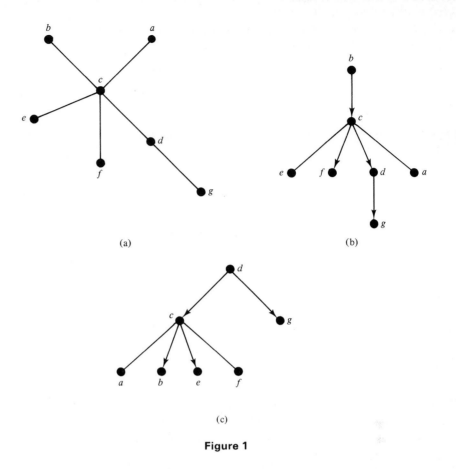

(a)   (b)

(c)

Figure 1

Example 2   Figure 2 shows the graph of a symmetric relation R. The path a, b, c, e, d is simple but the path f, e, d, c, d, a is not simple, since d, c and c, d correspond to the same undirected edge. Also, f, e, a, d, b, a, f and d, a, b, d are simple cycles, but f, e, d, c, e, f is not a simple cycle, since f, e and e, f correspond to the same undirected edge.

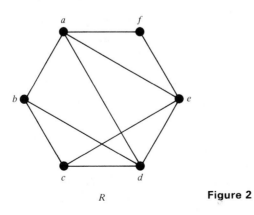

R **Figure 2**

We will say that a symmetric relation R is **acylic** if it contains no simple cycles. It can be shown that if R contains any cycles, then it contains a simple cycle. Recall (see Section 2.4) that a symmetric relation R is connected if there is a path in R from any vertex to any other vertex.

The following theorem provides a useful equivalent statement to the foregoing definition of an undirected tree.

Theorem 1   Let R be a symmetric relation on a set A. Then the following statements are equivalent.

    (a) R is an undirected tree.
    (b) R is connected and acyclic.

Proof. We will prove that (a) implies (b), and we will omit the proof that (b) implies (a). We suppose that R is an undirected tree, which means that R is the symmetric closure of some tree T on A. Note first that if $(a, b) \in R$, we must have either $(a, b) \in T$ or $(b, a) \in T$. In geometric terms, this means that every undirected edge in the graph of R appears in the digraph of T, directed one way or the other.

We will show by contradiction that R has no simple cycles. Suppose that R has a simple cycle $p = v_1, v_2, \ldots, v_n, v_1$. For each edge (v_i, v_j) in p, choose whichever pair (v_i, v_j) or (v_j, v_i) is in T. The result is a closed figure with edges in T, where each edge may be pointing in either direction. Now there are three possibilities. Either all arrows point "clockwise," as in Fig. 3(a), all point "counterclockwise," or some pair must be as in Fig. 3(b). Figure 3(b) is impossible, since in a tree T every vertex has in-degree 1 (see Theorem 1 in Section 5.1). But either of the other two cases would mean that T contains a cycle, which is also impossible. Thus the existence of the cycle p in R leads to a contradiction, and so is impossible.

We must also show that R is connected. Let v_0 be the root of the tree T. Then if a and b are any vertices in A, there must be paths p from v_0 to a, and q from v_0 to b, as shown in Fig. 3(c). Now all paths in T are reversible in R, so the path $p^{-1} \circ q$, shown in Fig. 3(d), connects a with b in R. Since a and b are arbitrary, R is connected, and (b) is proved.

There are other useful characterizations of undirected trees. We state two of these without proof in the following theorem.

Theorem 2   Let R be a symmetric relation on a set A. Then R is an undirected tree if and only if either of the following statements is true.

    (a) R is acyclic, and if any undirected edge is added to R, the new relation will not be acyclic.

    (b) R is connected, and if any undirected edge is removed from R, the new relation will not be connected.

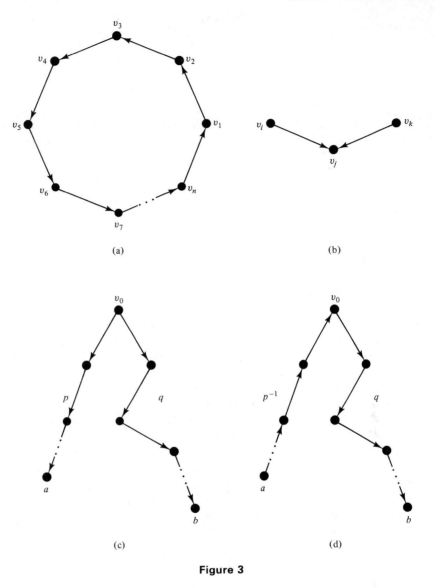

Figure 3

Spanning Trees of Connected Relations

If R is a symmetric, connected relation on a set A, we say that a tree T on A is a **spanning tree** for R if $T \subseteq R$. This simply says that T is a tree with exactly the same vertices as R and which can be obtained from R by deleting some edges of R.

Example 3   The symmetric relation R whose graph is shown in Fig. 4(a) has the tree T', whose digraph is shown in Fig. 4(b), as a spanning tree. Also, the tree T'', whose digraph is shown in Fig. 4(c), is a spanning tree for R. Since R, T', and T'' are all relations on the same set A, we have labeled the vertices to show the correspondence of elements. As this example illustrates, spanning trees are not unique.

Sometimes there is interest in an **undirected spanning tree** for a symmetric, connected relation R. This is just the symmetric closure of a spanning tree. Figure 4(d) shows an undirected spanning tree for R that is derived from the spanning tree of Fig. 4(c). If R is a complicated relation that is symmetric and connected, it might be difficult to devise a scheme for searching R, that is, for visiting each of its vertices once in some systematic manner. If R is reduced to a spanning tree, the searching algorithms discussed in Section 5.5 can be used.

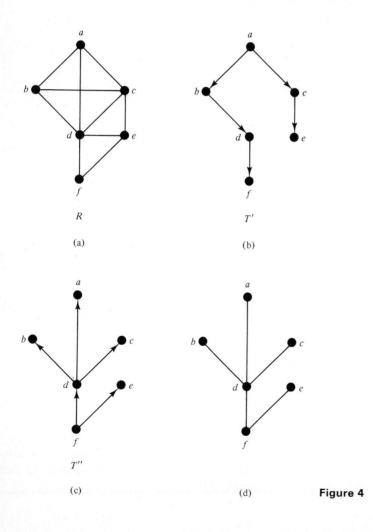

(a)

(b)

(c)

(d)   **Figure 4**

Theorem 2(b) suggests an algorithm for finding a undirected spanning tree for a relation R. Simply remove undirected edges from R until we reach a point where removal of one more undirected edge will result in a relation that is not connected. The result will be an undirected spanning tree.

Example 4 In Fig. 5(a) we repeat the graph of Fig. 4(a). We then show the result of successive removal of undirected edges, culminating in Fig. 5(f), the undirected spanning tree, which agrees with Fig. 4(d).

This algorithm is fine for small relations whose graphs are easily drawn. For large relations, perhaps stored in a computer, it is inefficient because at each stage one must check for connectedness, and this in itself requires a complicated algorithm. We now introduce a more efficient method, which also yields a spanning tree rather than an undirected spanning tree.

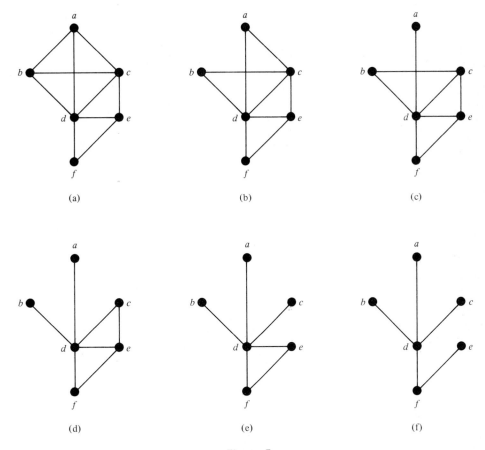

Figure 5

Let R be a relation on a set A, and let $a, b \in A$. Let $A_0 = A - \{a, b\}$, and $A' = A_0 \cup \{a'\}$, where a' is some new element not in A. Define a relation R' on A' as follows. Let $u, v \in A'$, $u \neq a'$, $v \neq a'$. Let $(a', u) \in R'$ if and only if $(a, u) \in R$ or $(b, u) \in R$. Let $(u, a') \in R'$ if and only if $(u, a) \in R$ or $(u, b) \in R$. Finally, let $(u, v) \in R'$ if and only if $(u, v) \in R$. We say that R' is a result of merging the vertices a and b.

Imagine, in the digraph of R, that the vertices are pins, and the edges are elastic bands that can be shrunk to zero length. Now physically move pins a and b together, shrinking the edge between them, if there is one, to zero length. The resulting digraph is the digraph of R'. If R is symmetric, we may perform the same operation on the graph of R.

Example 5   Figure 6(a) shows the graph of a symmetric relation R. In Fig. 6(b) we show the result of merging vertices v_0 and v_1 into a new vertex v_0'. In Fig. 6(c) we show the result of merging vertices v_0' and v_2, of the relation whose graph is shown in Fig. 6(b), into a new vertex v_0''. Notice in Fig. 6(c) that the undirected edges that were previously present between v_0' and v_5, and between v_2 and v_5, have been combined into one undirected edge.

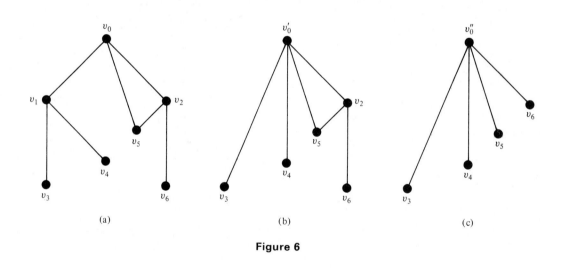

(a)   (b)   (c)

Figure 6

The algebraic form of this merging process is also very important. Let us restrict our attention to symmetric relations and their graphs. We know from Section 2.2 how to construct the matrix of a relation R.

If R is a relation on A, we will temporarily refer to elements of A as vertices of R. This will facilitate the discussion.

Suppose now that vertices a and b of a relation R are merged into a new vertex a' that replaces a and b, to obtain the relation R'. To determine the matrix of R' we proceed as follows.

Step 1. Let row i represent vertex a and row j represent vertex b. Replace row i by the join of rows i and j. The join of two n-tuples of 0's and 1's has a 1 in some position exactly when either of those two n-tuples has a 1 in that position.

Step 2. Replace column i by the join of columns i and j.

Step 3. Restore the main diagonal to its original values in R.

Step 4. Delete row j and column j.

We make the following observation regarding step 3. If $e = (a, b) \in R$ and we merge a and b, then e would become a cycle of length 1 at a'. We do not want to create this situation, since it does not correspond to "shrinking (a, b) to zero." Step 3 corrects for this occurrence.

Example 6   Figure 7 gives the matrices for the corresponding symmetric relations whose graphs were given in Fig. 6. In Fig. 7(b) we have merged vertices v_0 and v_1 into v_0'. Note that this is done by taking the join of the first two rows and entering the result in row 1, doing the same for the columns, then restoring the diagonal, and removing row 2 and column 2. If vertices v_0' and v_2 in the graph whose matrix is given by Fig. 7(b) are merged, the resulting graph has the matrix given by Fig. 7(c).

$$
\begin{array}{c}
\begin{array}{ccccccc}
v_0 & v_1 & v_2 & v_3 & v_4 & v_5 & v_6
\end{array} \\
\begin{array}{c}
v_0 \\ v_1 \\ v_2 \\ v_3 \\ v_4 \\ v_5 \\ v_6
\end{array}
\left[
\begin{array}{ccccccc}
0 & 1 & 1 & 0 & 0 & 1 & 0 \\
1 & 0 & 0 & 1 & 1 & 0 & 0 \\
1 & 0 & 0 & 0 & 0 & 1 & 1 \\
0 & 1 & 0 & 0 & 0 & 0 & 0 \\
0 & 1 & 0 & 0 & 0 & 0 & 0 \\
1 & 0 & 1 & 0 & 0 & 0 & 0 \\
0 & 0 & 1 & 0 & 0 & 0 & 0
\end{array}
\right]
\end{array}
$$

(a)

$$
\begin{array}{c}
\begin{array}{cccccc}
v_0' & v_2 & v_3 & v_4 & v_5 & v_6
\end{array} \\
\begin{array}{c}
v_0' \\ v_2 \\ v_3 \\ v_4 \\ v_5 \\ v_6
\end{array}
\left[
\begin{array}{cccccc}
0 & 1 & 1 & 1 & 1 & 0 \\
1 & 0 & 0 & 0 & 1 & 1 \\
1 & 0 & 0 & 0 & 0 & 0 \\
1 & 0 & 0 & 0 & 0 & 0 \\
1 & 1 & 0 & 0 & 0 & 0 \\
0 & 1 & 0 & 0 & 0 & 0
\end{array}
\right]
\end{array}
$$

(b)

$$
\begin{array}{c}
\begin{array}{ccccc}
v_0'' & v_3 & v_4 & v_5 & v_6
\end{array} \\
\begin{array}{c}
v_0'' \\ v_3 \\ v_4 \\ v_5 \\ v_6
\end{array}
\left[
\begin{array}{ccccc}
0 & 1 & 1 & 1 & 1 \\
1 & 0 & 0 & 0 & 0 \\
1 & 0 & 0 & 0 & 0 \\
1 & 0 & 0 & 0 & 0 \\
1 & 0 & 0 & 0 & 0
\end{array}
\right]
\end{array}
$$

(c)

Figure 7

We can now give an algorithm for finding a spanning tree for a symmetric, connected relation R on the set $A = \{v_1, v_2, \ldots, v_n\}$. The method is a special case of an algorithm called **Prim's algorithm**. The steps are as follows.

Step 1. Choose a vertex v_1 of R, and arrange the matrix of R so that the first row corresponds to v_1.

Step 2. Choose a vertex v_2 of R such that $(v_1, v_2) \in R$, merge v_1 and v_2 into a new vertex v_1', representing $\{v_1, v_2\}$, and replace v_1 by v_1'. Compute the matrix of the resulting relation R'. Call the vertex v_1' a merged vertex.

Step 3. Repeat steps 1 and 2 on R' and on all subsequent relations until a relation with a single vertex is obtained. At each step, keep a record of the set of original vertices that is represented by each merged vertex.

Step 4. Construct the spanning tree as follows. At each stage, when merging vertices a and b, select an edge in R from one of the original vertices represented by a to one of the original vertices represented by b.

Example 7   We apply Prim's algorithm to the symmetric relation whose graph is shown in Fig. 8.

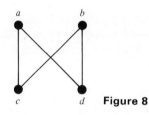

a b

c d **Figure 8**

We show below the matrices that are obtained when the original set of vertices is reduced by merging until a single vertex is obtained, and at each stage we keep track of the set of original vertices represented by each merged vertex, as well as of the new vertex that is about to be merged.

Matrix				Original vertices represented by merged vertices	New vertex to be merged (with first row)
$\begin{array}{c c c c c} & a & b & c & d \\ a & 0 & 0 & 1 & 1 \\ b & 0 & 0 & 1 & 1 \\ c & 1 & 1 & 0 & 0 \\ d & 1 & 1 & 0 & 0 \end{array}$				—	c
$\begin{array}{c c c c} & a' & b & d \\ a' & 0 & 1 & 1 \\ b & 1 & 0 & 1 \\ d & 1 & 1 & 0 \end{array}$				$a' \leftrightarrow \{a, c\}$	b
$\begin{array}{c c c} & a'' & d \\ a'' & 0 & 1 \\ d & 1 & 0 \end{array}$				$a'' \leftrightarrow \{a, c, b\}$	d
$\begin{array}{c c} & a''' \\ a''' & [0] \end{array}$				$a''' \leftrightarrow \{a, c, d, b\}$	—

The first vertex chosen is a and we choose c as the vertex to be merged with a, since there is a 1 at vertex c in row 1. We also select the edge (a, c) from the original graph. At the second stage, there is a 1 at vertex b in row 1, so we merge b with vertex a'. We select an edge in the original relation R from a vertex of $\{a, c\}$ to b, say (c, b). At the third stage, we have to merge d with vertex a". Again, we need an edge in R from a vertex of {a, c, b} to d, say (a, d). The selected edges (a, c), (c, b), and (a, d) form the spanning tree for R, which is shown in Fig. 9.

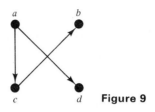

Figure 9

EXERCISE SET 5.6

In Exercises 1–5, use Prim's algorithm to construct a spanning tree for the connected graph shown. Use the indicated vertex as the root of the tree and draw the digraph of the spanning tree produced.

1. Use e as the root.

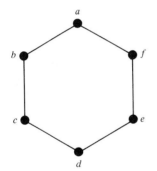

2. Use 5 as the root.

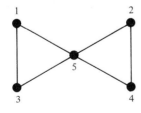

3. Use c as the root.

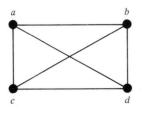

4. Use 4 as the root.

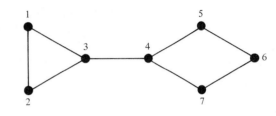

5. Use e as the root.

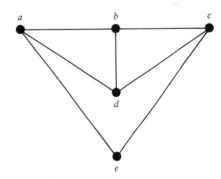

In Exercises 6–10, construct an undirected spanning tree for the connected graph G by removing edges in succession. Show the graph of the resulting undirected tree.

6. Let G be the graph of Exercise 1.
7. Let G be the graph of Exercise 2.
8. Let G be the graph of Exercise 3.
9. Let C be the graph of Exercise 4.
10. Let G be the graph of Exercise 5.

11. Consider the following connected graph. Show the graphs of three different undirected spanning trees.

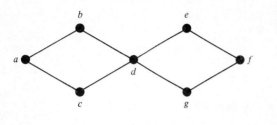

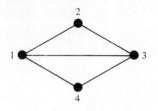

12. For the following connected graph, show the graphs of all undirected spanning trees.

13. For the following undirected tree, show the digraphs of all spanning trees. How many are there?

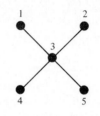

KEY IDEAS FOR REVIEW

☐ Tree: relation on a finite set A such that there exists a vertex $v_0 \in A$ with the property that there is a unique path from v_0 to any other vertex in A, and no path from v_0 to v_0.

☐ Root of tree: vertex v_0 in the definition of tree above.

☐ Rooted tree (T, v_0): tree T with root v_0.

☐ Theorem. Let (T, v_0) be a rooted tree. Then
   (a) There are no cycles in T.
   (b) v_0 is the only root of T.
   (c) Each vertex in T, other than v_0, has in-degree one, and v_0 has in-degree zero.

☐ Level: see page 22.

☐ Leaves: vertices having no offspring.

☐ Theorem. Let T be a rooted tree on a set A. Then
   (a) T is irreflexive.
   (b) T is asymmetric.
   (c) If $(a, b) \in T$ and $(b, c) \in T$, then $(a, c) \notin T$, for all a, b, and c in A.

☐ n-tree: tree where every vertex has at most n offspring.

☐ Binary tree: 2-tree.

☐ Theorem. If (T, v_0) is a rooted tree and $v \in T$, then $T(v)$ is also a rooted tree with root v.

☐ $T(v)$: subtree of T beginning at v.

- [ ] Binary positional tree: see page 231.
- [ ] Phrase structure grammar: see page 233.
- [ ] Production: a statement $w \mapsto w'$, where $(w, w') \in \mapsto$.
- [ ] Direct derivability: see page 233.
- [ ] Terminal symbols: the elements of S.
- [ ] Nonterminal symbols: the elements of $V - S$.
- [ ] Derivation of a sentence: substitution process that produces a valid sentence.
- [ ] Language of a grammar G: set of all properly constructed sentences that can be produced from G.
- [ ] Derivation tree for a sentence: see page 235.
- [ ] Types 0, 1, 2, 3 phrase structure grammars: see page 239.
- [ ] Context-free grammar: type 2 grammar.
- [ ] Regular grammar: type 3 grammar.
- [ ] Parsing: process of obtaining a derivation tree that will produce a given sentence.
- [ ] BNF notation: see page 242.
- [ ] Syntax diagram: see page 244.
- [ ] Theorem. Let S be a finite set, and $L \subseteq S^*$. Then L is a regular set if and only if $L = L(G)$ for some regular grammar $G = (V, S, v_0, \mapsto)$.
- [ ] Preorder search: see page 254.
- [ ] Inorder search: see page 257.
- [ ] Postorder search: see page 257.
- [ ] Polish notation: see page 257.
- [ ] Reverse Polish notation: see page 258.
- [ ] Computer representation of trees: see page 259.
- [ ] Searching general trees: see page 262.
- [ ] Linked-list representation of a tree: see page 264.
- [ ] Undirected tree: symmetric closure of a tree.
- [ ] Simple path: no two edges correspond to the same undirected edge.
- [ ] Connected symmetric relation R: There is a path in R from any vertex to any other vertex.
- [ ] Spanning tree for symmetric connected relation R: tree reaching all the vertices of R, and whose edges are all edges of R.
- [ ] Undirected spanning tree: symmetric closure of a spanning tree.
- [ ] Prim's algorithm: see page 273.

FURTHER READING

AHO, ALFRED V. and J. D. ULLMAN, *Principles of Compiler Design*, Addison-Wesley, Reading, Mass., 1977.

DENNING, PETER J., JACK B. DENNIS and JOSEPH L. QUALITZ, *Machines, Languages, and Computation*, Prentice-Hall, Englewood Cliffs, 1978.

GINSBURG, SEYMOUR, *The Mathematical Theory of Context-Free Languages*, McGraw-Hill, New York, 1966.

HOPCROFT, J. E. and J. D. ULLMAN, *Formal Languages and Their Relation to Automata*, Addison-Wesley, Reading, Mass., 1969.

KNUTH, D. E., *The Art of Computer Programming*, v. 1, Addison-Wesley, Reading, Mass., 1969.

LEWIS, P. M., II, D. J. ROSENKRANTZ, and R. E. STEARNS, *Compiler Design Theory*, Addison-Wesley, Reading, Mass., 1976.

POLLACK, S. V. ed., *Studies in Computer Science*, MAA Studies in Mathematics, v. 22, MAA, Washington, D.C., 1982.

RADER, ROBERT J., *Advanced Software Design Techniques*, Petrocelli, New York, 1978.

Semigroups and Groups

Prerequisites: Chapters 1, 2, and 3.

We are all familiar from high school algebra with the idea of a binary operation on a set of elements, for example, addition and multiplication on the set of all integers. In this chapter we deal with binary operations on sets from a more abstract point of view. This will help to develop the notion of semigroup, which will be used in our study of finite-state machines in Chapter 7. We shall also develop the basic ideas of group theory, which we apply to coding theory in Chapter 8.

6.1 Binary Operations

Although we have all dealt with binary operations on sets for many years, it will be important for our work in this chapter to provide a precise definition of this fundamental idea.

A **binary operation** on a set A is a function $f: A \times A \to A$. Observe the following properties that a binary operation must satisfy:

1. Since Dom $(f) = A \times A$, f assigns an element $f(a, b)$ of A to each ordered pair (a, b) of elements of A. That is, the binary operation must be defined for each ordered pair of elements of A.
2. Since a binary operation is a function, only one element of A is assigned to each ordered pair (a, b).

Thus we can say that a binary operation is a rule that assigns to each ordered pair of elements of A a unique element of A. We shall now turn to a number of examples.

It is customary to denote binary operations by a symbol such as $*$, instead of f, and to denote the element assigned to (a, b) by $a * b$ [instead of $*(a, b)$]. It should be emphasized that if a and b are elements in A, then $a * b \in A$; and this property is often described by saying that A is **closed** under the operation $*$.

Example 1   Let $A = Z$. Define $a * b$ as $a + b$. Then $*$ is a binary operation on Z.

Example 2   Let $A = \mathbb{R}$. Define $a * b$ as a/b. Then $*$ is not a binary operation, since it is not defined for every ordered pair of elements of A. For example, $3 * 0$ is not defined, since we cannot divide by zero.

Example 3   Let $A = Z^+$. Define $a * b$ as $a - b$. Then $*$ is not a binary operation since it does assign an element of A to every ordered pair of elements of A; for example, $2 * 5 \notin A$.

Example 4   Let $A = Z$. Define $a * b$ as a number less than both a and b. Then $*$ is not a binary operation, since it does not assign a *unique* element of A to each ordered pair of elements of A; for example, $8 * 6$ could be 5, 4, 3, 1, and so on. Thus, in this case, $*$ would be a relation from $A \times A$ to A, but not a function.

Example 5   Let $A = Z$. Define $a * b$ as $\max \{a, b\}$. Then $*$ is a binary operation; for example, $2 * 4 = 4$, $-3 * (-5) = -3$.

Example 6   Let $A = P(S)$, for some set S. If V and W are subsets of S, define $V * W$ as $V \cup W$. Then $*$ is a binary operation on A. Moreover, if we define $V * W$ as $V \cap W$, then $*$ is another binary operation on A.

As Example 6 shows, it is possible to define many binary operations on the same set.

Example 7   Let M be the set of all Boolean matrices. Define $A * B$ as $A \vee B$ (see Section 1.8). Then $*$ is a binary operation. This is also true of $A \wedge B$.

Example 8†   Let L be a lattice. Define $a * b$ as $a \wedge b$ (the greatest lower bound of a and b). Then $*$ is a binary operation on L. This is also true of $a \vee b$ (the least upper bound of a and b).

---

†Uses material from Chapter 4.

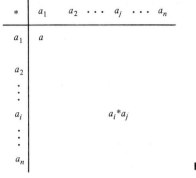

Figure 1

Tables

If $A = \{a_1, a_2, \ldots, a_n\}$ is a finite set, we can define a binary operation on A by means of a table as shown in Fig. 1. The entry in position i, j denotes the element $a_i * a_j$.

Example 9   Let $A = \{0, 1\}$. Recall from Section 1.8 that we defined the binary operations $\vee$ and $\wedge$ by the following tables

$\vee$	0	1
0	0	1
1	1	1

$\wedge$	0	1
0	0	0
1	0	1

If $A = \{a, b\}$, we shall now determine the number of binary operations that can be defined on A. Every binary operation $*$ on A can be described by the table

Since every blank can be filled in with the element a or b, we conclude that there are $2 \cdot 2 \cdot 2 \cdot 2 = 2^4 = 16$ ways to complete the table. Thus there are 16 binary operations on A.

Properties of Binary Operations

A binary operation on a set A is said to be **commutative** if

$$a * b = b * a$$

for all elements a and b in A.

Example 10   The binary operation of addition on Z (as discussed in Example 1) is commutative.

Example 11   The binary operation of subtraction on Z is not commutative, since

$$2 - 3 \neq 3 - 2$$

A binary operation that is described by a table is commutative if and only if the entries in the table are symmetric with respect to the main diagonal.

Example 12   Which of the following binary operations on $A = \{a, b, c, d\}$ are commutative?

$*$	a	b	c	d
a	a	c	b	d
b	b	c	b	a
c	c	d	b	c
d	a	a	b	b

(a)

$*$	a	b	c	d
a	a	c	b	d
b	c	d	b	a
c	b	b	a	c
d	d	a	c	d

(b)

Solution.
(a) Is not commutative since $a * b$ is c while $b * a$ is b.
(b) Is commutative since the entries in the table are symmetric with respect to the main diagonal.

A binary operation $*$ on a set A is said to be **associative** if

$$a * (b * c) = (a * b) * c$$

for all elements a, b, and c in A.

Example 13   The binary operation of addition on Z is associative.

Example 14   The binary operation of subtraction on Z is not associative since

$$2 - (3 - 5) \neq (2 - 3) - 5$$

Example 15†   Let L be a lattice. The binary operation defined by $a * b = a \wedge b$ (see Example 8) is commutative and associative. It also satisfies the **idempotent** property $a \wedge a = a$. A partial converse of this example is also true, as is shown in Example 16.

---

†Uses material from Chapter 4.

Example 16†   Let * be a binary operation on a set A, and suppose tha * satisfies the following properties for any a, b, and c in A.

1. $a = a * a$ (idempotent property)
2. $a * b = b * a$ (commutative property)
3. $a * (b * c) = (a * b) * c$ (associative property)

Define a relation $\leq$ on A by

$$a \leq b \quad \text{if and only if} \quad a = a * b$$

Show that $(A, \leq)$ is a poset, and for all a, b, glb $(a, b) = a * b$.

Solution. We must show that $\leq$ is reflexive, antisymmetric, and transitive. Since $a = a * a$, $a \leq a$ for all a in A, and $\leq$ is reflexive.

Now suppose that $a \leq b$ and $b \leq a$. Then, by definition and property 2, $a = a * b = b * a = b$, so $a = b$. Thus, $\leq$ is antisymmetric.

If $a \leq b$ and $b \leq c$, then $a = a * b = a * (b * c) = (a * b) * c = a * c$, sc $a \leq c$ and $\leq$ is transitive.

Finally, we must show that for all a and b in A, $a * b = a \wedge b$ (the greatest lower bound of a and b with respect to $\leq$). We have $a * b = a * (b * b) = (a * b) * b$, so $a * b \leq b$. In a similar way, we can show that $a * b \leq a$, so $a * b$ is a lower bound for a and b. Now if $c \leq a$ and $c \leq b$, then $c = c * a$ and $c = c * b$ by definition. Thus $c = (c * a) * b = c * (a * b)$, so $c \leq a * b$. This shows that $a * b$ is the greatest lower bound of a and b.

---

†Uses material from Chapter 4.

EXERCISE SET 6.1

In Exercises 1–8, determine whether the description of * is a valid definition of a binary operation on the set.

1. On $\mathbb{R}$, where $a * b$ is ab (ordinary multiplication).
2. On Z^+, where $a * b$ is a/b.
3. On Z, where $a * b$ is a^b.
4. On Z^+, where $a * b$ is a^b.
5. On Z^+, where $a * b$ is $a - b$.
6. On $\mathbb{R}$, where $a * b$ is $a\sqrt{b}$.
7. On $\mathbb{R}$, where $a * b$ is the largest rational number which is less than ab.
8. On Z, where $a * b$ is $2a + b$.

In Exercises 9–18, determine whether the binary operation * is commutative and whether it is associative on the set.

9. On Z^+, where $a * b$ is $a + b + 2$.
10. On Z, where $a * b$ is ab.
11. On $\mathbb{R}$, where $a * b$ is $a \times |b|$.
12. On the set of nonzero real numbers, where $a * b$ is a/b.
13. On $\mathbb{R}$, where $a * b$ is the minimum of a and b.
14. On the set of all $n \times n$ Boolean matrices, where $\mathbf{A} * \mathbf{B}$ is $\mathbf{A} \odot \mathbf{B}$ (see Section 1.8).
15. On $\mathbb{R}$, where $a * b$ is $ab/3$.
16. On $\mathbb{R}$, where $a * b$ is $ab + 2b$.
17. On a lattice A, where $a * b$ is $a \vee b$.
18. On a lattice A, where $a * b$ is $a \wedge b$.
19. Fill in the following table so that the binary operation * is commutative.

*	a	b	c
a	b		
b	c	b	a
c	a		c

20. Consider the binary operation * defined on the set $A = \{a, b, c, d\}$ by the following table.

*	a	b	c	d
a	a	c	b	d
b	d	a	b	c
c	c	d	a	a
d	d	b	a	c

Compute

(a) $c * d$ and $d * c$

(b) $b * d$ and $d * b$

(c) $a * (b * c)$ and $(a * b) * c$

(d) Is * commutative; associative?

In Exercises 21 and 22, complete the given table so that the binary operation * is associative.

21.

*	a	b	c	d
a	a	b	c	d
b	b	a	d	c
c	c	d	a	b
d				

22.

*	a	b	c	d
a	a	b	c	d
b	b	a	c	d
c				
d	d	c	c	d

23. Let A be a set with n elements.

(a) How many binary operations can be defined on A?

(b) How many commutative binary operations can be defined on A?

24. Let $A = \{a, b\}$.

(a) Determine tables for each of the 16 binary operations that can be defined on A.

(b) Determine the binary operations on A that are commutative.

(c) Determine the binary operations on A that are associative.

25. Let * be a binary operation on a set A, and suppose that * satisfies the idempotent, commutative, and associative properties, as discussed in Example 16. Define a relation $\leq$ on A by: $a \leq b$ if and only if $b = a * b$. Show that $(A, \leq)$ is a poset, and for all a and b, lub $(a, b) = a * b$.

6.2 Semigroups

In this section we define a simple algebraic system consisting of a set together with a binary operation that has many important applications.

A **semigroup** is a nonempty set S together with an associative binary operation * defined on S. We shall denote the semigroup by $(S, *)$ or, when it is clear what the operation * is, simply by S. We also refer to $a * b$ as the **product** of a and b. The semigroup $(S, *)$ is said to be commutative if * is a commutative operation.

Example 1   It follows from Section 6.1 that $(Z, +)$ is a commutative semigroup.

Example 2   The set $P(S)$, where S is a set, together with the operation of union is a commutative semigroup.

Example 3   The set Z with the binary operation of subtraction is not a semigroup, since subtraction is not associative.

Example 4   Let S be a fixed nonempty set, and let S^S be the set of all functions $f: S \to S$. If f and g are elements of S^S, we define $f * g$ as $f \circ g$, the composite function. Then $*$ is a binary operation on S^S and it follows from Section 2.6 that $*$ is associative. Hence $(S^S, *)$ is a semigroup. The semigroup S^S is not commutative.

Example 5†   Let $(L, \leq)$ be a lattice. Define a binary operation on L by $a * b = a \vee b$. Then L is a semigroup.

Example 6   Let $A = \{a_1, a_2, \ldots, a_n\}$ be a nonempty set. Recall from Section 1.2 that A^* is the set of all finite sequences of elements of A. That is, A^* consists of all words that can be formed from the alphabet A. Let α and β be elements of A^*. Observe that catenation is a binary operation $\cdot$ on A^*. Recall that if $\alpha = a_1 a_2 \cdots a_n$ and $\beta = b_1 b_2 \cdots b_k$, then $\alpha \cdot \beta = a_1 a_2 \cdots a_n b_1 b_2 \cdots b_k$. It is easy to see that if α, β, and γ are any elements of A^*, then

$$a \cdot (\beta \cdot \gamma) = (\alpha \cdot \beta) \cdot \gamma$$

so that $\cdot$ is an associative binary operation, and $(A^*, \cdot)$ is a semigroup. The semigroup $(A^*, \cdot)$ is called the **free semigroup generated by** A.

In a semigroup $(S, *)$ we can establish the following generalization of the associative property; we omit the proof.

Theorem 1   If $a_1, a_2, \ldots, a_n$, $(n \geq 3)$, are arbitrary elements of a semigroup, then all products of the elements $a_1, a_2, \ldots, a_n$ that can be formed by inserting arbitrary parentheses are equal.

If $a_1, a_2, \ldots, a_n$ are elements in a semigroup $(S, *)$ we shall write their product as

$$a_1 * a_2 * \cdots * a_n$$

Example 7   Theorem 1 shows that the products

$$((a_1 * a_2) * a_3) * a_4, \qquad a_1 * (a_2 * (a_3 * a_4)), \qquad (a_1 * (a_2 * a_3)) * a_4$$

are all equal.

---

†Uses material from Chapter 4.

An element e in a semigroup $(S, *)$ is called an **identity** element if

$$e * a = a * e = a$$

for all $a \in S$.

Example 8   The number 0 is an identity in the semigroup $(Z, +)$.

Example 9   The semigroup $(Z^+, +)$ has no identity element.

Theorem 2   If a semigroup $(S, *)$ has an identity element, it is unique.

   Proof. Suppose that e and e' are identity elements in S. Then since e is an identity,

$$e * e' = e'$$

Also, since e' is an identity,

$$e * e' = e$$

Hence

$$e = e'$$

A **monoid** is a semigroup $(S, *)$ that has an identity.

Example 10   The semigroup $P(S)$ defined in Example 2 has the identity $\varnothing$, since

$$\varnothing * A = \varnothing \cup A = A = A \cup \varnothing$$

for any element $A \in P(S)$. Hence $P(S)$ is a monoid.

Example 11   The semigroup S^S defined in Example 4 has the identity 1_S, since

$$1_S * f = 1_S \circ f = f = f \circ 1_S$$

for any element $f \in S^S$. Hence S^S is a monoid.

Example 12   The semigroup A^* defined in Example 6 is actually a monoid with identity Λ, the empty sequence, since $\alpha \cdot \Lambda = \Lambda \cdot \alpha = \alpha$ for all $\alpha \in A^*$.

Example 13   The set of all relations on a set A is a monoid under the operation of composition. The identity element is the equality relation Δ (see Section 2.6).

Let $(S, *)$ be a semigroup and let T be a subset of S. If T is closed under the operation $*$ (that is, $a * b \in T$ whenever a and b are elements of T), then $(T, *)$ is called a **subsemigroup** of $(S, *)$. Similarly, let $(S, *)$ be a monoid with identity e, and let T be a nonempty subset of S. If T is closed under the operation $*$, and $e \in T$, then $(T, *)$ is called a **submonoid** of $(S, *)$.

Observe that the associative property holds in any subset of a semigroup, so that a subsemigroup $(T, *)$ of a semigroup $(S, *)$ is itself a semigroup. Similarly, a submonoid of a monoid is itself a monoid.

Example 14   If $(S, *)$ is a semigroup, then $(S, *)$ is a subsemigroup of $(S, *)$. Similarly, let $(S, *)$ be a monoid. Then $(S, *)$ is a submonoid of $(S, *)$, and if $T = \{e\}$, then $(T, *)$ is also a submonoid of $(S, *)$.

Suppose that $(S, *)$ is a semigroup, and let $a \in S$. For $n \in Z^+$, we define the powers a^n recursively as follows:

$$a^1 = a, \qquad a^n = a^{n-1} * a, \qquad n \ge 2$$

Moreover, if $(S, *)$ is a monoid, we also define

$$a^0 = e$$

It can then be shown that if m and n are nonnegative integers, then

$$a^m * a^n = a^{m+n}$$

Example 15

(a) If $(S, *)$ is a semigroup, $a \in S$, and

$$T = \{a^i \mid i \in Z^+\}$$

then $(T, *)$ is a subsemigroup of $(S, *)$.

(b) If $(S, *)$ is a monoid, $a \in S$, and

$$T = \{a^i \mid i \in Z^+ \text{ or } i = 0\}$$

then $(T, *)$ is a submonoid of $(S, *)$.

Example 16   If T is the set of all even integers, then $(T, \times)$ is a subsemigroup of the monoid $(Z, \times)$ where $\times$ is ordinary multiplication, but it is not a submonoid since the identity of Z, the number 1, does not belong to T.

Isomorphism and Homomorphism

Let $(S, *)$ and $(T, *')$ be two semigroups. A function $f: S \rightarrow T$ is called an **isomorphism** from $(S, *)$ to $(T, *')$ if it is one to one, onto, and if

$$f(a * b) = f(a) *' f(b)$$

for all a and b in S.

If f is an isomorphism from $(S, *)$ to $(T, *')$, then, since f is one to one and onto, it follows from Theorem 1 in Section 3.1 that f^{-1} exists and is one to one and onto. We now show that f^{-1} is an isomorphism from $(T, *')$ to $(S, *)$. Let a' and b' be any elements of T. Since f is onto, we can find elements a and b in S such that $f(a) = a'$ and $f(b) = b'$. Then $a = f^{-1}(a')$ and $b = f^{-1}(b')$. Now

$$f^{-1}(a' *' b') = f^{-1}(f(a) *' f(b))$$
$$= f^{-1}(f(a * b))$$
$$= (f \circ f^{-1})(a * b)$$
$$= a * b = f^{-1}(a') * f^{-1}(b')$$

Hence f^{-1} is an isomorphism.

We now merely say that the semigroups $(S, *)$ and $(T, *')$ are **isomorphic** and we write $S \simeq T$.

To show that the semigroups $(S, *)$ and $(T, *')$ are isomorphic, we must use the following procedure:

Step 1. Define a function $f: S \rightarrow T$.
Step 2. Show that f is one to one.
Step 3. Show that f is onto.
Step 4. Show that $f(a * b) = f(a) *' f(b)$.

Example 17   Let T be the set of all even integers. Show that the semigroups $(Z, +)$ and $(T, +)$ are isomorphic.

Solution.

Step 1. We define the function $f: Z \rightarrow T$ by

$$f(a) = 2a$$

Step 2. We now show that f is one to one as follows. Suppose that

$$f(a_1) = f(a_2)$$

Then

$$2a_1 = 2a_2$$

so

$$a_1 = a_2$$

Hence f is one to one.

Step 3. We next show that f is onto. Suppose that b is any even integer. Then $a = b/2 \in Z$ and

$$f(a) = f(b/2) = 2(b/2) = b$$

so f is onto.

Step 4. We have

$$f(a + b) = 2(a + b)$$
$$= 2a + 2b = f(a) + f(b)$$

Hence $(Z, +)$ and $(T, +)$ are isomorphic semigroups.

In general, it is rather straightforward to verify that a given function $f: S \to T$ is or is not an isomorphism. However, it is generally difficult to show that two semigroups are isomorphic, because one has to create the isomorphism f.

As in the case of lattice isomorphisms, when two semigroups $(S, *)$ and $(T, *')$ are isomorphic, they can differ only in the nature of their elements; their semigroup structures are identical. If S and T are finite semigroups, their respective binary operations are given by multiplication tables. Then S and T are isomorphic if we can rearrange and relabel the elements of S so that its table is identical with that of T.

Example 18 Let $S = \{a, b, c\}$ and $T = \{x, y, z\}$. It is easy to verify that the following multiplication tables give semigroup structures for S and T, respectively.

$*$	a	b	c
a	a	b	c
b	b	c	a
c	c	a	b

$*'$	x	y	z
x	z	x	y
y	x	y	z
z	y	z	x

Let

$$f(a) = y$$
$$f(b) = x$$
$$f(c) = z$$

Replacing the elements in S by their images and rearranging the table, we obtain exactly the table for T. Thus S and T are isomorphic.

Theorem 3   Let $(S, *)$ and $(T, *')$ be monoids with respective identities e and e'. Let $f : S \to T$ be an isomorphism. Then $f(e) = e'$.

Proof. Let b be any element of T. Since f is onto, there is an element a in S such that $f(a) = b$. Then

$$a = a * e$$
$$b = f(a) = f(a * e) = f(a) *' f(e)$$
$$= b *' f(e)$$

Similarly, since $a = e * a$, $b = f(e) *' b$.
Thus for any $b \in T$,

$$b = b *' f(e) = f(e) *' b$$

which means that $f(e)$ is an identity for T. By Theorem 2, it follows that $f(e) = e'$.

If $(S, *)$ and $(T, *')$ are semigroups such that S has an identity and T does not, it then follows from Theorem 3 that $(S, *)$ and $(T, *')$ cannot be isomorphic.

Example 19   Let T be the set of all even integers and let $\times$ be ordinary multiplication. Then the semigroups $(Z, \times)$ and $(T, \times)$ are not isomorphic, since Z has an identity and T does not.

By dropping the condition of one to one and onto in the definition of an isomorphism of two semigroups, we get another important method for comparing the algebraic structures of the two semigroups.

Let $(S, *)$ and $(T, *')$ be two semigroups. A function $f : S \to T$ is called a **homomorphism** from $(S, *)$ to $(T, *')$ if

$$f(a * b) = f(a) *' f(b)$$

for all a and b in S. If f is also onto, say that T is the **homomorphic image** of S.

Example 20   Let $A = \{0, 1\}$ and consider the semigroups $(A^*, \cdot)$ and $(A, +)$, where $\cdot$ is the catenation operation and $+$ is defined by the multiplication table

+	0	1
0	0	1
1	1	0

Define the function $f : A^* \to A$ by

$$f(\alpha) = \begin{cases} 1 & \text{if } \alpha \text{ has an odd number of 1's} \\ 0 & \text{if } \alpha \text{ has an even number of 1's} \end{cases}$$

It is easy to verify that if α and β are any elements of A^*, then

$$f(\alpha \cdot \beta) = f(\alpha) + f(\beta)$$

Thus f is a homomorphism. The function f is onto since

$$f(0) = 0$$
$$f(1) = 1$$

but f is not an isomorphism, since it is not one to one.

The difference between an isomorphism and a homomorphism is that an isomorphism must be one to one and onto. For both an isomorphism and a homomorphism, the image of a product is the product of the images.

The proof of the following theorem, which is left as an exercise to the reader, is completely analogous to the proof of Theorem 3.

Theorem 4   Let $(S, *)$ and $(T, *')$ be monoids with respective identities e and e'. Let $f : S \to T$ be a homomorphism from $(S, *)$ onto $(T, *')$. Then $f(e) = e'$

Theorem 4, together with the following two theorems, shows that if a semigroup $(T, *')$ is the homomorphic image of the semigroup $(S, *)$, then $(T, *')$ has a strong algebraic resemblance to $(S, *)$.

Theorem 5   Let f be a homomorphism from a semigroup $(S, *)$ to semigroup $(T, *')$. If S' is a subsemigroup of $(S, *)$, then

$$f(S') = \{t \in T \mid t = f(s) \text{ for some } s \in S'\}$$

the image of S' under f, is a subsemigroup of $(T, *')$.

Proof. If t_1 and t_2 are any elements of $f(S')$, then

$$t_1 = f(s_1) \qquad \text{and} \qquad t_2 = f(s_2)$$

for some elements s_1 and s_2 in S'. Then

$$
\begin{aligned}
t_1 *' t_2 &= f(s_1) *' f(s_2) \\
&= f(s_1 * s_2) \\
&= f(s_3)
\end{aligned}
$$

where $s_3 = s_1 * s_2 \in S'$. Hence $t_1 *' t_2 \in f(S')$.

Thus $f(S')$ is closed under the operation $*'$. Since the associative property holds in T, it holds in $f(S')$, so $f(S')$ is a subsemigroup of $(T, *')$.

Theorem 6   If f is a homomorphism from a commutative semigroup $(S, *)$ onto a semigroup $(T, *')$, then $(T, *')$ is also commutative.

Proof. Let t_1 and t_2 be any elements of T. Then

$$t_1 = f(s_1) \qquad \text{and} \qquad t_2 = f(s_2)$$

for some elements s_1 and s_2 in S. Therefore

$$
\begin{aligned}
t_1 *' t_2 &= f(s_1) *' f(s_2) \\
&= f(s_1 * s_2) \\
&= f(s_2 * s_1) \\
&= f(s_2) *' f(s_1) \\
&= t_2 *' t_1
\end{aligned}
$$

Hence $(T, *')$ is also commutative.

EXERCISE SET 6.2

1. Let $A = \{a, b\}$. Which of the following multiplication tables define a semigroup on A? Which define a monoid on A?

$*$	a	b
a	a	b
b	a	a

(a)

$*$	a	b
a	a	b
b	b	b

(b)

$*$	a	b
a	b	a
b	a	b

(c)

$*$	a	b
a	a	b
b	b	a

(d)

$*$	a	b
a	a	a
b	b	b

(e)

$*$	a	b
a	b	b
b	a	a

(f)

In Exercises 2–11, determine whether the set together with the binary operation is a semigroup, a monoid, or neither. If it is a monoid, specify the identity.

If it is a semigroup or a monoid, determine if it is commutative.

2. Z^+, where $*$ is defined as ordinary multiplication.

3. Z^+, where $a * b$ is defined as max $\{a, b\}$.

4. Z^+, where $a * b$ is defined as GCD $\{a, b\}$.

5. Z^+, where $a * b$ is defined as a.

6. The nonzero real numbers, where $*$ is ordinary multiplication.

7. $P(S)$, with S a set, where $*$ is defined as intersection.

8. A Boolean algebra B, where $a * b$ is defined as $a \wedge b$.

9. $S = \{1, 2, 3, 6, 12\}$, where $a * b$ is defined as GCD (a, b).

10. $S = \{1, 2, 3, 6, 9, 18\}$, where $a * b$ is defined as LCM (a, b).

11. Z, where $a * b = a + b - ab$.

12. Which of the following multiplication tables gives a semigroup?

*	a	b	c
a	c	b	a
b	b	c	b
c	a	b	c

(a)

*	a	b	c
a	a	c	b
b	c	b	a
c	b	a	c

(b)

13. Complete the blanks in the following table to obtain a semigroup.

*	a	b	c
a	c	a	b
b	a	b	c
c			a

14. Let $S = \{a, b\}$. Determine the multiplication table for the semigroup S^S. Is the semigroup commutative?

15. Let $S = \{a, b\}$. Determine the multiplication table for the semigroup $(P(S), \cup)$.

16. Let $A = \{a, b, c\}$ and consider the semigroup $(A^*, \cdot)$, where $\cdot$ is the operation of catenation. If $\alpha = abac$, $\beta = cba$, and $\gamma = babc$, compute (a) $(\alpha \cdot \beta) \cdot \gamma$ (b) $\gamma \cdot (\alpha \cdot \alpha)$ (c) $(\gamma \cdot \beta) \cdot \alpha$.

17. Prove that the intersection of two subsemigroups of a semigroup $(S, *)$ is a subsemigroup of $(S, *)$.

18. Prove that the intersection of two submonoids of a monoid $(S, *)$ is a submonoid of $(S, *)$.

19. Let $A = \{0, 1\}$, and consider the semigroup $(A^*, \cdot)$, where $\cdot$ is the operation of catenation. Let T be the subset of A^* consisting of all sequences having an odd number of 1's. Is $(T, \cdot)$ a subsemigroup of $(A, \cdot)$?

20. Let $A = \{a, b\}$. Are there two semigroups $(A, *)$ and $(A, *')$ which are not isomorphic?

21. An element x in a monoid is called an **idempotent** if $x^2 = x * x = x$. Show that the set of all idempotents in a commutative monoid S is a submonoid of S.

22. Let $(S_1, *)$, $(S_2, *')$, and $(S_3, *'')$ be semigroups, and let $f: S_1 \to S_2$ and $g: S_2 \to S_3$ be isomorphisms. Show that $f \circ g: S_1 \to S_3$ is an isomorphism.

23. Let R^+ be the set of all positive real numbers. Show that the function $f: R^+ \to R$ defined by $f(x) = \ln x$ is an isomorphism of the semigroup $(R^+, \times)$ to the semigroup $(R, +)$, where $\times$ and $+$ are ordinary multiplication and addition, respectively.

6.3 Products and Quotients of Semigroups

In this section we shall obtain new semigroups from existing semigroups.

Theorem 1   If $(S, *)$ and $(T, *')$ are semigroups, then $(S \times T, *'')$ is a semigroup, where $*''$ is defined by $(s_1, t_1) *'' (s_2, t_2) = (s_1 * s_2, t_1 *' t_2)$.

Proof. The proof is left as an exercise.

It follows at once from Theorem 1 that if S and T are monoids with identities e_S and e_T, respectively, then $S \times T$ is a monoid with identity (e_S, e_T).

We now turn to a discussion of equivalence relations on a semigroup $(S, *)$. Since a semigroup is not merely a set, we shall find that certain equivalence relations on a semigroup give additional information about the structure of the semigroup.

An equivalence relation R on the semigroup $(S, *)$ is called a **congruence relation** if

$$a \, R \, a' \quad \text{and} \quad b \, R \, b' \quad \text{imply} \quad (a * b) \, R \, (a' * b')$$

Example 1 Consider the semigroup $(Z, +)$ and the equivalence relation R on Z defined by

$$a \, R \, b \quad \text{if and only if} \quad 2 \text{ divides } a - b.$$

Recall that we discussed this equivalence relation in Section 2.4, where we denoted $a \, R \, b$ by

$$a \equiv b \pmod{2}$$

We now show that this relation is a congruence relation as follows.

If

$$a \equiv b \pmod{2} \quad \text{and} \quad c \equiv d \pmod{2}$$

then 2 divides $a - b$ and 2 divides $c - d$, so

$$a - b = 2m \quad \text{and} \quad c - d = 2n$$

where m and n are in Z. Adding these equations, we have

$$(a - b) + (c - d) = 2m + 2n$$

or

$$(a + c) - (b + d) = 2(m + n)$$

so

$$a + c \equiv b + d \pmod{2}$$

Hence, the relation is a congruence relation.

Example 2   Let $A = \{0, 1\}$ and consider the free semigroup $(A^*, \cdot)$ generated by A. Define the following relation on A:

$$\alpha \; R \; \beta \quad \text{if and only if} \quad \alpha \text{ and } \beta \text{ have the same number of 1's.}$$

Show that R is a congruence relation on $(A^*, \cdot)$.

    Solution. We first show that R is an equivalence relation. We have

1. $\alpha \; R \; \alpha$ for any $\alpha \in A^*$.
2. If $\alpha \; R \; \beta$, then α and β have the same number of 1's, so $\beta \; R \; \alpha$.
3. If $\alpha \; R \; \beta$ and $\beta \; R \; \gamma$, then α and β have the same number of 1's, β and γ have the same number of 1's, so α and γ have the same number of 1's. Hence $\alpha \; R \; \gamma$.

    We next show that R is a congruence relation. Suppose that $\alpha \; R \; \alpha'$ and $\beta \; R \; \beta'$. Then α and α' have the same number of 1's and β and β' have the same number of 1's. Since the number of 1's in $\alpha \cdot \beta$ is the sum of the number of 1's in α and the number of 1's in β, we conclude that the number of 1's in $\alpha \cdot \beta$ is the same as the number of 1's in $\alpha' \cdot \beta'$. Hence

$$(\alpha \cdot \beta) \; R \; (\alpha' \cdot \beta')$$

and thus R is a congruence relation.

Example 3   Consider the semigroup $(Z, +)$, where $+$ is ordinary addition. Let $f(x) = x^2 - x - 2$. We now define the following relation on Z:

$$a \; R \; b \quad \text{if and only if} \quad f(a) = f(b)$$

It is straightforward to verify that R is an equivalence relation on Z. However, R is not a congruence relation since we have

$$-1 \; R \; 2 \qquad (f(-1) = f(2) = 0)$$

and

$$-2 \; R \; 3 \qquad (f(-2) = f(3) = 4)$$

but

$$-3 \; \cancel{R} \; 5$$

Since $f(-3) = 10$ and $f(5) = 18$.

Recall from Section 2.4 that the equivalence relation R on the semigroup $(S, *)$ determines a partition of S. As before, we let $[a]$ denote the equivalence class containing a, and S/R denote the set of all equivalence classes.

Theorem 2 Let R be a congruence relation on the semigroup $(S, *)$. Consider the relation $\circledast$ from $S/R \times S/R$ to S/R in which the ordered pair $([a], [b])$ is, for a and b in S, related to $[a * b]$.

(a) $\circledast$ is a function from $S/R \times S/R$ to S/R, and as usual we denote $\circledast$ $([a], [b])$ by $[a] \circledast [b]$. Thus $[a] \circledast [b] = [a * b]$.

(b) $(S/R, \circledast)$ is a semigroup.

Proof. Suppose that $([a], [b]) = ([a'], [b'])$. Then $a R a'$ and $b R b'$, so we must have $a * b R a' * b'$, since R is a congruence relation. Thus $[a * b] = [a' * b']$; that is, $\circledast$ is a function. This means that $\circledast$ is a binary operation on S/R.

Next, we must verify that $\circledast$ is an associative operation. We have

$$[a] \circledast ([b] \circledast [c]) = [a] \circledast [b * c]$$
$$= [a * (b * c)]$$
$$= [(a * b) * c] \qquad \text{(by associative property of } * \text{ in } S\text{)}$$
$$= [a * b] \circledast [c]$$
$$= ([a] \circledast [b]) \circledast [c]$$

Hence S/R is a semigroup. We call S/R the **quotient semigroup** or **factor semigroup**. Observe that $\circledast$ is a type of "quotient binary relation" on S/R that is constructed from the original binary relation $*$ on S by using the congruence relation R.

Corollary 1 Let R be a congruence relation on the monoid $(S, *)$. If we define the operation $\circledast$ in S/R by $[a] \circledast [b] = [a * b]$ then $(S/R, \circledast)$ is a monoid.

Proof. If e is the identity in $(S, *)$, then it is easy to verify that $[e]$ is the identity in $(S/R, \circledast)$.

Example 4 Consider the situation in Example 2. Since R is a congruence relation on the monoid $S = (A^*, \cdot)$ we conclude that $(S/R, \odot)$ is a monoid, where

$$[\alpha] \odot [\beta] = [\alpha \cdot \beta]$$

Example 5 As has already been pointed out in Section 2.4, we can repeat Example 14 of that section with the positive integer n instead of 2. That is, we define the following relation on the semigroup $(Z, +)$.

$$a R b \quad \text{if and only if} \quad n \text{ divides } a - b$$

Using exactly the same method as in Example 12, we show that R is an equivalence relation and as in the case of $n = 2$, we write R as $\equiv$ (mod n). Thus if $n = 4$, then

$$2 \equiv 6 \quad (\text{mod } 4)$$

Since 4 divides $(2 - 6)$. We also leave it for the reader to show that $\equiv$ (mod n) is a congruence relation on Z.

We now let $n = 4$ and we compute the equivalence classes determined by the congruence relation $\equiv$ (mod 4) on Z. We obtain

$$[0] = \{\ldots, -8, -4, 0, 4, 8, 12, \ldots\} = [4] = [8] = \cdots$$
$$[1] = \{\ldots, -7, -3, 1, 5, 9, 13, \ldots\} = [5] = [9] = \cdots$$
$$[2] = \{\ldots, -6, -2, 2, 6, 10, 14, \ldots\} = [6] = [10] = \cdots$$
$$[3] = \{\ldots, -5, -1, 3, 7, 11, 15, \ldots\} = [7] = [11] = \cdots$$

These are all the distinct equivalence classes which form the quotient set $Z/\equiv$ (mod 4). It is customary to denote the quotient set $Z/\equiv$ (mod n) by Z_n; Z_n is a monoid with operation $\oplus$ and identity $[0]$. We now determine the addition table for the semigroup Z_4 with operation $\oplus$.

$\oplus$	$[0]$	$[1]$	$[2]$	$[3]$
$[0]$	$[0]$	$[1]$	$[2]$	$[3]$
$[1]$	$[1]$	$[2]$	$[3]$	$[0]$
$[2]$	$[2]$	$[3]$	$[0]$	$[1]$
$[3]$	$[3]$	$[0]$	$[1]$	$[2]$

The entries in this table are obtained from the equation

$$[a] \oplus [b] = [a + b]$$

Thus

$$[1] \oplus [2] = [1 + 2] = [3]$$
$$[1] \oplus [3] = [1 + 3] = [4] = [0]$$
$$[2] \oplus [3] = [2 + 3] = [5] = [1]$$
$$[3] \oplus [3] = [3 + 3] = [6] = [2]$$

It can be shown that Z_n has the n equivalence classes

$$[0], [1], [2], \ldots, [n - 1]$$

and that

$$[a] \oplus [b] = [r]$$

where r is the remainder when $a + b$ is divided by n. Thus if $n = 6$,

$$[2] \oplus [3] = [5]$$
$$[3] \oplus [5] = [2]$$
$$[3] \oplus [3] = [0]$$

We shall now examine the connection between the structure of the semigroup $(S, *)$ and the quotient semigroup $(S/R, \circledast)$, where R is a congruence relation on $(S, *)$.

Theorem 3   Let R be a congruence relation on a semigroup $(S, *)$, and let $(S/R, \circledast)$ be the corresponding quotient semigroup. Then the function $f_R : S \rightarrow S/R$ defined by

$$f_R(a) = [a]$$

is an onto homomorphism, called the **natural homomorphism**.

Proof. If $[a] \in S/R$, then $f_R(a) = [a]$, so f_R is an onto function. Moreover, if a and b are elements of S, then

$$f_R(a * b) = [a * b]$$
$$= [a] \circledast [b]$$
$$= f_R(a) \circledast f_R(b)$$

so f_R is a homomorphism.

Theorem 4   (**Fundamental Homomorphism Theorem**). Let $f : S \rightarrow T$ be a homomorphism of the semigroup $(S, *)$ onto the semigroup $(T, *')$. Let R be the relation on S defined by $a \ R \ b$ if and only if $f(a) = f(b)$, for a and b in S. Then

(a) R is a congruence relation.
(b) $(T, *')$ and the quotient semigroup $(S/R, \circledast)$ are isomorphic.

Proof.
(a) We show that R is an equivalence relation. First, $a \ R \ a$ for every $a \in S$, since $f(a) = f(a)$. Next, if $a \ R \ b$, then $f(a) = f(b)$, so $b \ R \ a$. Finally, if $a \ R \ b$ and $b \ R \ c$, then $f(a) = f(b)$ and $f(b) = f(c)$, so $f(a) = f(c)$ and $a \ R \ c$. Hence R is an equivalence relation. Now suppose that $a \ R \ a_1$ and $b \ R \ b_1$. Then

$$f(a) = f(a_1) \qquad \text{and} \qquad f(b) = f(b_1)$$

Multiplying the left sides and right sides of these equations in T, we obtain

$$f(a) *' f(b) = f(a_1) *' f(b_1)$$

Since f is a homomorphism, this last equation can be rewritten as

$$f(a * b) = f(a_1 * b_1)$$

Hence

$$(a * b) \ R \ (a_1 * b_1)$$

and R is a congruence relation.

(b) We now consider the relation $\bar{f}$ from S/R to T defined as follows:

$$\bar{f} = \{([a], f(a)) \,|\, [a] \in S/R\}$$

We first show that $\bar{f}$ is a function. Suppose that $[a] = [a']$. Then $a \ R \ a'$, so $f(a) = f(a')$, which implies that $\bar{f}$ is a function. We may now write $\bar{f} : S/R \rightarrow T$, where $\bar{f}([a]) = f(a)$ for $[a] \in S/R$.

We next show that $\bar{f}$ is one to one. Suppose that $\bar{f}([a]) = \bar{f}([a'])$. Then

$$f(a) = f(a')$$

so $a \ R \ a'$, which implies that $[a] = [a']$. Hence $\bar{f}$ is one to one.

Now we show that $\bar{f}$ is onto. Suppose that $b \in T$. Since f is onto, $f(a) = b$ for some element a in S. Then

$$\bar{f}([a]) = f(a) = \text{b}$$

so $\bar{f}$ is onto.

Finally,

$$\bar{f}([a] \circledast [b]) = \bar{f}([a * b])$$
$$= f(a * b) = f(a) *' f(b)$$
$$= \bar{f}([a]) *' \bar{f}([b])$$

Hence $\bar{f}$ is an isomorphism.

Example 6   Let $A = \{0, 1\}$, and consider the free semigroup A^* generated by A under the operation of catenation. Note that A^* is a monoid with the empty string Λ as its identity. Let N be the set of all nonnegative integers. Then N is a semigroup under the operation of ordinary addition, denoted by $(N, +)$. The function $f : A^* \rightarrow N$

defined by

$$f(\alpha) = \text{the number of 1's in } \alpha$$

is readily checked to be a homomorphism. Let R be the following relation on A^*:

$$\alpha \ R \ \beta \quad \text{if and only if} \quad f(\alpha) = f(\beta)$$

That is, $\alpha \ R \ \beta$ if and only if α and β have the same number of 1's. Theorem 4 implies that $A^*/R \simeq N$, under the isomorphism $\bar{f}: A^*/R \rightarrow N$ defined by

$$\bar{f}([\alpha]) = f(\alpha) = \text{the number of 1's in } \alpha$$

Part (b) of Theorem 4 can be described by the diagram shown in Fig. 1. Here f_R is the natural homomorphism. It follows from the definitions of f_R and $\bar{f}$ that

$$f_R \circ \bar{f} = f$$

since

$$(f_R \circ \bar{f})(a) = \bar{f}(f_R(a))$$
$$= \bar{f}([a]) = f(a)$$

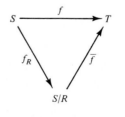

Figure 1

EXERCISE SET 6.3

1. Let $(S, *)$ and $(T, *')$ be commutative semigroups. Show that $S \times T$ (see Theorem 1) is also a commutative semigroup.

2. Let $(S, *)$ and $(T, *')$ be monoids. Show that $S \times T$ is also a monoid. Show that the identity of $S \times T$ is (e_S, e_T).

3. Let $(S, *)$ and $(T, *')$ be semigroups. Show that the function $f: S \times T \rightarrow S$ defined by

$$f(s, t) = s$$

is a homomorphism of the semigroup $S \times T$ onto the semigroup S.

4. Let S and T be semigroups. Show that $S \times T$ and $T \times S$ are isomorphic semigroups.

In Exercises 5–14, determine whether the relation R on the semigroup S is a congruence relation.

5. $S = Z$ under the operation of ordinary addition; $a\ R\ b$ if and only if 2 does not divide $a - b$.

6. $S = Z$ under the operation of ordinary addition; $a\ R\ b$ if and only if $a + b$ is even.

7. $S =$ any semigroup; $a\ R\ b$ if and only if $a = b$.

8. $S =$ the set of all rational numbers under the operation of addition; $a/b\ R\ c/d$ if and only if $ad = bc$.

9. $S =$ the set of all rational numbers under the operation of multiplication; $a/b\ R\ c/d$ if and only if $ad = bc$.

10. $S = Z$ under the operation of ordinary addition; $a\ R\ b$ if and only if 3 divides $a - b$.

11. $S = Z$ under the operation of ordinary addition; $a\ R\ b$ if and only if a and b are both even or a and b are both odd.

12. $S = Z^+$, under the operation of ordinary multiplication; $a\ R\ b$ if and only if $|a - b| \le 2$.

13. $A = \{0, 1\}$ and $S = A^*$, the free semigroup generated by A under the operation of catenation, $\alpha\ R\ \beta$ if and only if α and β both have an even number of 1's or both have an odd number of 1's.

14. $S = \{0, 1\}$ under the operation $*$ defined by the table

*	0	1
0	0	1
1	1	0

$a\ R\ b$ if and only if $a * a = b * b$. (*Hint*: Observe that if x is any element in S, then $x * x = 0$.)

15. Show that the intersection of two congruence relations on a semigroup is a congruence relation.

16. Show that the composite of two congruence relations on a semigroup need not be a congruence relation.

17. Consider the semigroup $S = \{a, b, c, d\}$ with the following multiplication table.

*	a	b	c	d
a	a	b	c	d
b	b	a	d	c
c	c	d	a	b
d	d	c	b	a

Consider the congruence relation $R = \{(a, a), (a, b), (b, a), (b, b), (c, c), (c, d), (d, c), (d, d)\}$ on S.

(a) Determine the multiplication table of the quotient semigroup S/R.

(b) Determine the natural homomorphism $f_R : S \to S/R$.

18. Consider the monoid $S = \{e, a, b, c\}$ with the following multiplication table.

*	e	a	b	c
e	e	a	b	c
a	a	e	b	c
b	b	c	b	c
c	c	b	b	c

Consider the congruence relation $R = \{(e, e), (e, a), (a, e), (a, a), (b, b), (b, c), (c, b), (c, c)\}$ on S.

(a) Determine the multiplication table of the quotient monoid S/R.

(b) Determine the natural homomorphism $f_R : S \to S/R$.

19. Let $A = \{0, 1\}$ and consider the free semigroup A^* generated by A under the operation of catenation. Let N be the semigroup of all nonnegative integers under the operation of ordinary addition.

(a) Verify that the function $f : A^* \to N$, defined by $f(\alpha) =$ the number of digits in α, is a homomorphism.

(b) Let R be the following relation on A^*: $\alpha\ R\ \beta$ if and only if $f(\alpha) = f(\beta)$. Show that R is a congruence relation on A^*.

(c) Show that A^*/R and N are isomorphic.

6.4 Groups

In this section we examine a special type of monoid, called a group, which has applications in every area where symmetry occurs. Applications of groups can be found in mathematics, physics, and chemistry, as well as in less obvious areas such as sociology. Very recent and exciting applications of group theory have arisen in quarks (astrophysics) and in the solutions of puzzles such as Rubik's cube. In this book, we shall present an important application of group theory to binary codes in Section 8.2.

A **group** $(G, *)$ is a monoid, with identity e, which has the additional property that for every element $a \in G$, there exists an element $a' \in G$ such that $a * a' = a' * a = e$. Thus a group is a set G together with a binary operation $*$ on G such that

1. $(a * b) * c = a * (b * c)$ for any elements a, b, and c in G.
2. There is a unique element e in G such that

$$a * e = e * a \qquad \text{for any } a \in G$$

3. For every $a \in G$ there is an element $a' \in G$, called an **inverse** of a, such that

$$a * a' = a' * a = e$$

Observe that if $(G, *)$ is a group, then $*$ is a binary operation so G must be closed under $*$; that is,

$$a * b \in G \qquad \text{for any elements } a \text{ and } b \text{ in } G$$

To simplify our notation, from now on when only one group $(G, *)$ is under consideration and there is no possibility of confusion, we shall write the product $a * b$ of the elements a and b in the group $(G, *)$ simply as ab, and we shall also refer to $(G, *)$ simply as G.

A group G is said to be **abelian** if $ab = ba$ for all elements a and b in G.

Example 1   The set of all integers Z with the operation of ordinary addition is an abelian group. If $a \in Z$, then an inverse of a is its negative $-a$.

Example 2   The set Z^+ under the operation of ordinary multiplication is not a group since the element 2 in Z^+ has no inverse. However, this set together with the given operation is a monoid.

Example 3   The set of all nonzero real numbers under the operation of ordinary multiplication is a group. An inverse of $a \neq 0$ is $1/a$.

Example 4   Let G be the set of all nonzero real numbers and let

$$a * b = \frac{ab}{2}$$

Show that $(G, *)$ is an abelian group.

Solution. We first verify that $*$ is a binary operation. If a and b are elements of G, then $a * b$ $(= ab/2)$ is a nonzero real number and hence is in G. We next verify associativity. Since

$$(a * b) * c = \left(\frac{ab}{2}\right) * c = \frac{(ab)c}{4}$$

and

$$a * (b * c) = a * \left(\frac{bc}{2}\right) = \frac{a(bc)}{4} = \frac{(ab)c}{4}$$

the operation $*$ is associative.

The number 2 is the identity in G, for if $a \in G$, then

$$a * 2 = \frac{(a)(2)}{2} = a = \frac{(2)(a)}{2} = 2 * a$$

Finally, if $a \in G$, then $a' = 4/a$ is an inverse of a, since

$$a * a' = a * \frac{4}{a} = \frac{a(4/a)}{2} = 2 = \frac{(4/a)(a)}{2} = \frac{4}{a} * a = a' * a$$

Since $a * b = b * a$ for all a and b in G, we conclude that G is an abelian group.

Before proceeding with additional examples of groups, we develop several important properties that are satisfied in any group G.

Theorem 1   Let G be a group. Each element a in G has only one inverse in G.

Proof. Let a' and a'' be inverses of a. Then

$$a'(aa'') = a'e = a'$$

and

$$(a'a)a'' = ea'' = a''$$

Hence, by associativity,

$$a' = a''$$

From now on we shall denote the inverse of a by a^{-1}. Thus in a group G we have

$$aa^{-1} = a^{-1}a = e$$

Theorem 2 Let G be a group and let a, b, and c be elements of G. Then

(a) $ab = ac$ implies that $b = c$ (**left cancellation property**).

(b) $ba = ca$ implies that $b = c$ (**right cancellation property**).

Proof.
(a) Suppose that

$$ab = ac$$

Multiplying both sides of this equation by a^{-1}, we obtain

$$
\begin{aligned}
a^{-1}(ab) &= a^{-1}(ac) \\
(a^{-1}a)b &= (a^{-1}a)c && \text{(by associativity)} \\
eb &= ec && \text{(by the definition of an inverse)} \\
b &= c && \text{(by the definition of an identity)}
\end{aligned}
$$

(b) The proof is similar to that of (a).

Theorem 3 Let G be a group and let a and b be elements of G. Then

(a) $(a^{-1})^{-1} = a$

(b) $(ab)^{-1} = b^{-1}a^{-1}$

Proof.
(a) We have

$$aa^{-1} = a^{-1}a = e$$

and since the inverse of an element is unique, we conclude that $(a^{-1})^{-1} = a$.

(b) We easily verify that

$$(ab)(b^{-1}a^{-1}) = a(b(b^{-1}a^{-1})) = a((bb^{-1})a^{-1}) = a(ea^{-1}) = aa^{-1} = e$$

and similarly,

$$(b^{-1}a^{-1})(ab) = e$$

so

$$(ab)^{-1} = b^{-1}a^{-1}$$

Theorem 4   Let G be a group, and let a and b be elements of G. Then:

(a) The equation $ax = b$ has a unique solution in G.
(b) The equation $ya = b$ has a unique solution in G.

Proof.
(a) The element $x = a^{-1}b$ is a solution of the equation $ax = b$, since

$$a(a^{-1}b) = (aa^{-1})b = eb = b$$

Suppose now that x_1 and x_2 are two solutions of the equation $ax = b$. Then

$$ax_1 = b \quad \text{and} \quad ax_2 = b$$

Hence

$$ax_1 = ax_2$$

Theorem 2 implies that $x_1 = x_2$.
(b) The proof is similar to that of (a).

From our discussion of monoids, we know that if a group G has a finite number of elements, then its binary operation can be given by a multiplication table. The multiplication table of a group $G = \{a_1, a_2, \ldots, a_n\}$ under the binary operation $*$ must satisfy the following properties:

1. The row labeled by e must contain the elements

$$a_1, a_2, \ldots, a_n$$

and the column labeled by e must contain the elements

$$a_1$$

$$a_2$$

$$\vdots$$

$$a_n$$

2. From Theorem 4 it follows that each element b of the group must appear exactly once in each row and column of the table. Thus each row and column is a permutation of the elements $a_1, a_2, \ldots, a_n$ of G, and each row (and each column) determines a different permutation.

If G is a group that has a finite number of elements, we say that G is a **finite group**, and the **order** of G is the number of elements $|G|$ in G. We shall now determine the multiplication tables of all groups of orders 1, 2, 3, and 4.

If G is a group of order 1, then $G = \{e\}$, and we have $ee = e$. Now let $G = \{e, a\}$ be a group of order 2. Then we obtain the multiplication table shown in Table 1, where we need to fill in the blank.

Table 1

	e	a
e	e	a
a	a	

The blank can be filled in by e or by a. Since there can be no repeats in any row or column, we must write e in the blank. The multiplication table shown in Table 2 satisfies the associative property and the other properties of a group, so it is the multiplication table of a group of order 2.

Table 2

	e	a
e	e	a
a	a	e

Next, let $G = \{e, a, b\}$ be a group of order 3. We have the multiplication table shown in Table 3, where we must fill in the four blanks.

Table 3

	e	a	b
e	e	a	b
a	a		
b	b		

Table 4

	e	a	b
e	e	a	b
a	a	b	e
b	b	e	a

A little experimentation shows that we can only complete the table as shown in Table 4. It can be shown (a tedious task) that Table 4 satisfies the associative property and the other properties of a group. Thus it is the multiplication table of a group of order 3. Observe that the groups of order 1, 2, and 3 are also abelian and that there is just one group of each order for a fixed labeling of the elements.

We next come to a group $G = \{e, a, b, c\}$ of order 4. It is not difficult to show that the possible multiplication table for G can be completed as shown in Tables 5 to 8. It can be shown that each of these tables satisfies the associative property and the other properties of a group. Thus there are four possible multiplication tables for a group of order 4. Again, observe that a group of order 4 is abelian. We shall return to groups of order 4 toward the end of this section, where we shall see that there are only two and not four different groups of order 4.

Table 5	e	a	b	c
e	e	a	b	c
a	a	e	c	b
b	b	c	e	a
c	c	b	a	e

Table 6	e	a	b	c
e	e	a	b	c
a	a	e	c	b
b	b	c	a	e
c	c	b	e	a

Table 7	e	a	b	c
e	e	a	b	c
a	a	b	c	e
b	b	c	e	a
c	c	e	a	b

Table 8	e	a	b	c
e	e	a	b	c
a	a	c	e	b
b	b	e	c	a
c	c	b	a	e

Example 5   Let $B = \{0, 1\}$, and let $+$ be the operation defined on B as follows:

+	0	1
0	0	1
1	1	0

Then B is a group. In this group, every element is its own inverse.

We next turn to an important example of a group.

Example 6   Consider the equilateral triangle shown in Fig. 1 with vertices 1, 2, and 3. A **symmetry** of the triangle (or of any geometrical figure) is a one-to-one function from the set of points forming the triangle (the geometrical figure) to itself which preserves the distance between adjacent points. Since the triangle is determined by its vertices, a symmetry of the triangle is merely a permutation of the vertices that preserves the distance between adjacent points. Let l_1, l_2, and l_3 be the angle bisectors of the corresponding angles as shown in Fig. 1, and let O be their point of intersection. We now describe the symmetries of this triangle. First, there is a counterclockwise rotation f_2 of the triangle about O through $120°$. Then f_2 can be written (see Section

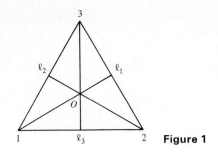

Figure 1

3.2) as the permutation

$$f_2 = \begin{pmatrix} 1 & 2 & 3 \\ 2 & 3 & 1 \end{pmatrix}$$

We next obtain a counterclockwise rotation f_3 about O through 240°, which can be written as the permutation

$$f_3 = \begin{pmatrix} 1 & 2 & 3 \\ 3 & 1 & 2 \end{pmatrix}$$

Finally, there is a counterclockwise rotation f_1 about O through 360° which can be written as the permutation

$$f_1 = \begin{pmatrix} 1 & 2 & 3 \\ 1 & 2 & 3 \end{pmatrix}$$

Of course, f_1 can also be viewed as the result of rotating the triangle about O through 0°.

We may also obtain three additional symmetries of the triangle, g_1, g_2, and g_3, by reflecting about the lines l_1, l_2, and l_3, respectively. Thus we may denote these reflections as the following permutations

$$g_1 = \begin{pmatrix} 1 & 2 & 3 \\ 1 & 3 & 2 \end{pmatrix}, \qquad g_2 = \begin{pmatrix} 1 & 2 & 3 \\ 3 & 2 & 1 \end{pmatrix}, \qquad g_3 = \begin{pmatrix} 1 & 2 & 3 \\ 2 & 1 & 3 \end{pmatrix}$$

Observe that the set of all symmetries of the triangle is the set of permutations of the set $\{1, 2, 3\}$, which has been considered in Section 3.2 and has been denoted by S_3. Thus

$$S_3 = \{f_1, f_2, f_3, g_1, g_2, g_3\}$$

We now introduce the operation of composition on the set S_3 and we obtain the multiplication table shown in Table 9.

Table 9

∘	f_1	f_2	f_3	g_1	g_2	g_3
f_1	f_1	f_2	f_3	g_1	g_2	g_3
f_2	f_2	f_3	f_1	g_2	g_3	g_1
f_3	f_3	f_1	f_2	g_3	g_1	g_2
g_1	g_1	g_3	g_2	f_1	f_3	f_2
g_2	g_2	g_1	g_3	f_2	f_1	f_3
g_3	g_3	g_2	g_1	f_3	f_2	f_1

Each of the entries in this table can be obtained in one of two ways: algebraically or geometrically. For example, suppose that we want to compute $f_2 \circ g_2$. Proceeding algebraically, we have

$$f_2 \circ g_2 = \begin{pmatrix} 1 & 2 & 3 \\ 2 & 3 & 1 \end{pmatrix} \circ \begin{pmatrix} 1 & 2 & 3 \\ 3 & 2 & 1 \end{pmatrix} = \begin{pmatrix} 1 & 2 & 3 \\ 2 & 1 & 3 \end{pmatrix} = g_3$$

Geometrically, we proceed as in Fig. 2. Since composition of functions is always associative, we see that $\circ$ is an associative operation on S_3. Observe that f_1 is the identity in S_3 and that every element of S_3 has a unique inverse in S_3. For example, $f_2^{-1} = f_3$. Hence S_3 is a group called the **group of symmetries of the triangle**. Observe that S_3 is the first example that we have given of a group that is not abelian.

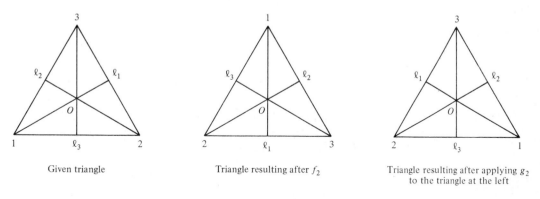

Given triangle   Triangle resulting after f_2 Triangle resulting after applying g_2 to the triangle at the left

Figure 2

Example 7   The set of all permutations of n elements is a group of order $n!$ under the operation of composition. This group is called the **symmetric group on n letters** and is denoted by S_n. It agrees with the group of symmetries of the equilateral triangle for $n = 3$.

As in Example 6, we can also consider the group of symmetries of a square. However, it turns out that this group is of order 8, so it does not agree with the group S_4, whose order is $4! = 24$.

Example 8   In Section 6.3 we discussed the monoid Z_n. We now show that Z_n is a group as follows. Let $[a] \in Z_n$. Then we may assume that $0 \le a < n$. Moreover, $[n - a] \in Z_n$ and since

$$[a] \oplus [n - a] = [a + n - a] = [n] = [0]$$

we conclude that $[n - a]$ is the inverse of $[a]$. Thus if $n = 6$, then $[2]$ is the inverse of $[4]$. Observe that Z_n is an abelian group.

We next turn to a discussion of certain important subsets of a group.

Let H be a subset of a group G such that:

(a) The identify e of G belongs to H.
(b) If a and b belong to H, then $ab \in H$.
(c) If $a \in H$, then $a^{-1} \in H$.

Then H is called a **subgroup** of G. Part (b) above says that H is a subsemigroup of G. Thus a subgroup of G can be viewed as a subsemigroup having properties (a) and (c) above.

Observe that if G is a group and H is a subgroup of G, then H is also a group with respect to the operation in G, since the associative property in G also holds in H.

Example 9   Let G be a group. Then G and $H = \{e\}$ are subgroups of G, called the *trivial* subgroups of G.

Example 10   Consider S_3, the group of symmetries of the equilateral triangle, whose multiplication table is shown in Table 9. It is easy to verify that $H = \{f_1, f_2, f_3\}$ is a subgroup of S_3.

Example 11   Let A_n be the set of all even permutations (see Section 3.2) in the group S_n. It follows from the definition of even permutation that A_n is a subgroup of S_n, called the **alternating group on *n* letters**.

Example 12   Let G be a group and let $a \in G$. Since a group is a monoid, we have already defined, in Section 6.2, a^n for $n \in Z^+$ as $aa \cdots a$ (n factors), and a^0 as e. If n is a negative integer, we now define a^{-n} as $a^{-1}a^{-1} \cdots a^{-1}$ (n factors). It then follows that if n and m are any integers, we have

$$a^n a^m = a^{n+m}$$

It is then easy to show that

$$H = \{a^i \,|\, i \in Z\}$$

is a subgroup of G.

Let $(G, *)$ and $(G', *')$ be two groups. Since groups are also semigroups, we can consider isomorphisms and homomorphisms from $(G, *)$ to $(G', *')$.

Since an isomorphism must be a one-to-one and onto function, it follows that two groups whose orders are unequal cannot possibly be isomorphic.

Example 13 Let G be the group of real numbers under addition, and let G' be the group of positive real numbers under multiplication. Let $f : G \to G'$ be defined by $f(x) = e^x$. We now show that f is an isomorphism.

If $f(a) = f(b)$, so that $e^a = e^b$, then $a = b$. Thus f is one-to-one.

If $c \in G'$, then $\ln c \in G$ and

$$f(\ln c) = e^{\ln c} = c$$

so f is onto.

Finally,

$$f(a + b) = e^{a+b} = e^a e^b = f(a)f(b)$$

Hence, f is an isomorphism.

Example 14 Let G be the symmetric group on n letters and let G' be the group B defined in Example 5. Let $f : G \to G'$ be defined as follows: for $p \in G$,

$$f(p) = \begin{cases} 0 & \text{if } p \in A_n \\ 1 & \text{if } p \notin A_n \end{cases} \quad \text{(the subgroup of all even permutations in } G\text{)}$$

Then f is a homomorphism.

Example 15 Let G be the group of integers under addition and let G' be the group Z_n as discussed in Example 8. Let $f : G \to G'$ be defined as follows: If $m \in G$, then $f(m) = [r]$, where r is the remainder when m is divided by n. We now show that f is a homomorphism of G onto G'.

Let $[r] \in Z_n$. Then we may assume that $0 \leqslant r < n$, so

$$r = 0 \cdot n + r$$

which means that the remainder when r is divided by n is r. Hence

$$f(r) = [r]$$

and thus f is onto.

Next, let a and b be elements of G expressed as

$$a = q_1 n + r_1, \quad \text{where } 0 \leqslant r_1 < n, \text{ and } r_1 \text{ and } q_1 \text{ are integers} \qquad (1)$$

$$b = q_2 n + r_2, \quad \text{where } 0 \leqslant r_2 < n, \text{ and } r_2 \text{ and } q_2 \text{ are integers} \qquad (2)$$

so that

$$f(a) = [r_1] \quad \text{and} \quad f(b) = [r_2]$$

Then

$$f(a) + f(b) = [r_1] + [r_2] = [r_1 + r_2]$$

To find $[r_1 + r_2]$, we need the remainder when $r_1 + r_2$ is divided by n: Write

$$r_1 + r_2 = q_3 n + r_3, \quad \text{where } 0 \leqslant r_3 < n, \text{ and } r_3 \text{ and } q_3 \text{ are integers}$$

Thus

$$f(a) + f(b) = [r_3]$$

Adding equations (1) and (2), we have

$$a + b = q_1 n + q_2 n + r_1 + r_2$$
$$= (q_1 + q_2 + q_3)n + r_3$$

so

$$f(a + b) = [r_1 + r_2] = [r_3]$$

Hence

$$f(a + b) = f(a) + f(b)$$

which implies that f is a homomorphism.

When $n = 2$, f assigns each even integer to $[0]$ and each odd integer to $[1]$.

Theorem 5   Let $(G, *)$ and $(G', *')$ be two groups, and let $f : G \to G'$ be a homomorphism from G to G'.

(a) If e is the identity in G and e' is the identity in G', then $f(e) = e'$.

(b) If $a \in G$, then $f(a^{-1}) = (f(a))^{-1}$.

(c) If H is a subgroup of G, then

$$f(H) = \{f(h) \mid h \in H\}$$

is a subgroup of G'.

Proof.

(a) Let $x = f(e)$. Then

$$x *' x = f(e) *' f(e) = f(e * e) = f(e) = x$$

so $x *' x = x$. Multiplying both sides by x^{-1}, we obtain

$$x = x *' x *' x^{-1} = x *' x^{-1} = e'$$

Thus, $f(e) = e'$.

(b)

$$a * a^{-1} = e$$

so

$$f(a * a^{-1}) = f(e) = e' \qquad \text{[by part (a)]}$$

or

$$f(a) *' f(a^{-1}) = e' \qquad (\text{since } f \text{ is a homomorphism})$$

Similarly,

$$f(a^{-1}) *' f(a) = e'$$

Hence $f(a^{-1}) = (f(a))^{-1}$.

(c) This follows from Theorem 5 of Section 6.2, and parts (a) and (b) above.

Example 16   The groups S_3 and Z_6 are both of order 6. However, S_3 is not abelian and Z_6 is abelian. Hence they are not isomorphic.

Example 17   Earlier in this section we found four possible multiplication tables (Tables 5 to 8) for a group of order 4. We now show that the groups with multiplication tables 6, 7, and 8 are isomorphic as follows. Let $G = \{e, a, b, c\}$ be the group whose multiplication table is Table 6, and let $G' = \{e', a', b', c'\}$ be the group whose multiplication table is Table 7, where we put primes on every entry in this last table. Let $f : G \rightarrow G'$ be defined by $f(e) = e', f(a) = b', f(b) = a', f(c) = c'$. We can then verify that under this renaming of elements, the two tables become identical, so the corresponding groups are isomorphic. Similarly, let $G'' = \{e'', a'', b'', c''\}$ be the group whose multiplication table is Table 8, where we put double primes on every entry in this last table. Let $g : G \rightarrow G''$ be defined by $g(e) = e'', g(a) = c'', g(b) = b'', g(c) = a''$. We can then verify that under this renaming of elements, the two tables become identical, so the corresponding groups are isomorphic. That is, the groups given by Tables 6, 7, and 8 are isomorphic.

Now, how can we be sure that Tables 5 and 6 do not yield isomorphic groups? Observe that if x is any element in the group determined by Table 5, then $x^2 = e$. If the groups were isomorphic, then the group determined by Table 6 would have the same property. Since it does not, we conclude that these groups are not isomorphic. Thus there are exactly two nonisomorphic groups of order 4.

The group with multiplication Table 5 is called the **Klein 4 group** and it is denoted by V. The one with multiplication Table 6, 7, or 8 is denoted by Z_4 since a relabeling of the elements of Z_4 results in this multiplication table.

EXERCISE SET 6.4

In Exercises 1–11, determine whether the set together with the binary operation is a group. If it is a group, determine if it is abelian; specify the identity and the inverse of an element a.

1. Z, where $*$ is ordinary multiplication.

2. Z, where $*$ is subtraction.

3. Q, the set of all rational numbers under the operation of addition.

4. Q, the set of all rational numbers under the operation of multiplication.

5. $\mathbb{R}$, under the operation of multiplication.

6. $\mathbb{R}$, where $a * b = a + b + 2$.

7. Z^+, under the operation of addition.

8. The real numbers that are not equal to -1, where $a * b = a + b + ab$.

9. The set of odd integers under the operation of multiplication.

10. The set of all $m \times n$ matrices under the operation of matrix addition.

11. If S is a nonempty set, the set $P(S)$, where $A * B = A \oplus B$. That is, if A and B are elements of $P(S)$, then $A \oplus B$ is the symmetric difference (see Section 1.3) of A and B.

12. Let $S = \{x \mid x$ is a real number and $x \neq 0, x \neq -1\}$. Consider the following functions $f_i : S \rightarrow S$, $i = 1, 2, \ldots, 6$:

$$f_1(x) = x, \qquad f_2(x) = 1 - x, \qquad f_3(x) = \frac{1}{x},$$

$$f_4(x) = \frac{1}{1-x}, \qquad f_5(x) = 1 - \frac{1}{x}, \qquad f_6(x) = \frac{x}{x-1}.$$

Show that $G = \{f_1, f_2, f_3, f_4, f_5, f_6\}$ is a group under the operation of composition. Determine the multiplication table of G.

13. Let G be a group with identity e. Show that if $x^2 = x$ for some x in G, then $x = e$.

14. Show that a group G is abelian if and only if $(ab)^2 = a^2b^2$ for all elements a and b in G.

15. Let G be the group defined in Example 4. Solve the following equations: (a) $3 * x = 4$; (b) $y * 5 = -2$.

16. Let G be a group with identity e. Show that if $a^2 = e$ for all a in G, then G is abelian.

17. Consider the square shown below.

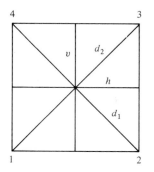

The symmetries of the square are:

Rotations f_1, f_2, f_3, and f_4 through $0°$, $90°$, $180°$, and $270°$, respectively.

f_5 and f_6, reflections about the lines v and h, respectively.

f_7 and f_8, reflections about the diagonals d_1 and d_2, respectively.

Write the multiplication table of D_4, the group of symmetries of the square.

18. Let G be a group. Show by mathematical induction that if $ab = ba$, then $(ab)^n = a^n b^n$ for $n \in Z^+$.

19. Let G be a finite group with identity e and let a be an arbitrary element of G. Prove that there exists a nonnegative integer n such that $a^n = e$.

20. Let G be the group of integers under the operation of addition. Which of the following subsets of G are subgroups of G? (a) the set of all even integers; (b) the set of all odd integers.

21. Is the set of positive rationals a subgroup of the group of real numbers under the operation of addition?

22. Let G be the nonzero integers under the operation of multiplication and let $H = \{3^n \mid n \in Z\}$. Is H a subgroup of G?

23. Let G be the group of integers under the operation of addition and let $H = \{3k \mid k \in Z\}$. Is H a subgroup of G?

24. Let G be an abelian group with identity e and let $H = \{x \mid x^2 = e\}$. Show that H is a subgroup of G.

25. Let G be a group and let $H = \{x \mid x \in G$ and $xy = yx$ for all $y \in G\}$. Prove that H is a subgroup of G.

26. Let G be a group and let $a \in G$. Define $H_a = \{x \mid x \in G$ and $xa = ax\}$. Prove that H_a is a subgroup of G.

27. Let A_n be the set of all even permutations in S_n. Show that A_n is a subgroup of S_n.

28. Let H and K be subgroups of a group G.
   (a) Prove that $H \cap K$ is a subgroup of G.
   (b) Show that $H \cup K$ need not be a subgroup of G.

29. Find all subgroups of D_4 (see Exercise 17).

30. Let G be an abelian group and n a fixed integer. Prove that the function $f : G \to G$ defined by $f(a) = a^n$, for $a \in G$, is a homomorphism.

31. Prove that the function $f(x) = |x|$ is a homomorphism from the group G, of nonzero real numbers under multiplication, to the group G', of positive real numbers under multiplication.

32. Let G be a group with identity e. Show that the function $f : G \to G$ defined by $f(a) = e$ for all $a \in G$ is a homomorphism.

33. Let G be a group. Show that the function $f : G \to G$ defined by $f(a) = a^2$ is a homomorphism if and only if G is abelian.

34. Show that the group in Exercise 12 is isomorphic to S_3.

35. Show that if $f : G \to G'$ is an isomorphism, then $f^{-1} : G' \to G$ is also an isomorphism.

36. Let S be the set of all finite groups and define the following relation R on S: G R G' if and only if G and G' are isomorphic. Prove that R is an equivalence relation (*Hint*: Use Exercise 35).

37. Let G be the group of integers under the operation of addition and let G' be the group of all even integers under the operation of addition. Show that the function $f : G \to G'$ defined by $f(a) = 2a$ is an isomorphism.

38. Let G be a group. Show that the function $f : G \to G$ defined by $f(a) = a^{-1}$ is an isomorphism if and only if G is abelian.

39. Let G be a group and let a be a fixed element of G. Show that the function $f_a : G \to G$ defined by $f_a(x) = axa^{-1}$, for $x \in G$, is an isomorphism.

40. Let $G = \{e, a, a^2, a^3, a^4, a^5\}$ be a group under the operation $a^i a^j = a^r$, where $i + j \equiv r$ (mod 6). Prove that G and Z_6 are isomorphic.

6.5 Products and Quotients of Groups

In this section we shall obtain new groups from other groups by using the ideas of product and quotient. Since a group has more structure than a semigroup, our results will be deeper than the analogous results for semigroups as discussed in Section 6.3.

Theorem 1 If G_1 and G_2 are groups, then $G = G_1 \times G_2$ is a group with operation defined by

$$(a_1, b_1)(a_2, b_2) = (a_1 a_2, b_1 b_2)$$

Example 1 Let G_1 and G_2 both be the group Z_2. For simplicity of notation, we shall write the elements of Z_2 as $\bar{0}$ and $\bar{1}$, respectively, instead of $[0]$ and $[1]$. Then the multiplication table of $G = G_1 \times G_2$ is given in Table 1.

Table 1 Multiplication Table of $Z_2 \times Z_2$

	$(\bar{0}, \bar{0})$	$(\bar{1}, \bar{0})$	$(\bar{0}, \bar{1})$	$(\bar{1}, \bar{1})$
$(\bar{0}, \bar{0})$	$(\bar{0}, \bar{0})$	$(\bar{1}, \bar{0})$	$(\bar{0}, \bar{1})$	$(\bar{1}, \bar{1})$
$(\bar{1}, \bar{0})$	$(\bar{1}, \bar{0})$	$(\bar{0}, \bar{0})$	$(\bar{1}, \bar{1})$	$(\bar{0}, \bar{1})$
$(\bar{0}, \bar{1})$	$(\bar{0}, \bar{1})$	$(\bar{1}, \bar{1})$	$(\bar{0}, \bar{0})$	$(\bar{1}, \bar{0})$
$(\bar{1}, \bar{1})$	$(\bar{1}, \bar{1})$	$(\bar{0}, \bar{1})$	$(\bar{1}, \bar{0})$	$(\bar{0}, \bar{0})$

Since G is a group of order 4, it must be isomorphic to V or to Z_4 (see Section 6.4), the only groups of order 4. By looking at the multiplication tables we see that the function $f: V \to Z_2 \times Z_2$ defined by $f(e) = (0, 0), f(a) = (1, 0), f(b) = (0, 1)$, and $f(c) = (1, 1)$ is an isomorphism.

If we repeat Example 1 with Z_2 and Z_3, we find that $Z_2 \times Z_3 \simeq Z_6$. It can be shown, in general, that $Z_m \times Z_n \simeq Z_{mn}$ if and only if GCD $(m, n) = 1$, that is, if and only if m and n are relatively prime.

Theorem 1 can obviously be extended to show that if $G_1, G_2, \ldots, G_n$ are groups, then $G = G_1 \times G_2 \times \cdots \times G_n$ is also a group.

Example 2 Let $B = \{0, 1\}$ be the group defined in Example 5 of Section 6.4, where $+$ is defined as follows:

+	0	1
0	0	1
1	1	0

Then $B^n = B \times B \times \cdots \times B$ (n factors) is a group with operation $\oplus$ defined by

$$(x_1, x_2, \ldots, x_n) \oplus (y_1, y_2, \ldots, y_n) = (x_1 + y_1, x_2 + y_2, \ldots, x_n + y_n)$$

The identity of B is $(0, 0, \ldots, 0)$ and every element is its own inverse.

A congruence relation on a group is simply a congruence relation on the group when it is viewed as a semigroup. We now discuss quotient structures determined by a congruence relation on a group.

Theorem 2   Let R be a congruence relation on the group $(G, *)$. Then the semigroup $(G/R, \circledast)$ is a group, where the operation $\circledast$ is defined on G/R by

$$[a] \circledast [b] = [a * b] \qquad \text{(see Section 6.3)}$$

Proof. Since a group is a monoid, we know from Corollary 1 in Section 6.3 that G/R is a monoid. We need to show that each element of G/R has an inverse. Let $[a] \in G/R$. Then $[a^{-1}] \in G/R$, and

$$[a] \circledast [a^{-1}] = [a * a^{-1}] = [e]$$

so $[a]^{-1} = [a^{-1}]$. Hence $(G/R, \circledast)$ is a group.

Since the definitions of homomorphism, isomorphism, and congruence for groups involve only the semigroup and monoid structure of groups, the following corollary is an immediate consequence of Theorems 3 and 4 of Section 6.3.

Corollary 1   (a) If R is a congruence relation on a group G, then the function $f_R : G \to G/R$, given by $f_R(a) = [a]$, is a group homomorphism.

(b) If $f : G \to G'$ is a homomorphism from the group $(G, *)$ onto the group $(G', *')$, and R is the relation defined on G by $a\,R\,b$ if and only if $f(a) = f(b)$, for a and b in G, then:

(1) R is a congruence relation.

(2) The function $\bar{f} : G/R \to G'$, given by $\bar{f}([a]) = f(a)$, is an isomorphism from the group $(G/R, \circledast)$ onto the group $(G', *')$.

Congruence relations on groups have a very special form which we will now develop. Let H be a subgroup of a group G, and let $a \in G$. The **left coset** of H in G determined by a is the set $aH = \{ah \,|\, h \in H\}$. The **right coset** of H in G determined by a is the set $Ha = \{ha \,|\, h \in H\}$. Finally, we will say that a subgroup H of G is **normal** if $aH = Ha$ for all a in G.

Warning. If $Ha = aH$, it does *not* follow that for $h \in H$ and $a \in G$, $ha = ah$. It does follow that $ha = ah'$, where h' is some element in H.

If H is subgroup of G, we shall need to compute all the left cosets of H in G.

First, suppose that $a \in H$. Then $aH \subseteq H$, since H is a subgroup of G; moreover, if $h \in H$, then $h = ah'$, where $h' = a^{-1}h \in H$, so that $H \subseteq aH$. Thus if $a \in H$, then $aH = H$. This means that when finding all the cosets of H, we need not compute aH for $a \in H$, since it will always be H.

Example 3   Let G be the symmetric group S_3 discussed in Example 6 of Section 6.4. The subset $H = \{f_1, g_2\}$ is a subgroup of G. Compute all the distinct left cosets of H in G.

Solution. If $a \in H$, then $aH = H$. Thus

$$f_1 H = g_2 H = H$$

Also,

$$f_2 H = \{f_2, g_3\}$$
$$f_3 H = \{f_3, g_1\}$$
$$g_1 H = \{g_1, f_3\} = f_3 H$$
$$g_3 H = \{g_3, f_2\} = f_2 H$$

The distinct left cosets of H in G are $H, f_2 H,$ and $f_3 H$.

Example 4   Let G and H be as in Example 3. Then the right coset $Hf_2 = \{f_2, g_1\}$. In Example 3 we saw that $f_2 H = \{f_2, g_3\}$. It follows that H is not a normal subgroup of G.

Example 5   Show that if G is an abelian group, every subgroup of G is a normal subgroup.

Solution. Let H be a subgroup of G and let $a \in G$ and $h \in H$. Then $ha = ah$, so $Ha = aH$, which implies that H is a normal subgroup of G.

Theorem 3   Let R be a congruence relation on a group G, and let $H = [e]$, the equivalence class containing the identity. Then H is a normal subgroup of G and for each $a \in G$, $[a] = aH = Ha$.

Proof. Let a and b be any elements in G. Since R is an equivalence relation, $b \in [a]$ if and only if $[b] = [a]$. Also G/R is a group by Theorem 2. Therefore, $[b] = [a]$ if and only if $[e] = [a]^{-1}[b] = [a^{-1}b]$. Thus $b \in [a]$ if and only if $H = [e] = [a^{-1}b]$. That is, $b \in [a]$ if and only if $a^{-1}b \in H$ or $b \in aH$. This proves that $[a] = aH$ for every $a \in G$. We can show similarly that $b \in [a]$ if and only if $H = [e] = [b][a]^{-1} = [ba^{-1}]$. This is equivalent to the statement $[a] = Ha$. Thus $[a] = aH = Ha$, and H is normal.

Combining Theorem 3 with Corollary 1, we see that in this case the quotient group G/R consists of all the left cosets of $N = [e]$. The operation in G/R is given by

$$(aN)(bN) = [a] \circledast [b] = [ab] = abN$$

and the function $f_R: G \to G/R$, defined by $f_R(a) = aN$, is a homomorphism from G onto G/R. For this reason, we will often write G/R as G/N.

We next consider the question of whether every normal subgroup of a group G is the equivalence class of the identity of G for some congruence relation.

Theorem 4   Let N be a normal subgroup of a group G, and let R be the following relation on G:

$$a \ R \ b \quad \text{if and only if} \quad a^{-1}b \in N.$$

Then:

(a) R is a congruence relation on G.
(b) N is the equivalence class $[e]$ relative to R, where e is the identity of G.

Proof.
(a) Let $a \in G$. Then $a \ R \ a$, since $a^{-1}a = e \in N$, so R is reflexive. Next, suppose that $a \ R \ b$, so that $a^{-1}b \in N$. Then $(a^{-1}b)^{-1} = b^{-1}a \in N$, so $b \ R \ a$. Hence R is symmetric. Finally, suppose that $a \ R \ b$ and $b \ R \ c$. Then $a^{-1}b \in N$ and $b^{-1}c \in N$. Then $(a^{-1}b)(b^{-1}c) = a^{-1}c \in N$, so $a \ R \ c$. Hence R is transitive. Thus R is an equivalence relation on G.

Next we show that R is a congruence relation on G. Suppose that $a \ R \ b$ and $c \ R \ d$. Then $a^{-1}b \in N$ and $c^{-1}d \in N$. Since N is normal, $Nd = dN$; that is, for any $n_1 \in N$, $n_1 d = n_2 d$ for some $n_2 \in N$. In particular, since $a^{-1}b \in N$, we have $a^{-1}bd = dn_2$ for some $n_2 \in N$. Then $(ac)^{-1}bd = (c^{-1}a^{-1})(bd) = c^{-1}(a^{-1}b)d = (c^{-1}d)n_2 \in N$, so $ac \ R \ bd$. Hence R is a congruence relation on G.

(b) Suppose that $x \in N$. Then $x^{-1}e = x^{-1} \in N$ since N is a subgroup, so $x \ R \ e$ and therefore $x \in [e]$. Thus $N \subseteq [e]$. Conversely, if $x \in [e]$, then $x \ R \ e$, so $x^{-1}e = x^{-1} \in N$. Then $x \in N$ and $[e] \subseteq N$. Hence $N = [e]$.

We see, thanks to Theorems 3 and 4, that if G is any group, then the equivalence classes with respect to a congruence relation on G are always the cosets of some normal subgroup of G. Conversely, the cosets of any normal subgroup of G are just the equivalence classes with respect to some congruence relation on G. We may now, therefore, translate part (b) of Corollary 1 as follows: Let f be a homomorphism from a group $(G, *)$ onto a group $(G', *')$, and let the **kernel** of f, ker (f), be defined by

$$\ker (f) = \{a \in G \mid f(a) = e'\}$$

Then:

(a) ker (f) is a normal subgroup of G.

(b) The quotient group $G/\text{ker}(f)$ is isomorphic to G'.

This follows from Corollary 1 and Theorem 3, since if R is the congruence relation on G given by

$$a \, R \, b \quad \text{if and only if} \quad f(a) = f(b)$$

then it is easy to show that ker $(f) = [e]$.

Example 6 Consider the homomorphism f from Z onto Z_n defined by

$$f(m) = [r]$$

where r is the remainder when m is divided by n. (See Example 15 in Section 6.4.) Find ker (f).

Solution. The integer m in Z belongs to ker (f) if and only if $f(m) = [0]$, that is, if and only if m is a multiple of n. Hence ker $(f) = nZ$.

EXERCISE SET 6.5

1. Write out the multiplication table of the group $Z_2 \times Z_3$.

2. Prove that if G and G' are abelian groups then $G \times G'$ is an abelian group.

3. Let G_1 and G_2 be groups. Prove that $G_1 \times G_2$ and $G_2 \times G_1$ are isomorphic.

4. Let G_1 and G_2 be groups. Show that the function $f: G_1 \times G_2 \to G_1$ defined by $f(a, b) = a$, for $a \in G_1$ and $b \in G_2$, is a homomorphism.

5. Determine the multiplication table of the quotient group $Z/3Z$, where Z has operation $+$.

6. Let Z be the group of integers under the operation of addition. Prove that the function $f: Z \times Z \to Z$ defined by $f(a, b) = a + b$ is a homomorphism.

7. Let $G = Z_4$. For each of the following subgroups H of G, determine all the left cosets of H in G.

   (a) $H = \{[0]\}$ (b) $H = \{[0], [2]\}$
   (c) $H = \{[0], [1], [2], [3]\}$

8. Let $G = S_3$. For each of the following subgroups H of G, determine all the left cosets of H in G.

   (a) $H = \{f_1, g_1\}$ (b) $H = \{f_1, g_3\}$

   (c) $H = \{f_1, f_2, f_3\}$ (d) $H = \{f_1\}$

   (e) $H = \{f_1, f_2, f_3, g_1, g_2, g_3\}$

9. Let $G = Z_8$. For each of the following subgroups H of G, determine all the left cosets of H in G.

   (a) $H = \{[0], [4]\}$
   (b) $H = \{[0], [2], [4], [6]\}$

10. Let G be the group all nonzero real numbers under the operation of multiplication, and consider the subgroup $H = \{3^n \mid n \in Z\}$ of G. Determine all the left cosets of H in G.

11. Let Z be the group of integers under the operation of addition and let $G = Z \times Z$. Consider the subgroup $H = \{(x, y) \mid x \in Z, y \in Z \text{ and } x = y\}$ of G. Describe the left cosets of H in G.

12. Let N be a subgroup of a group G, and let $a \in G$. Define

$$a^{-1}Na = \{a^{-1}na \mid n \in N\}$$

Prove that N is a normal subgroup of G if and only if $a^{-1}Na = N$ for all $a \in N$.

13. Let N be a subgroup of a group G. Prove that N is a normal subgroup of G if and only if $a^{-1}Na \subseteq N$ for all $a \in G$.

14. Find all the normal subgroups of S_3.

15. Find all the normal subgroups of D_4. (See Exercise 17 of Section 6.4.)

16. Let G be a group and let $H = \{x \mid x \in G \text{ and } xa = ax \text{ for all } a \in G\}$. Show that H is a normal subgroup of G.

17. Let H be a subgroup of a group G. Prove that every left coset aH of H has the same number of elements as H by showing that the function $f_a : H \rightarrow aH$ defined by $f_a(h) = ah$, for $h \in H$, is one to one and onto.

18. Let H and K be normal subgroups of G. Show that $H \cap K$ is a normal subgroup of G.

19. Let G be a group and H a subgroup of G. Let S be the set of all left cosets of H in G and let T be the set of all right cosets of H in G. Prove that the function $f : S \rightarrow T$ defined by $f(aH) = Ha^{-1}$ is one-to-one and onto.

20. Let G_1 and G_2 be groups. Let $f : G_1 \times G_2 \rightarrow G_2$ be the homomorphism from $G_1 \times G_2$ onto G_2 given by $f((g_1, g_2)) = g_2$. Compute ker (f).

21. Let f be a homomorphism from a group G_1 onto a group G_2, and suppose that G_2 is abelian. Show that ker (f) contains all elements of G_1 of the form $aba^{-1}b^{-1}$, where a and b are arbitrary in G_1.

22. Let G be an abelian group and N a subgroup of G. Prove that G/N is an abelian group.

23. Let H be a subgroup of the finite group G and suppose that there are only two left cosets of H in G. Prove that H is a normal subgroup of G.

24. Let H and N be subgroups of the group G. Prove that if N is a normal subgroup of G, then $H \cap N$ is a normal subgroup of H.

KEY IDEAS FOR REVIEW

☐ Binary operation on A: function $f : A \times A \rightarrow A$.

☐ Commutative binary operation: $a * b = b * a$.

☐ Associative binary operation: $a * (b * c) = (a * b) * c$.

☐ Semigroup: nonempty set S together with an associative operation $*$ defined on S.

☐ Theorem: If a semigroup $(S, *)$ has an identity element, it is unique.

☐ Monoid: semigroup that has an identity.

☐ Subsemigroup $(T, *)$ of semigroup $(S, *)$: T is a nonempty subset of S and $a * b \in T$ whenever a and b are in T.

☐ Submonoid $(T, *)$ of monoid $(S, *)$: T is a nonempty subset of S, $e \in T$, and $a * b \in T$ whenever a and b are in T.

☐ Isomorphism: see pages 288 and 311.

☐ Homomorphism: see pages 290 and 311.

☐ Theorem: Let $(S, *)$ and $(T, *')$ be monoids with respective identities e and e' and suppose $f : S \rightarrow T$ is an isomorphism. Then $f(e) = e'$.

☐ Theorem: If $(S, *)$ and $(T, *')$ are semigroups, then $(S \times T, *'')$ is a semigroup, where $*''$ is defined by

$$(s_1, t_1) *'' (s_2, t_2) = (s_1 * s_2, t_1 *' t_2)$$

☐ Congruence relation R on semigroup $(S, *)$: Equivalance relation R such that $a \, R \, a'$ and $b \, R \, b'$ imply that $(a * b) \, R \, (a' * b')$.

☐ Theorem: Let R be a congruence relation on the semigroup $(S, *)$. Define the operation $\circledast$ in S/R as follows:

$$[a] \circledast [b] = [a * b]$$

Then $(S/R, \circledast)$ is a semigroup.

☐ Quotient semigroup or factor semigroup S/R: see page 296.

☐ Z_n: see page 297.

☐ Theorem (Fundamental Homomorphism Theorem): Let $f: S \to T$ be a homomorphism of the semigroup $(S, *)$ onto the semigroup $(T, *')$. Let R be the relation on S defined by $a \, R \, b$ if and only if $f(a) = f(b)$, for a and b in S. Then:

    (a) R is a congruence relation.

    (b) T is isomorphic to S/R.

☐ Group $(G, *)$: monoid with identity e such that for every $a \in G$, there exists $a' \in G$ with the property that $a * a' = a' * a = e$.

☐ Theorem: Let G be a group and let a, b, and c be elements of G. Then:

    (a) $ab = ac$ implies that $b = c$ (left cancellation property).

    (b) $ba = ca$ implies that $b = c$ (right cancellation property).

☐ Theorem: Let G be a group and let a and b be elements of G. Then

    (a) $(a^{-1})^{-1} = a$

    (b) $(ab)^{-1} = b^{-1}a^{-1}$

☐ Order of a group G: $|G|$, the number of elements in G.

☐ S_n: the symmetric group on n letters.

☐ Subgroup: see page 310.

☐ Left coset aH of H in G determined by a: $\{ah \, | \, h \in H\}$.

☐ Theorem: Let R be a congruence relation on the group $(G, *)$. Then the semigroup $(G/R, \circledast)$ is a group, where the operation $\circledast$ is defined in G/R by

$$[a] \circledast [b] = [a * b]$$

☐ Theorem: Let R be a congruence relation on a group G, and let $H = [e]$, the equivalence class containing the identity. Then H is a normal subgroup of G and for each $a \, \varepsilon \, G$, $[a] = aH = Ha$.

☐ Theorem: Let N be a normal subgroup of a group G, and let R be the following relation on G:

$$a \, R \, b \quad \text{if and only if} \quad a^{-1} b \in N$$

Then:

(a) R is a congruence relation on G.

(b) N is the equivalence class $[e]$ relative to R, where e is the identity of G.

FURTHER READING

CONNELL, IAN, *Modern Algebra, A Constructive Introduction*, North-Holland, New York, 1982.

DORNHOFF, LARRY L., and FRANZ E. HOHN, *Applied Modern Algebra*, Macmillan, New York, 1978.

FRALEIGH, JOHN B., *A First Course in Abstract Algebra*, 3rd ed., Addison-Wesley, Reading, Mass., 1982.

GILBERT, WILLIAM J., *Modern Algebra with Applications*, Wiley, New York, 1976.

GOLDSTEIN, LARRY J., *Abstract Algebra: A First Course*, Prentice-Hall, Englewood Cliffs, N.J., 1973.

LIPSON, JOHN D., *Elements of Algebra and Algebraic Computing*, Addison-Wesley, Reading, Mass., 1981.

Finite-State Machines and Languages

Prerequisites: Chapters 1, 2, 3, 5, and 6.

We think of a machine as a system that can accept **input**, possibly produce **output**, and have some sort of internal memory that can keep track of certain information about previous inputs. The complete internal condition of the machine and all of its memory, at any particular time, is said to constitute the **state** of the machine at that time. The state in which a machine finds itself at any instant, summarizes its "memory" of past inputs, and determines how it will react to subsequent input. When more input arrives, the given state of the machine determines (with the input) the next state to be occupied, and any output that may be produced. If the number of states is finite, the machine is a finite-state machine.

A finite-state machine is a model that will include devices ranging from simple "flip-flops" all the way to entire computers. They can also be used to describe the effect of certain computer programs and other "nonhardware" systems. Finite-state machines are useful in the study of formal languages, and are often found in compilers and interpreters for various computer programming languages. There are several types and extensions of finite-state machines, and in this chapter we will introduce the most elementary versions.

7.1 Finite-State Machines

Suppose that we have a finite set $S = \{s_0, s_1, \ldots, s_m\}$, a finite set I, and for each $x \in I$, a function $f_x: S \to S$. Let $\mathscr{F} = \{f_x \mid x \in I\}$. The triple $(S, I, \mathscr{F})$ is called a **finite-state machine**, S is called the **state set** of the machine and the elements of S are called **states**. The set I is called the **input set** of the machine. For any input $x \in I$, the function f_x describes the effect that this input has on the states of the machine, and is called a **state transition function**. Thus if the machine is in state s_i and input x occurs, the next state of the machine will be $f_x(s_i)$.

Since the next state $f_x(s_i)$ is uniquely determined by the pair (x, s_i), there is a function $F: I \times S \rightarrow S$ given by

$$F(x, s_i) = f_x(s_i)$$

The individual functions f_x can all be recovered from a knowledge of F. Many authors will use a function $F: I \times S \rightarrow S$, instead of a set $\{f_x \mid x \in I\}$ to define a finite-state machine. The definitions are completely equivalent.

Example 1   Let $S = \{s_0, s_1\}$, and $I = \{0, 1\}$. Define f_0 and f_1 as follows:

$$f_0(s_0) = s_0, \qquad f_1(s_0) = s_1,$$
$$f_0(s_1) = s_1, \qquad f_1(s_1) = s_0$$

This finite-state machine has two states, s_0 and s_1, and accepts two possible inputs, 0 and 1. The input 0 leaves each state fixed, and the input 1 reverses states. One can think of this machine as a model for a circuit (or logical) device and visualize such a device as in Fig. 1. The output signals will, at any given time, consist of two voltages, one higher than the other. Either line 1 will be at the higher voltage and line 2 at the lower, or the reverse. The first set of output conditions will be denoted s_0, and the second will be denoted s_1. An input pulse, represented by the symbol 1, will reverse output voltages. The symbol 0 represents the absence of an input pulse, and so results in no change of output. This device is often called a **T flip-flop**, and is a concrete realization of the machine in this example.

We summarize this machine in Fig. 2. The table shown there lists the states down the side and inputs across the top. The column under each input gives the values of the function corresponding to that input, at each state shown on the left.

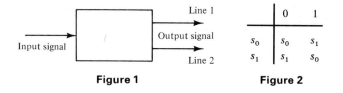

| Figure 1 | | Figure 2 |

	0	1
s_0	s_0	s_1
s_1	s_1	s_0

The arrangement illustrated in Fig. 2 for summarizing the effect of inputs on states, is called the **state transition table** of the finite-state machine. It can be used with any machine of reasonable size, and is a convenient method of specifying the machine.

Example 2   Consider the state transition table shown in Fig. 3. Here a and b are the possible inputs, and there are three states, s_0, s_1, and s_2. The table shows us that

$$f_a(s_0) = s_0, \qquad f_a(s_1) = s_2, \qquad f_a(s_2) = s_1$$

and

$$f_b(s_0) = s_1, \qquad f_b(s_1) = s_0, \qquad f_b(s_2) = s_2$$

If M is a finite-state machine with states S, inputs I, and state transition functions $\{f_x \mid x \in I\}$, we can determine a relation R_M on S in a natural way. If s_i, $s_j \in S$, we say that $s_i \, R_M \, s_j$ if there is an input x so that $f_x(s_i) = s_j$.

Thus $s_i \, R_M \, s_j$ means that if the machine is in state s_i there is some input $x \in I$ which, if received next, will put the machine in state s_j. The relation R_M permits us to describe the machine M as a labeled digraph of the relation R_M on S, where each edge is labeled by the set of all inputs which cause the machine to change states as indicated by that edge.

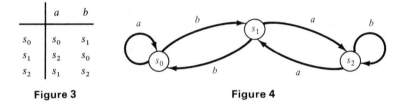

	a	b
s_0	s_0	s_1
s_1	s_2	s_0
s_2	s_1	s_2

Figure 3   Figure 4

Example 3   Consider the machine of Example 2. Fig. 4 shows the digraph of the relation R_M, with each edge labeled appropriately. Notice that the entire structure of M can be recovered from this digraph, since edges and their labels indicate where each input sends each state.

Example 4   Consider the machine M whose state table is shown in Fig. 5(a). The digraph of R_M is then shown in Fig. 5(b), with edges labeled appropriately.

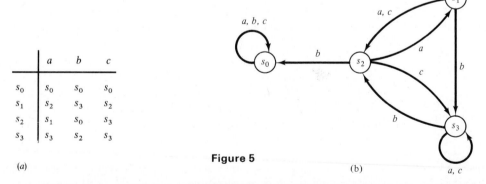

	a	b	c
s_0	s_0	s_0	s_0
s_1	s_2	s_3	s_2
s_2	s_1	s_0	s_3
s_3	s_3	s_2	s_3

(a)

Figure 5

(b)

Note that an edge may be labeled by more than one input, since several inputs may cause the same change of state. The reader will observe that every input must be part of the label of exactly one edge out of each state. This is a general property which holds for the labeled digraphs of all finite-state machines. For brevity, we will refer to the labeled digraph of a machine M simply as the **digraph** of M.

It is possible to add a variety of extra features to a finite-state machine in order to increase the utility of the concept. A simple, yet very useful extension results in what is often called a **Moore machine**, or **recognition machine**, which is defined as a sequence $(S, I, \mathscr{F}, s_0, T)$, where $(S, I, \mathscr{F})$ constitutes a finite-state machine, $s_0 \in S$ and $T \subseteq S$. The state s_0 is called the **starting state** of M, and it will be used to represent the condition of the machine before it receives any input. The set T is called the set of **acceptance states** of M. These states will be used in Section 7.2 in connection with language recognition.

When the digraph of a Moore machine is drawn, the acceptance states are indicated with two concentric circles, instead of one. No special notation will be used on these digraphs for the starting state, but unless otherwise specified, this state will be named s_0.

Example 5   Let M be the Moore machine $(S, I, \mathscr{F}, s_0, T)$, where $(S, I, \mathscr{F})$ is the finite-state machine of Fig. 5, and $T = \{s_1, s_3\}$. Figure 6 shows the digraph of M.

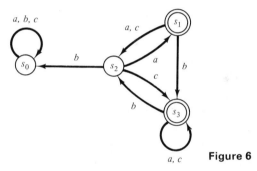

Figure 6

Machine Congruence and Quotient Machines

Let $M = (S, I, \mathscr{F})$ be a finite-state machine, and suppose that R is an equivalence relation on S. We say that R is a **machine congruence** on M if, for any $s, t \in S$, $s \, R \, t$ implies that $f_x(s) \, R \, f_x(t)$ for all $x \in I$. In other words, R is a machine congruence if R-equivalent pairs of states are always taken into R-equivalent pairs of states, by every input in I. If R is a machine congruence on $M = (S, I, \mathscr{F})$, we let $\bar{S} = S/R$ be the partition of S corresponding to R (see Section 2.4). Then $\bar{S} = \{[s] \,|\, s \in S\}$.

For any input $x \in I$, consider the relation $\bar{f}_x$ on $\bar{S}$ defined by

$$\bar{f}_x = \{([s], [f_x(s)])\}$$

If $[s] = [t]$, then $s \, R \, t$; therefore, $f_x(s) \, R \, f_x(t)$, so $[f_x(s)] = [f_x(t)]$. This shows that the relation $\bar{f}_x$ is a function from $\bar{S}$ to $\bar{S}$, and we may write $\bar{f}_x([s]) = [f_x(s)]$ for all equivalence classes $[s]$ in $\bar{S}$. If we let $\bar{\mathscr{F}} = \{\bar{f}_x \mid x \in I\}$, then the triple $\bar{M} = (\bar{S}, I, \bar{\mathscr{F}})$ is a finite-state machine called the **quotient of** M **corresponding to** R. We will also denote $\bar{M}$ by M/R.

Generally, a quotient machine will be simpler than the original machine. We will show in Section 7.4 that it is often possible to find a simpler quotient machine which will replace the original machine.

Example 6   Let M be the finite-state machine whose state transition table is shown in Fig. 7.

Then $S = \{s_0, s_1, s_2, s_3, s_4, s_5\}$. Let R be the equivalence relation on S whose matrix is

$$\mathbf{M}_R = \begin{bmatrix} 1 & 0 & 1 & 0 & 0 & 0 \\ 0 & 1 & 0 & 1 & 0 & 1 \\ 1 & 0 & 1 & 0 & 0 & 0 \\ 0 & 1 & 0 & 1 & 0 & 1 \\ 0 & 0 & 0 & 0 & 1 & 0 \\ 0 & 1 & 0 & 1 & 0 & 1 \end{bmatrix}$$

Then we have $S/R = \{[s_0], [s_1], [s_4]\}$, where $[s_0] = \{s_0, s_2\} = [s_2]$, $[s_1] = \{s_1, s_3, s_5\} = [s_3] = [s_5]$, and $[s_4] = \{s_4\}$. We check that R is a machine congruence. The state transition table in Fig. 7 shows that f_a takes each element of $[s_i]$ to an element of $[s_i]$ for $i = 0, 1, 4$. Also, f_b takes each element of $[s_0]$ to an element of $[s_4]$, each element of $[s_1]$ to an element of $[s_0]$, and each element of $[s_4]$ to an element of $[s_1]$. These observations show that R is a machine congruence, and that the state transition table of the quotient machine M/R is as shown in Fig. 8.

	a	b
s_0	s_0	s_4
s_1	s_1	s_0
s_2	s_2	s_4
s_3	s_5	s_2
s_4	s_4	s_3
s_5	s_3	s_2

Figure 7

	a	b
$[s_0]$	$[s_0]$	$[s_4]$
$[s_1]$	$[s_1]$	$[s_0]$
$[s_4]$	$[s_4]$	$[s_1]$

Figure 8

Example 7   Let $I = \{0, 1\}$, $S = \{s_0, s_1, s_2, s_3, s_4, s_5, s_6, s_7\}$, and let $M = (S, I, \mathscr{F})$ be the finite-state machine whose digraph is shown in Fig. 9.

Suppose that R is the equivalence relation whose corresponding partition of S, S/R, is $\{\{s_0, s_4\}, \{s_1, s_2, s_5\}, \{s_6\}, \{s_3, s_7\}\}$. Then it is easily checked, from the digraph of Fig. 9, that R is a machine congruence. To obtain the digraph of the quotient machine, $\bar{M}$, draw a vertex for each equivalence class, $[s_0] = \{s_0, s_4\}$, $[s_1] = \{s_1, s_2, s_5\}$, $[s_6] = \{s_6\}$, $[s_3] = \{s_3, s_7\}$ and construct an edge from $[s_i]$ to $[s_j]$ if there is, in the original digraph, an edge from some vertex in $[s_i]$ to some

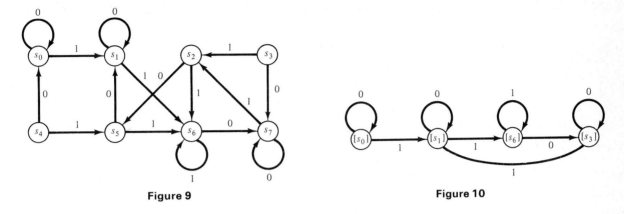

Figure 9   **Figure 10**

vertex in $[s_j]$. In this case, the constructed edge is labeled with all inputs that take some vertex in $[s_i]$ to some vertex in $[s_j]$. Figure 10 shows the result. The procedure illustrated in this example works in general.

If $M = (S, I, \mathscr{F}, s_0, T)$ is a Moore machine, and R is a machine congruence on M, then we may let $\bar{T} = \{[t] \mid t \in T\}$. In this case, the sequence $\bar{M} = (\bar{S}, I, \bar{\mathscr{F}}, [s_0], \bar{T})$ is a Moore machine. In other words, we compute the usual quotient machine M/R; then we designate $[s_0]$ as a starting state, and let $\bar{T}$ be the set of equivalence classes of acceptance states. The resulting Moore machine $\bar{M}$, constructed as above, will be called the **quotient Moore machine** of M.

Example 8   Consider the Moore machine $(S, I, \mathscr{F}, s_0, T)$, where $(S, I, \mathscr{F})$ is the finite-state machine of Example 6, and T is the set $\{s_1, s_3, s_4\}$. The digraph of the resulting quotient Moore machine is shown in Fig. 11.

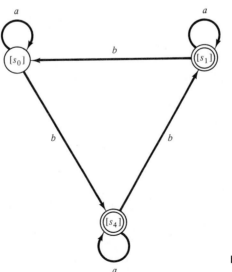

Figure 11

EXERCISE SET 7.1

In Exercises 1–4, draw the digraph of the machine whose state transition table is shown. Remember to label the edges with the appropriate inputs.

1.

	0	1
s_0	s_0	s_1
s_1	s_1	s_2
s_2	s_2	s_0

2.

	0	1	2
s_0	s_1	s_0	s_2
s_1	s_0	s_0	s_1
s_2	s_2	s_0	s_2

3.

	a	b
s_0	s_1	s_0
s_1	s_2	s_0
s_2	s_2	s_0

4.

	a	b
s_0	s_1	s_0
s_1	s_2	s_1
s_2	s_3	s_2
s_3	s_3	s_3

In Exercises 5–7, construct the state transition table of the finite-state machine whose digraph is shown.

5.

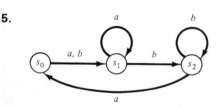

6.

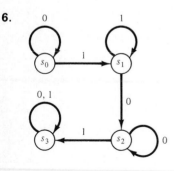

7.

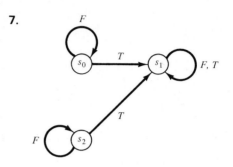

8. Let $M = (S, I, \mathscr{F})$ be a finite-state machine. Define a relation R on I as follows: $x_1 \ R \ x_2$ if and only if $f_{x_1}(s) = f_{x_2}(s)$ for every s in S. Show that R is an equivalence relation on I.

9. Let $(S, *)$ be a finite semigroup. Then we may consider the machine $(S, S, \mathscr{F})$ where $\mathscr{F} = \{f_x \mid x \in S\}$, and $f_x(y) = x * y$ for all $x, y \in S$. Thus we have a finite-state machine in which the state set and the input set are the same. Define a relation R on S as follows: $x \ R \ y$ if and only if there is some $z \in S$ such that $f_z(x) = y$. Show that R is transitive.

10. Consider a finite group $(S, *)$ and let $(S, S, \mathscr{F})$ be the finite-state machine constructed in Exercise 9. Show that if R is the relation defined in Exercise 9, then R is an equivalence relation.

11. Let $I = \{0, 1\}$, $S = \{a, b\}$. Construct all possible state transition tables of finite-state machines that have S as state set and I as input set.

12. Consider the machine whose state transition table is shown below.

	0	1
1	1	4
2	3	2
3	2	3
4	4	1

Here $S = \{1, 2, 3, 4\}$. Show that $R = \{(1, 1), (1, 4), (4, 1), (4, 4), (2, 2), (2, 3), (3, 2), (3, 3)\}$ is a machine congruence and construct the state transition table for the corresponding quotient machine.

13. Consider the Moore machine whose digraph is shown below.

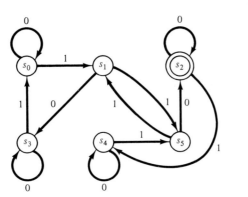

Show that the relation R on S whose matrix is

$$\mathbf{M}_R = \begin{bmatrix} 1 & 0 & 0 & 0 & 1 & 0 \\ 0 & 1 & 0 & 0 & 0 & 1 \\ 0 & 0 & 1 & 1 & 0 & 0 \\ 0 & 0 & 1 & 1 & 0 & 0 \\ 1 & 0 & 0 & 0 & 1 & 0 \\ 0 & 1 & 0 & 0 & 0 & 1 \end{bmatrix}$$

is a machine congruence. Draw the digraph of the corresponding quotient Moore machine.

7.2 Semigroups, Machines, and Languages

Let $M = (S, I, \mathscr{F})$ be a finite-state machine with state set $S = \{s_0, s_1, \ldots, s_n\}$, input set I, and state transition functions $\mathscr{T} = \{f_x \mid x \in I\}$.

We will associate with M two monoids, whose construction we recall from Section 6.2. First, there is the free monoid I^* on the input set I. This monoid consists of all finite sequences (or "strings" or "words") from I, with catenation as its binary operation. The identity is the empty string Λ. Second, we have the monoid S^S, which consists of all functions from S to S, and which has function composition as its binary operation. The identity in S^S is the function 1_S defined by $1_S(s) = s$, for all s in S.

If $w = x_1 x_2 \cdots x_n \in I^*$, we let $f_w = f_{x_1} \circ f_{x_2} \circ \cdots \circ f_{x_n}$, the composition of the functions $f_{x_1}, f_{x_2}, \ldots, f_{x_n}$. Also define f_Λ to be 1_S. In this way we assign an element f_w of S^S to each element w of I^*. If we think of each f_x as the "effect" of the input x on the states of the machine M, then f_w represents the combined effect of all the input letters in the word w, received in the sequence specified by w. We call f_w the **state transition function corresponding to** w.

Example 1 Let $M = (S, I, \mathscr{F})$, where $S = \{s_0, s_1, s_2\}$, $I = \{0, 1\}$, and $\mathscr{F}$ is given by the following state transition table.

	0	1
s_0	s_0	s_1
s_1	s_2	s_2
s_2	s_1	s_0

Let $w = 011 \in I^*$. Then

$$f_w(s_0) = (f_0 \circ f_1 \circ f_1)(s_0) = f_1(f_1(f_0(s_0)))$$
$$= f_1(f_1(s_0)) = f_1(s_1) = s_2$$

Similarly,

$$f_w(s_1) = f_1(f_1(f_0(s_1))) = f_1(f_1(s_2)) = f_1(s_0) = s_1$$

and

$$f_w(s_2) = f_1(f_1(f_0(s_2))) = f_1(f_1(s_1)) = f_1(s_2) = s_0$$

Example 2   Let us consider the same machine M as in Example 1, and examine the problem of computing f_w a little differently. In Example 1 we used the definition directly, and for a large machine we would program an algorithm to compute the values of f_w in just that way. However, if the machine is of moderate size, human beings may find another procedure to be preferable.

We begin by drawing the digraph of the machine M as shown in Fig. 1. We may use this digraph to compute word transition functions by just following the edges corresponding to successive input letter transitions. Thus to compute $f_w(s_0)$, we start at state s_0 and see that input 0 takes us to state s_0. The input 1 which follows takes us on to state s_1, and the final input of 1 takes us to s_2. Thus $f_w(s_0) = s_2$, as before.

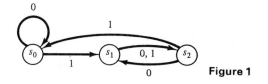

Figure 1

Let us compute $f_{w'}$ where $w' = 01011$. The successive transitions of s_0 are

$$s_0 \xrightarrow{0} s_0 \xrightarrow{1} s_1 \xrightarrow{0} s_2 \xrightarrow{1} s_0 \xrightarrow{1} s_1$$

so $f_{w'}(s_0) = s_1$. Similar displays show that $f_{w'}(s_1) = s_2$ and $f_{w'}(s_2) = s_0$.

This method of interpreting word transition functions such as f_w and $f_{w'}$ is useful in designing machines that have word transitions possessing certain desired properties. This is a crucial step in the practical application of the theory, and we will consider it in the next section.

Let $M = (S, I, \mathcal{F})$ be a finite-state machine. We define a function T from I^* to S^S. If w is a string in I^*, let $T(w) = f_w$ as defined above. Then we have the following result.

Theorem 1   (a) T is a monoid homomorphism from I^* to S^S.

(b) If $\mathcal{M} = T(I^*)$, then $\mathcal{M}$ is a submonoid of S^S.

Proof.

(a) Let $w_1 = x_1 x_2 \cdots x_k$ and $w_2 = y_1 y_2 \cdots y_m$ be two strings in I^*. Then $T(w_1 \cdot w_2) = T(x_1 x_2 \cdots x_k y_1 y_2 \cdots y_m) = (f_{x_1} \circ f_{x_2} \circ \cdots \circ f_{x_k}) \circ (f_{y_1} \circ f_{y_2} \circ \cdots \circ f_{y_m}) = T(w_1) \circ T(w_2)$. Also, $T(\Lambda) = 1_S$ by definition. Thus T is a monoid homomorphism.

(b) Follows from Theorem 5 of Section 6.2 and the fact that $1_S \in T(I^*)$. The monoid $\mathcal{M}$ is called the **monoid of the machine M**.

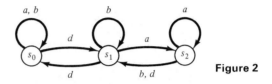

Figure 2

Example 3   Let $S = \{s_0, s_1, s_2\}$, $I = \{a, b, d\}$, and consider the finite-state machine $M = (S, I, \mathscr{F})$ defined by the digraph shown in Fig. 2. Compute the functions f_{bad}, f_{add}, and f_{badadd}, and verify that

$$f_{bad} \circ f_{add} = f_{badadd}$$

Solution. f_{bad} is computed by the following sequence of transitions:

$$s_0 \xrightarrow{b} s_0 \xrightarrow{a} s_0 \xrightarrow{d} s_1$$

$$s_1 \xrightarrow{b} s_1 \xrightarrow{a} s_2 \xrightarrow{d} s_1$$

$$s_2 \xrightarrow{b} s_1 \xrightarrow{a} s_2 \xrightarrow{d} s_1$$

Thus $f_{bad}(s_0) = s_1, f_{bad}(s_1) = s_1$, and $f_{bad}(s_2) = s_1$.

Similarly, for f_{add},

$$s_0 \xrightarrow{a} s_0 \xrightarrow{d} s_1 \xrightarrow{d} s_0$$

$$s_1 \xrightarrow{a} s_2 \xrightarrow{d} s_1 \xrightarrow{d} s_0$$

$$s_2 \xrightarrow{a} s_2 \xrightarrow{d} s_1 \xrightarrow{d} s_0$$

so $f_{add}(s_i) = s_0$ for $i = 0, 1, 2$. A similar computation shows that

$$f_{badadd}(s_0) = s_0, \quad f_{badadd}(s_1) = s_0, \quad f_{badadd}(s_2) = s_0$$

and the same formulas hold for $f_{bad} \circ f_{add}$. In fact,

$$(f_{bad} \circ f_{add})(s_0) = f_{add}(f_{bad}(s_0)) = f_{add}(s_1) = s_0$$
$$(f_{bad} \circ f_{add})(s_1) = f_{add}(f_{bad}(s_1)) = f_{add}(s_1) = s_0$$
$$(f_{bad} \circ f_{add})(s_2) = f_{add}(f_{bad}(s_2)) = f_{add}(s_1) = s_0$$

Example 4   Consider the machine whose digraph is shown in Fig. 3. Show that $f_w(s_0) = s_0$ if and only if w has $3k$ 1's for some $k \geq 0$.

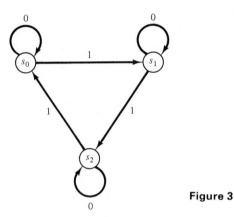

Figure 3

Solution.   From Fig. 3 we see that $f_0 = 1_S$, so that the 0's in a string $w \in I^*$ have no effect on f_w. Thus if $\bar{w}$ is w with all 0's removed, then $f_w = f_{\bar{w}}$. Let $l(w)$ denote the **length** of w, that is, the number of digits in w. Then $l(\bar{w})$ is the number of 1's in w, for all $w \in I^*$. For each $k \geq 0$ consider the statement

$P(k)$: Let $w \in I^*$ and let $l(\bar{w}) = n$.

(a) If $n = 3k, f_w(s_0) = s_0$.

(b) If $n = 3k + 1, f_w(s_0) = s_1$.

(c) If $n = 3k + 2, f_w(s_0) = s_2$.

We prove by mathematical induction that $P(k)$ is true for all $k \geq 0$.

Basis Step.   Suppose that $k = 0$. In case (a) $n = 0$, therefore w has no 1's and $f_w(s_0) = 1_S(s_0) = s_0$. In case (b) $n = 1$, so $\bar{w} = 1$ and $f_w(s_0) = f_{\bar{w}}(s_0) = f_1(s_0) = s_1$. Finally, in case (c) $n = 2$, so $\bar{w} = 11$, and $f_w(s_0) = f_{\bar{w}}(s_0) = f_{11}(s_0) = f_1(s_1) = s_2$.

Induction Step.   Suppose that $P(k)$ is true for some $k \geq 0$. We must show that $P(k + 1)$ is true. Let $w \in I^*$, and denote $l(\bar{w})$ by n. In case (a) $n = 3(k + 1) = 3k + 3$; therefore, $\bar{w} = 111 \cdot w'$, where $l(w') = 3k$. Then $f_{w'}(s_0) = s_0$ by the induction hypothesis, and $f_{111}(s_0) = s_0$ by direct computation, so $f_{\bar{w}}(s_0) = f_{w'}(f_{111}(s_0)) = f_{w'}(s_0) = s_0$. Cases (b) and (c) are handled in the same way.

By mathematical induction, $P(k)$ is true for all $k \geq 0$, so $f_w(s_0) = s_0$ if and only if the number of 1's in w is a multiple of 3.

Suppose now that $(S, I, \mathscr{F}, s_0, T)$ is a Moore machine. As in Chapter 5, we may think of certain subsets of I^* as "languages" with "words" from I. Using M, we can define such a subset, which we will denote by $L(M)$, and call the **language of the machine** M. Define $L(M)$ to be the set of all $w \in I^*$ such that $f_w(s_0) \in T$. In other words, $L(M)$ consists of all strings which, when used as input to the machine, cause the starting state s_0 to move to an "acceptance state" in T. Thus, in this sense, M "accepts" the string. It is for this reason that the states in T were named acceptance states in Section 7.1.

Example 5   Let $M = (S, I, \mathscr{F}, s_0, T)$ be the Moore machine in which $(S, I, \mathscr{F})$ is the finite-state machine whose digraph is shown in Fig. 3, and $T = \{s_1\}$. The discussion of Example 4 shows that $f_w(s_0) = s_1$ if and only if the number of 1's in w is of the form $3k + 1$ for some $k \geq 0$. Thus $L(M)$ is exactly the set of all strings with $3k + 1$ 1's for some $k \geq 0$.

Example 6   Consider the Moore machine M whose digraph is shown in Fig. 4. Here state s_0 is the starting state, and $T = \{s_2\}$. What is $L(M)$? Clearly, the input set is $I = \{a, b\}$. Observe that, in order for a string w to cause a transition from s_0 to s_2, w must contain at least two b's. After reaching s_2, any additional letters have no effect. Thus $L(M)$ is the set of all strings having two or more b's. We see, for example, that $f_{aabaa}(s_0) = s_1$, so $aabaa$ is rejected. On the other hand, $f_{abaab}(s_0) = s_2$, so $abaab$ is accepted.

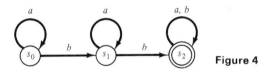

Figure 4

EXERCISE SET 7.2

In Exercises 1–5, we refer to the finite-state machine whose state transition table is given below.

	0	1
s_0	s_0	s_1
s_1	s_1	s_2
s_2	s_2	s_3
s_3	s_3	s_0

1. List the values of the transition function f_w for $w = 01001$.
2. List the values of the transition function f_w for $w = 11100$.
3. Describe the set of binary words (sequences of 0's and 1's) w having the property that $f_w(s_0) = s_0$.
4. Describe the set of binary words w having the property that $f_w = f_{010}$.
5. Describe the set of binary words w having the property that $f_w(s_0) = s_2$.

In Exercises 6–10, we refer to the finite-state machine whose digraph is given below.

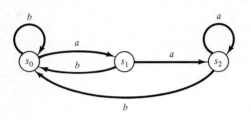

6. List the values of the transition function f_w for $w = abba$.

7. List the values of the transition function f_w for $w = babab$.

8. Describe the set of binary words w having the property that $f_w(s_0) = s_2$.

9. Describe the set of binary words w having the property that $f_w(s_0) = s_0$.

10. Describe the set of binary words w having the property that $f_w = f_{aba}$.

11. Describe the language recognized by the Moore machine whose digraph is shown below.

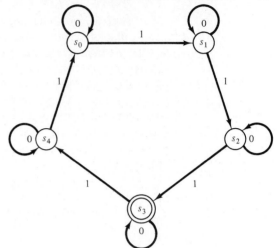

12. Describe the language recognized by the Moore machine whose state table is shown below, if $T = \{s_2\}$ and s_0 is the starting state.

	0	1
s_0	s_1	s_2
s_1	s_1	s_2
s_2	s_2	s_1

7.3 Machines and Regular Languages

Let $M = (S, I, \mathcal{F}, s_0, T)$ be a Moore machine. In Section 7.2 we defined the language, $L(M)$, of the machine M. It is natural to ask if there is a connection between such a language and the languages of phrase structure grammars, discussed in Section 5.3. The following theorem, due to S. Kleene, describes the connection.

Theorem 1 Let I be a set and let $L \subseteq I^*$. Then L is a type 3 language, that is, $L = L(G)$, where G is a type 3 grammar having I as its set of terminal symbols, if and only if $L = L(M)$ for some Moore machine $M = (S, I, \mathcal{F}, s_0, T)$.

We stated in Section 5.4 that a set $L \subseteq I^*$ is a type 3 language if and only if L is a regular set, that is, if and only if L corresponds to some regular expression over I. This leads to the following corollary of Theorem 1.

Corollary 1 Let I be a set and let $L \subseteq I^*$. Then $L = L(M)$, for some Moore machine $M = (S, I, \mathscr{F}, s_0, T)$, if and only if L is a regular set.

We will not give a complete and detailed proof of Theorem 1. However, it is easy to give a construction which produces a type 3 grammar from a given Moore machine. This is done in such a way that the grammar and the machine have the same language. Let $M = (S, I, \mathscr{F}, s_0, T)$ be a given Moore machine. We construct a type 3 grammar $G = (V, I, s_0, \mapsto)$ as follows. Let $V = I \cup S$; that is, I will be the set of terminal symbols for G, while S will be the set of nonterminal symbols. Let s_i and s_j be in S, and $x \in I$. We write $s_i \mapsto x s_j$, if $f_x(s_i) = s_j$, that is, if the input x takes state s_i to s_j. We also write $s_i \mapsto x$ if $f_x(s_i) \in T$, that is if the input x takes the state s_i to some acceptance state. Now let $\mapsto$ be the relation determined by the two conditions above, and take this relation as the production relation of G.

The grammar G constructed above has the same language as M. Suppose, for example, that $w = x_1 x_2 x_3 \in I^*$. The string w is in $L(M)$ if and only if $f_w(s_0) = f_{x_3}(f_{x_2}(f_{x_1}(s_0))) \in T$. Let $a = f_{x_1}(s_0)$, $b = f_{x_2}(a)$, and $c = f_{x_3}(b)$, where $c = f_w(s_0)$ is in T. Then the rules given above for constructing $\mapsto$ tell us that

1. $s_0 \mapsto x_1 a$
2. $a \mapsto x_2 b$
3. $b \mapsto x_3$

are all productions in G. The last one occurs because $c \in T$. If we begin with s_0, and substitute, using (1), (2), and (3) in succession, we see that $s_0 \Rightarrow^* x_1 x_2 x_3 = w$ (see Section 5.3), so $w \in L(G)$. A similar argument works for any string in $L(M)$, so $L(M) \subseteq L(G)$. If we reverse the argument given above, we can see that we also have $L(G) \subseteq L(M)$. Thus M and G have the same language.

Example 1 Consider the Moore machine M shown in Fig. 4 of Section 7.2. Construct a type 3 grammar G such that $L(G) = L(M)$. Also, find a regular expression over $I = \{a, b\}$ which corresponds to $L(M)$.

Solution. Let $I = \{a, b\}$, $S = \{s_0, s_1, s_2\}$, and $V = I \cup S$. We construct the grammar $(V, I, s_0, \mapsto)$, where $\mapsto$ is described below.

$$\mapsto: \quad \begin{array}{ll} s_0 \mapsto a s_0 & s_2 \mapsto b s_2 \\ s_0 \mapsto b s_1 & s_1 \mapsto b \\ s_1 \mapsto a s_1 & s_2 \mapsto a \\ s_1 \mapsto b s_2 & s_2 \mapsto b \\ s_2 \mapsto a s_2 & \end{array}$$

The production relation $\mapsto$ is constructed as we indicated previously; therefore, $L(M) = L(G)$.

If we consult Fig. 4 of Section 7.2, we see that a string $w \in L(M)$ has the following properties. Any number $n \geq 0$ of a's can occur at the beginning of w. At some point, a b must occur, in order to cause the transition from s_0 to s_1. After this b, any number $k \geq 0$ of a's may occur, followed by another b to cause transition to s_2. The remainder of w, if any, is completely arbitrary, since the machine cannot leave s_2 after once entering this state. From this description we can readily see that $L(M)$ corresponds to the regular expression

$$a*ba*b(a \lor b)*$$

Example 2 Consider the Moore machine whose digraph is shown in Fig. 1. Describe, in English, the language $L(M)$. Then construct the regular expression which corresponds to $L(M)$, and describe the productions of the corresponding grammar G in BNF form.

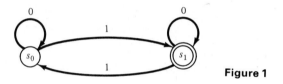

Figure 1

Solution. It is clear that 0's in the input string have no effect on the states. If an input string w has an odd number of 1's, then $f_w(s_0) = s_1$. If w has an even number of 1's, then $f_w(s_0) = s_0$. Since $T = \{s_1\}$, we see that $L(M)$ consists of all w in I^* which have an odd number of 1's as components.

We now find the regular expression corresponding to $L(M)$. Any input string corresponding to the expression 0*10* will be accepted, since it will have exactly one 1. If an input w begins in this way, but has more 1's, then the additional ones must come in pairs, with any number of 0's allowed between, or after each pair of 1's. The previous sentence describes the set of strings corresponding to the expression (10*10*)*. Thus $L(M)$ corresponds to the regular expression

$$0*10*(10*10*)*$$

Finally, the type 3 grammar constructed from M is $G = (V, I, s_0, \mapsto)$ with $V = I \cup S$. The BNF of the relation $\mapsto$ is

$$\langle s_0 \rangle ::= 0\langle s_0 \rangle \,|\, 1\langle s_1 \rangle \,|\, 1$$

$$\langle s_1 \rangle ::= 0\langle s_1 \rangle \,|\, 1\langle s_0 \rangle \,|\, 0$$

Occasionally, we may need to determine the function performed by a given Moore machine, as we did in the examples above. More commonly, however, it is necessary to construct a machine that will perform a given task. This task may be defined by giving an ordinary English description, a regular expression, or an equivalent type 3 grammar, perhaps in BNF or with a syntax diagram. There are systematic, almost mechanical ways to construct such a machine. Most of these use the concept of "nondeterministic machines," and employ a tedious translation process from such machines to the Moore machines that we have discussed. The reader may find discussions of these methods in the guide to further reading at the end of this chapter. If the task of the machine is not too complex, we may use simple reasoning to construct the machine in steps, usually in the form of its digraph. Whichever method is used, the resulting machine may be quite inefficient; for example, it may have unneeded states. We will give, in Section 7.4, a procedure for constructing an equivalent machine which may be much more efficient.

Example 3   Construct a Moore machine M that will accept exactly the string 001 from input strings of 0's and 1's. In other words, $I = \{0, 1\}$ and $L(M) = \{001\}$.

Solution. We must begin with a starting state s_0. If w is an input string of 0's and 1's, and if w begins with a 0, then w "may" be accepted (depending on the remainder of its components). Thus one step toward acceptance has been taken, and there needs to be a state s_1 that corresponds to this step. We therefore begin as in Fig. 2(a). If we next receive another 0, we have progressed one more step toward acceptance. We therefore construct another state s_2 and let 0 give a transition from s_1 to s_2. State s_1 represents the condition "first component of input is a 0," whereas state s_2 represents the condition "first two components of the input are respectively 00." This situation is shown in Fig. 2(b). Finally, if the third input component is a 1, we move to an acceptance state, as shown in Fig. 2(c). Any other beginning sequence of input digits, or any additional digits will move us to a "failure state" s_4 from which there is no escape. Thus Fig. 2(d) shows the completed machine.

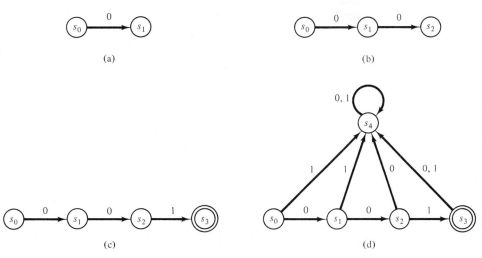

Figure 2

The process illustrated in Example 3 is difficult to precisely describe or generalize. We try to construct states representing each successive stage of input complexity, leading up to an acceptable string. There must also be states indicating the ways in which a promising input pattern may be destroyed when a certain component is received. If the machine is to recognize several, essentially different, types of input, then we will need to construct separate "branches" corresponding to each type of input. This process may result in some redundancy, but the machine can be simplified later.

Example 4   Let $I = \{0, 1\}$. Construct a Moore machine which accepts those input sequences w that contain the string 01 or the string 10 anywhere within them. In other words, we are to accept exactly those strings which do not consist entirely of 0's, or entirely of 1's.

Solution. This is a simple example in which, whatever input digit is received first, a string will be accepted if and only if the other digit is eventually received. There must be a starting state s_0, states s_1 and s_2 corresponding respectively to first receiving a 0 or 1, and (acceptance) states s_3 and s_4, which will be reached if and when the other digit is received. Having once reached an acceptance state, the machine stays in that state. Thus we construct the digraph of this machine as shown in Fig. 3.

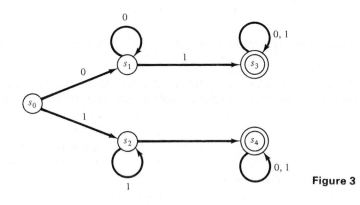

Figure 3

In Example 3, once an acceptance state is reached, any additional input will cause a permanent transition to a nonaccepting state. In Example 4, once an acceptance state is reached, any additional input will have no effect. Sometimes the situation is between these two extremes. As input is received, the machine may repeatedly enter and leave acceptance states. Consider the Moore machine M whose digraph is shown in Fig. 4. This machine is a slight modification of the finite-state machine given in Example 4 of Section 7.2. We know from that example that $w \in L(M)$ if and only if the number of 1's in w is of the form $3k$, $k \geq 0$. As input components are received, M may enter and leave s_0 repeatedly. The conceptual

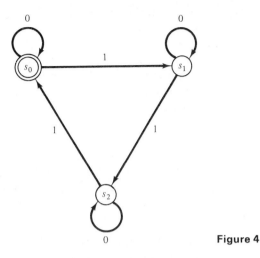

Figure 4

states "one 1 has been received" and "four 1's have been received" may both be represented by s_1. When constructing machines, one should keep in mind the fact that a state, previously defined to represent one conceptual input condition, may be used for a new input condition, if these two conditions represent the same "degree of progress" of the input stream toward acceptance. The next example illustrates this fact.

Example 5   Construct a Moore machine that accepts exactly those input strings of x's and y's that end in yy.

Solution. Again we need a starting state s_0. If the input string begins with a y, we progress one step to a new state s_1 ("last input component received is a y"). On the other hand, if the input begins with an x, we have made no progress toward acceptance. Thus we may suppose that M is again in state s_0. This situation is shown in Fig. 5(a). If, while in state s_1, a y is received, we progress to an acceptance state s_2 ("last two components of input received were y's"). If instead the input received is an x, we must again receive two y's in order to be in an acceptance state. Thus we may again regard this as a return to state s_0. The situation at this point is shown in Fig. 5(b). Having reached state s_2, an additional input of y will have no effect, but an input of x will necessitate two more y's for acceptance. Thus we can again regard M as being in state s_0. The final Moore machine is shown in Fig. 5(c).

We have not mentioned the question of implementation of finite-state machines. Indeed, many such machines, including all digital computers, are implemented as hardware devices, that is, as electronic circuitry. There are, however, many occasions when finite-state machines are simulated in software. This is frequently seen in compilers and interpreters, where Moore machines may be programmed to retrieve and interpret words and symbols in an input string. We provide just a hint

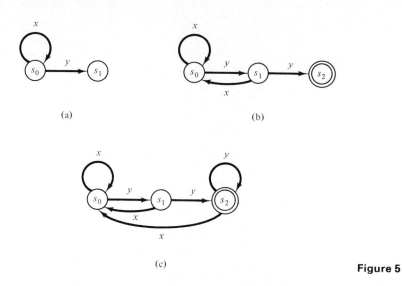

(a)

(b)

(c)

Figure 5

of the techniques available by simulating the machine of Example 2 in pseudocode. The reader should refer back to Example 2 and Fig. 1 for the details of the machine. Figure 6 gives a pseudocode program for this machine. This program uses a subroutine INPUT to get the next 0 or 1 in variable X and assumes that a logical variable EOI will be set true when no further input is available. The variable RESULT will be true if the input string contains an odd number of 1's; otherwise, it will be false.

```
SUBROUTINE ODDONES (RESULT)
1. EOI ← F
2. RESULT ← F
3. STATE ← 0
4. UNTIL (EOI)
    a. CALL INPUT (X, EOI)
        1. IF (EOI = F) THEN
            a. IF (STATE = 0) THEN
                1. IF (X = 1) THEN
                    a. RESULT ← T
                    b. STATE ← 1
            b. ELSE
                1. IF (X = 1) THEN
                    a. RESULT → F
                    b. STATE ← 0
5. RETURN
   END OF SUBROUTINE ODDONES
```

Figure 6

In this particular coding technique, a state is denoted by a variable which may be assigned different values depending on input, and whose values then determine other effects of the input. An alternative procedure is to represent a state by a particular location in code. This location then determines the effect of input, and the branch to a new location (subsequent state). Figure 7 shows the same subroutine

SUBROUTINE ODDONES (RESULT) — version 2
    1. RESULT ← F
S0: 2. CALL INPUT (X, EOI)
    3. IF (EOI) THEN
       a. RETURN
    4. ELSE
       a. IF (X = 1) THEN
          1. RESULT ← T
          2. GO TO S1
       b. ELSE
          1. GO TO S0
S1: 5. CALL INPUT (X, EOI)
    6. IF (EOI) THEN
       a. RETURN
    7. ELSE
       a. IF (X = 1) THEN
          1. RESULT ← F
          2. GO TO S0
       b. ELSE
          1. GO TO S1
END OF SUBROUTINE ODDONES — version 2   **Figure 7**

ODDONES coded in this alternative way. It is awkward to avoid GO TO statements in this approach, and we have used them. In languages with "multiple" GO
TO statements, such as FORTRAN's indexed GO TO or Pascal's CASE statement,
this method may be particularly efficient for finite-state machines with a fairly large
number of states. In such cases, the first method may become quite cumbersome.

EXERCISE SET 7.3

In Exercises 1–4, describe (in words) the language
recognized by the Moore machines whose digraphs are
given.

1.

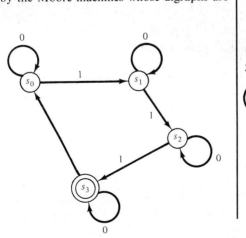

2.

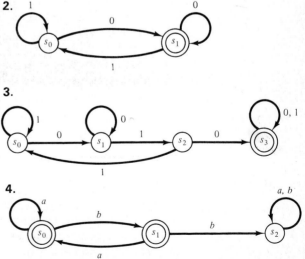

3.

4.

In Exercises 5 and 6, describe (in words) the language recognized by the Moore machines whose state tables are given. The starting state is s_0, and the set T, of acceptance states, is shown.

5.

	0	1
s_0	s_1	s_0
s_1	s_1	s_2
s_2	s_1	s_0

$T = \{s_2\}$

6.

	0	1
s_0	s_0	s_1
s_1	s_0	s_1

$T = \{s_1\}$

7. Let M be the Moore machine of Exercise 1. Construct a type 3 grammar $G = (V, I, s_0, \mapsto)$, such that $L(M) = L(G)$.

8. Let M be the Moore machine of Exercise 2. Give a regular expression over $I = \{0, 1\}$, which corresponds to the language $L(M)$.

9. Let M be the Moore machine of Exercise 6. Give a regular expression over $I = \{0, 1\}$, which corresponds to the language $L(M)$.

10. Let M be the Moore machine of Exercise 3. Construct a type 3 grammar $G = (V, I, s_0, \mapsto)$, such that $L(M) = L(G)$. Describe $\mapsto$ in BNF.

11. Let M be the Moore machine of Exercise 4. Construct a type 3 grammar $G = (V, I, s_0, \mapsto)$, such that $L(M) = L(G)$. Describe $\mapsto$ in BNF.

12. Let M be the Moore machine of Exercise 5. Construct a type 3 grammar $G = (V, I, s_0, \mapsto)$, such that $L(M) = L(G)$. Describe $\mapsto$ by a syntax diagram.

In Exercises 13–26, construct the digraph of a Moore machine that accepts the input strings described, and no others.

13. Inputs a, b: strings where the number of b's is divisible by 3.

14. Inputs a, b: strings where the number of a's is even and the number of b's is a multiple of 3.

15. Inputs x, y: strings that have an even number of y's.

16. Inputs 0, 1: strings that contain 0011.

17. Inputs 0, 1: strings that end with 0011.

18. Inputs $\square$, $\triangle$: strings that contain $\square\triangle$ or $\triangle\square$.

19. Inputs $+$, $\times$: strings that contain $+\times\times$ or $\times++$.

20. Inputs w, z: strings that contain wz or zzw.

21. Inputs a, b: strings that contain ab and end in bbb.

22. Inputs $+$, $\times$: strings that end in $+\times\times$.

23. Inputs w, z: strings that end in wz or zzw.

24. Inputs 0, 1, 2: string 0120 is the only string recognized.

25. Inputs a, b, c: strings aab or abc are to be recognized.

26. Inputs x, y, z: strings xzx or yx or zyx are to be recognized.

In Exercises 27–30, construct the state table of a Moore machine that recognizes the given input strings, and no others.

27. Inputs 0, 1: strings ending in 0101.

28. Inputs a, b: strings where the number of b's is divisible by 4.

29. Inputs x, y: strings having exactly two x's.

30. Inputs a, b: strings that do not have two successive b's.

31. Let $\mathcal{M} = (S, I, \mathcal{F}, s_0, T)$ be a Moore machine. Define a relation R on S as follows: $s_i \, R \, s_j$ if and only if $f_w(s_i)$ and $f_w(s_j)$ either both belong to T or neither does, for every $w \in I^*$. Show that R is an equivalence relation on S.

32. Construct a subroutine, in pseudocode, which simulates the Moore machine of Exercise 2.

33. Construct a subroutine, in pseudocode, which simulates the Moore machine of Exercise 4.

7.4 Simplification of Machines

As we have seen in the preceding section, the construction of a finite state machine to perform a given task is as much an art as a science. Generally, graphical methods are first used, and states are constructed for all "intermediate steps" in the process.

Not surprisingly, a machine constructed in this way may not be efficient, and we need to find a method for obtaining an equivalent, more efficient machine. Fortunately, there is a method available that is quite systematic (and can even be computerized), and that method will take any correct machine, however redundant it is, and produce an equivalent machine which is usually more efficient. Here we will use the number of states as our measure of efficiency. We will demonstrate this technique for Moore machines, but the principles extend, with small changes, to various other types of finite-state machines. Let $(S, I, \mathscr{F}, s_0, T)$ be a Moore machine, and k a nonnegative integer. We define a relation R on S as follows: For any $s, t \in S$ and $w \in I^*$, we say that s and t are w-**compatible** if $f_w(s)$ and $f_w(t)$ both belong to T, or neither does. Let $s \, R \, t$ mean that s and t are w-compatible for all $w \in I^*$.

Theorem 1   Let $(S, I, \mathscr{F}, s_0, T)$ be a Moore machine, and let R be the relation defined above.

      (a) R is an equivalence relation on S.

      (b) R is a machine congruence (see Section 7.1).

Proof.

(a) R is clearly reflexive and symmetric. Suppose now that $s \, R \, t$ and $t \, R \, u$ for $s, t,$ and u in S, and let $w \in I^*$. Then s and t are w-compatible, as are t and u, so if we consider $f_w(s), f_w(t), f_w(u)$, it follows that either all belong to T, or all belong to $\bar{T}$. Thus s and u are w-compatible, so R is transitive, and therefore R is an equivalence relation.

(b) We must show that if s and t are in S and $x \in I$, then $s \, R \, t$ implies that $f_x(s) \, R \, f_x(t)$. To show this, let $w \in I^*$, and let $w' = x \cdot w$ ($\cdot$ is the operation of catenation). Since $s \, R \, t$, $f_{w'}(s)$ and $f_{w'}(t)$ are both in T or both in $\bar{T}$. But $f_{w'}(s) = f_{x \cdot w}(s) = f_w(f_x(s))$ and $f_{w'}(t) = f_{x \cdot w}(t) = f_w(f_x(t))$, so that $f_x(s)$ and $f_x(t)$ are w-compatible. Since w is arbitrary in I^*, $f_x(s) \, R \, f_x(t)$.

Since R is a machine congruence, we may form the quotient Moore machine $\bar{M} = (S/R, I, \bar{\mathscr{F}}, [s_0], T/R)$ as in Section 7.1. The machine $\bar{M}$ is the efficient version of M which we have promised. We will show that $\bar{M}$ is **equivalent** to M, meaning that $L(\bar{M}) = L(M)$.

Example 1   Consider the Moore machine whose digraph is shown in Fig. 1.

In this machine $I = \{0, 1\}$. The starting state is s_0, and $T = \{s_2, s_3\}$. Let us compute the quotient machine $\bar{M}$. First, we see that $s_0 \, R \, s_1$. In fact, $f_w(s_0) \in T$ if and only if w contains at least one 1, and $f_w(s_1) \in T$ under precisely the same condition. Thus s_0 and s_1 are w-compatible for all $w \in I^*$. Now $s_2 \, \cancel{R} \, s_0$ and $s_3 \, \cancel{R} \, s_0$, since $f_0(s_2) \in T$, $f_0(s_3) \in T$, but $f_0(s_0) \notin T$. This implies that $\{s_0, s_1\}$ is one R-equivalence class. Also $s_2 \, R \, s_3$, since $f_w(s_2) \in T$ and $f_w(s_3) \in I$ for all $w \in I^*$. This proves that

$$S/R = \{\{s_0, s_1\}, \{s_2, s_3\}\} = \{[s_0], [s_2]\}.$$

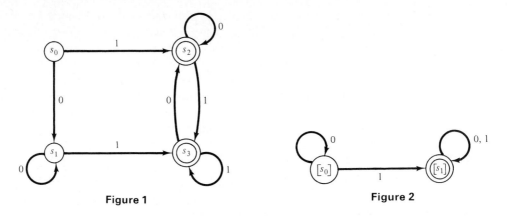

Figure 1   Figure 2

Also note that $T/R = \{[s_2]\}$. The resulting quotient Moore machine $\bar{M}$ is equivalent to M and its digraph is shown in Fig. 2.

In this case it is clear that M and $\bar{M}$ are equivalent since each accepts a word w if and only if w has at least one 1. In general, we have the following result.

Theorem 2   Let $M = (S, I, \bar{\mathscr{F}}, s_0, T)$ be a Moore machine, let R be the equivalence relation defined above, and let $\bar{M} = (S/R, I, \mathscr{F}, [s_0], T/R)$ be the corresponding quotient Moore machine. Then $L(\bar{M}) = L(M)$.

Proof. Suppose that w is accepted by M, so that $f_w(s_0) \in T$. Then $\bar{f}_w([s_0]) = [f_w(s_0)] \in T/R$; that is, $\bar{M}$ also accepts w.

Conversely, suppose that $\bar{M}$ accepts w, so that $\bar{f}_w([s_0]) = [f_w(s_0)]$ is in T/R. This means that $t \; R \; f_w(s_0)$ for some element t in T. By definition of R, we know that t and $f_w(s_0)$ are w' compatible, for every $w' \in I^*$. If $w' = \Lambda$, the empty string, then $f_{w'} = 1_S$, so $t = f_{w'}(t)$ and $f_w(s_0) = f_{w'}(f_w(s_0))$ are both in T or both in $\bar{T}$. Since $t \in T$, we must have $f_w(s_0) \in T$, so M accepts w.

Thus we see that after initially designing the Moore machine M, we may compute R and pass to the quotient machine $\bar{M} = M/R$, thereby obtaining an equivalent machine which may be considerably more efficient, in the sense that it may have many fewer states. Often the quotient machine is one that would have been difficult to discover at the outset.

We now need an algorithm for computing the relation R. In Example 1 we found R by direct analysis of input, but this example was chosen to be particularly simple. In general, a direct analysis will be very difficult. We now define and investigate a set of relations which provide an effective method for computing R.

If k is a nonnegative integer, we define a relation R_k on S, the state set of a Moore machine $(S, I, \mathscr{F}, s_0, T)$. If $w \in I^*$, recall that $l(w)$ is the length of the string w, that is, the number of symbols in w. Note that $l(\Lambda) = 0$. Then if s and $t \in S$, we let $s \; R_k \; t$ mean that s and t are w-compatible for all $w \in I^*$ with $l(w) \leq k$. The relations R_k are not machine congruences, but are successive approximations to the congruence R.

Theorem 3   (a) $R_k \supseteq R_{k+1}$ for all $k \geq 0$.

(b) Each R_k is an equivalence relation.

(c) $R_k \supseteq R$ for all $k \geq 0$.

Proof. If s, $t \in S$, and s and t are w-compatible for all $w \in I^*$ or for all w with $l(w) \leq k + 1$, then in either case, s and t are compatible for all w with $l(w) \leq k$. This proves (a) and (c). The proof of (b) is similar to the proof of Theorem 1(a), and we omit it.

The key result for computing the relations R_k recursively is the following theorem.

Theorem 4   (a) $S/R_0 = \{T, \bar{T}\}$

(b) Let k be a nonnegative integer and s, $t \in S$. Then $s\ R_{k+1}\ t$ if and only if:

    (i) $s\ R_k\ t$

    (ii) $f_x(s)\ R_k\ f_x(t)$ for all $x \in I$

Proof.

(a) Since only Λ has length 0, it follows that $s\ R_0\ t$ if and only if both s and t are in T, or both are in $\bar{T}$. This proves that $S/R_0 = \{T, \bar{T}\}$.

(b) Let $w \in I^*$ be such that $l(w) \leq k + 1$. Then $w = x \cdot w'$, for some $x \in I$ and for some $w' \in I^*$ with $l(w') \leq k$.

Conversely, if any $x \in I$ and $w' \in I^*$ with $l(w') \leq k$ are chosen, the resulting string $w = x \cdot w'$ has length less than or equal to $k + 1$.

Now $f_w(s) = f_{x \cdot w'}(s) = f_{w'}(f_x(s))$ and $f_w(t) = f_{w'}(f_x(t))$, for any s, t in S. This equation shows that s and t are w-compatible for any $w \in I^*$ with $l(w) \leq k + 1$ if and only if $f_x(s)$ and $f_x(t)$ are, for all $x \in I$, w'-compatible, for any w' with $l(w') \leq k$. That is, $s\ R_{k+1}\ t$ if and only if $f_x(s)\ R_k\ f_x(t)$ for all $x \in I$.

Now either of these equivalent conditions implies that $s\ R_k\ t$, since $R_{k+1} \subseteq R_k$, so we have proved the theorem.

This result says that we may find the partitions P_k, corresponding to the relations R_k, by the following recursive method:

Step 1. Begin with $P_0 = \{T, \bar{T}\}$.

Step 2. Having reached partition $P_k = \{A_1, A_2, \ldots, A_m\}$, examine each equivalence class A_i, and break it into pieces where two elements s and t of A_i fall into the same piece if all inputs x take both s and t into the same subset A_j (depending on x).

Step 3. The new partition of S, obtained by taking all pieces of all of the A_i, will be P_{k+1}.

The final step in the method above, telling us when to stop, is given by the following result.

Theorem 5   If $R_k = R_{k+1}$ for any nonnegative integer k, then $R_k = R$.

Proof. Suppose that $R_k = R_{k+1}$. Then by Theorem 4, $s\ R_{k+2}\ t$ if and only if $f_x(s)\ R_{k+1}\ f_x(t)$ for all $x \in I$, or (since $R_k = R_{k+1}$) if and only if $f_x(s)\ R_k\ f_x(t)$ for all $x \in I$. This happens if and only if $s\ R_{k+1}\ t$. Thus $R_{k+2} = R_{k+1} = R_k$. By induction, it follows that $R_k = R_n$ for all $n \geq k$. Now it is easy to see that $R = \bigcap_{n=0}^{\infty} R_n$, since every string w in I^* must have some finite length. Since $R_1 \supseteq R_2 \cdots \supseteq R_k = R_{k+1} = \ldots$, the intersection of the R_n's is exactly R_k, so $R = R_k$.

A procedure for reducing a given Moore machine to an equivalent machine is as follows.

Step 1. Start with the partition $P_0 = \{T, \bar{T}\}$ of S.

Step 2. Construct successive partitions $P_1, P_2, \ldots$ corresponding to the equivalence relations $R_1, R_2, \ldots$ by using the method outlined after Theorem 4.

Step 3. Whenever $P_k = P_{k+1}$, stop. The resulting partition $P = P_k$ corresponds to the relation R.

Step 4. The resulting quotient machine is equivalent to the given Moore machine.

Example 2   Consider the machine of Example 1. Here $S = \{s_0, s_1, s_2, s_3\}$, $T = \{s_2, s_3\}$. We use the method above to compute an equivalent quotient machine. First, $P_0 = \{\{s_0, s_1\}, \{s_2, s_3\}\}$. We must decompose this partition in order to find P_1. Consider first the set $\{s_0, s_1\}$. Input 0 takes each of these states into $\{s_0, s_1\}$. Input 1 takes both s_0 and s_1 into $\{s_2, s_3\}$. Thus the equivalence class $\{s_0, s_1\}$ does not decompose in passing to P_1. We also see that input 0 takes both s_2 and s_3 into $\{s_2, s_3\}$, and input 1 takes both s_2 and s_3 into $\{s_2, s_3\}$. Again, the equivalence class $\{s_2, s_3\}$ does not decompose in passing to P_1. This means that $P_1 = P_0$, so P_0 corresponds to the congruence R. We found this result directly in Example 1.

Example 3   Let M be the Moore machine shown in Fig. 3. Find the relation R and draw the digraph of the corresponding quotient machine $\bar{M}$.

Solution. The partition $P_0 = \{T, \bar{T}\} = \{\{s_0, s_5\}, \{s_1, s_2, s_3, s_4\}\}$. Consider first the set $\{s_0, s_5\}$. Input 0 carries both s_0 and s_5 into T, and input 1 carries both into $\bar{T}$. Thus $\{s_0, s_5\}$ does not decompose further in passing to P_1. Next consider the set $\bar{T} = \{s_1, s_2, s_3, s_4\}$. State s_1 is carried to $\bar{T}$ by input 0 and to T by input 1. This is also true for state s_4, but not for s_2 and s_3, so the equivalence class of s_1 in P_1 will be $\{s_1, s_4\}$. Since states s_2 and s_3 are carried into $\bar{T}$ by inputs 0 and 1, they will also form an equivalence class in P_1. Thus $\bar{T}$ has decomposed into the subsets $\{s_1, s_4\}$ and $\{s_2, s_3\}$ in passing to P_1, and $P_1 = \{\{s_0, s_5\}, \{s_1, s_4\}, \{s_2, s_3\}\}$.

In order to find P_2, we must examine each subset of P_1 in turn. Consider $\{s_0, s_5\}$. Input 0 takes s_0 and s_5 to $\{s_0, s_5\}$, and input 1 takes each of them to $\{s_1, s_4\}$. This means that $\{s_0, s_5\}$ does not further decompose in passing to P_2. A

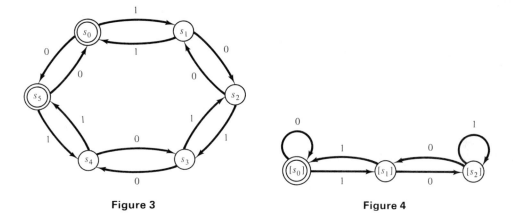

Figure 3   Figure 4

similar argument shows that neither of the sets $\{s_1, s_4\}$ and $\{s_2, s_3\}$ decomposes, so that $P_2 = P_1$. Hence P_1 corresponds to R. The resulting quotient machine is shown in Fig. 4. It can be shown (we omit the proof) that each of these machines will accept a string $w = b_1 b_2 \cdots b_n$ in $\{0, 1\}^*$ if and only if w is the binary representation of a number that is evenly divisible by 3.

EXERCISE SET 7.4

In Exercises 1–7, find the specified relation R_k for the Moore machine whose digraph is given.

1.   Find R_0.

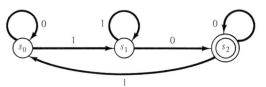

2.   In Exercise 1, find R_1.
3.   Find R_1.

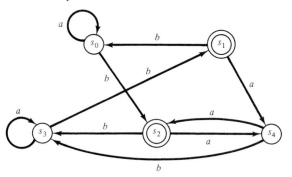

4.   In Exercise 3, find R_2.
5.   In Exercise 3, find R_{127}.
6.   Find R_1.

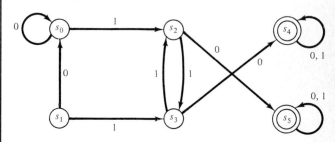

7.   In Exercise 6, find R_2.
8.   Find R for the machine of Exercise 1.
9.   Find R for the machine of Exercise 3.
10. Find R for the machine of Exercise 6.

In Exercises 11 and 12, find the partition corresponding to the relation R, and construct the state table of the corresponding quotient machine that is equivalent to the Moore machine whose state table is shown.

11.

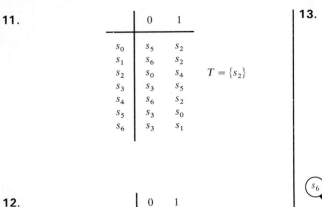

	0	1
s_0	s_5	s_2
s_1	s_6	s_2
s_2	s_0	s_4
s_3	s_3	s_5
s_4	s_6	s_2
s_5	s_3	s_0
s_6	s_3	s_1

$T = \{s_2\}$

12.

	0	1
a	a	c
b	g	d
c	f	e
d	a	d
e	a	d
f	g	f
g	g	c

$s_0 = a$

$T = \{d, e\}$

13. Find the relation R, and construct the digraph of the corresponding equivalent quotient machine, for the Moore machine whose digraph is shown.

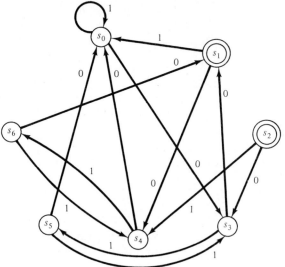

KEY IDEAS FOR REVIEW

☐ Finite-state machine: $(S, I, \mathscr{F})$, where S is a finite set of states, I is a set of inputs, and $\mathscr{F} = \{f_x \mid x \in I\}$.

☐ State table: see page 325.

☐ R_M: $s_i \; R_M \; s_j$ if there is an input x so that $f_x(s_i) = s_j$.

☐ Machine congruence R on M: For any $s, t \in S$, $s \; R \; t$ implies that $f_x(s) \; R \; f_x(t)$ for all $x \in I$.

☐ Quotient of M corresponding to R: see page 328.

☐ State transition function f_w, $w = x_1 x_2 \cdots x_n$: $f_w = f_{x_1} \circ f_{x_2} \circ \cdots \circ f_{x_n}, f_\Lambda = 1_S$.

☐ Theorem: Let $M = (S, I, \mathscr{F})$ be a finite state machine. Define $T: I^* \to S^S$ by the equations $T(w) = f_w$, $w \neq \Lambda$, and $T(\Lambda) = 1_S$. Then:
  (a) T is a monoid homomorphism from I^* to S^S.
  (b) If $\mathscr{M} = T(I^*)$, then $\mathscr{M}$ is a submonoid of S^S.

☐ Monoid of a machine: $\mathscr{M}$ in theorem above.

☐ Recognition or Moore machine: $M = (S, I, \mathscr{F}, s_0, T)$, where $s_0 \in S$ is the starting state and $T \subseteq S$ is the set of acceptance states.

☐ $l(w)$: length of the string w.

☐ Language accepted by M: $L(M) = \{w \in I^* \mid f_w(s_0) \in T\}$.

☐ Theorem: Let I be a set and $L \subseteq I^*$. Then L is a type 3 language, that is, $L = L(G)$

where G is a type 3 grammar having I as its set of terminal symbols, if and only if $L = L(M)$ for some Moore machine $M = (S, I, \mathscr{F}, s_0, T)$.

☐ w-compatible: see page 345.

☐ R: s R t if s and t are w-compatible for all $w \in I^*$.

☐ Theorem: Let $(S, I, \mathscr{F}, s_0, T)$ be a Moore machine, and let R be the relation defined above. Then (a) R is an equivalence relation on S.

(b) R is a machine congruence (see Section 7.1).

☐ Theorem: Let $M = (S, I, \mathscr{F}, s_0, T)$ be a Moore machine, let R be the equivalence relation defined above, and let $\bar{M} = (S/R, I, \mathscr{F}, [s_0], T/R)$ be the corresponding quotient Moore machine. Then $L(M) = L(\bar{M})$.

☐ R_k: s R_k t if s and t are w-compatible for all $w \in I^*$ with $l(w) \leq k$.

☐ Theorem: (a) $R_k \supseteq R_{k+1}$ for all $k \geq 0$.

(b) Each R_k is an equivalence relation.

(c) $R_k \supseteq R$ for all $k \geq 0$.

☐ Theorem: (a) $S/R_0 = \{T, \bar{T}\}$

(b) Let k be a nonnegative integer and $s, t \in S$. Then s R_{k+1} t if and only if:

(i) s R_k t

(ii) $f_x(s)$ R_k $f_x(t)$ for all $x \in I$

☐ Theorem: If $R_k = R_{k+1}$ for any nonnegative integer k, then $R_k = R$.

FURTHER READING

AHO, ALFRED V., and JEFFREY D. ULLMAN, *Principles of Compiler Design*, Addison-Wesley, Reading, Mass., 1977.

ARBIB, M. A., *Algebraic Theory of Machines, Languages, and Semigroups*, Academic Press, New York, 1969.

HOPCROFT, JOHN E., and JEFFREY D. ULLMAN, *Introduction to Automata Theory, Languages, and Computation*, Prentice-Hall, Englewood Cliffs, N.J., 1979.

KOHAVI, Z., *Switching and Finite Automata Theory*, McGraw-Hill, New York, 1970.

MINSKY, MARVIN L., *Computation: Finite and Infinite Machines*, Prentice-Hall, Englewood Cliffs, N.J., 1967.

Groups and Coding

Prerequisites: Chapters 1, 2, 3, and 6.

In today's modern world of communication, data items are constantly being transmitted from point to point. This transmission may range from the simple task of a computer terminal interacting with the mainframe computer located 200 feet away, to the more complicated task of sending a signal thousands of miles away via a satellite which is parked in an orbit 20,000 miles from the earth, or to a telephone call or letter to another part of the country. The basic problem in transmission of data is that of receiving the data as sent and not receiving a distorted piece of data. Distortion can be caused by a number of factors.

Coding theory has developed techniques for introducing redundant information in transmitted data which help in detecting, and sometimes in correcting, errors. Some of these techniques make use of group theory. We present a brief introduction to these ideas in this chapter.

8.1 Coding of Binary Information and Error Detection

The basic unit of information, called a **message**, is a finite sequence of characters from a finite alphabet. We shall choose our alphabet as the set $B = \{0, 1\}$. Every character or symbol that we want to transmit is now represented as a sequence of m elements from B. That is, every character or symbol is represented in binary form. Our basic unit of information, called a **word**, is a sequence of m 0's and 1's.

Table 1

+	0	1
0	0	1
1	1	0

The set B is a group under the binary operation $+$ whose multiplication table is shown in Table 1. (See Example 5 in Section 6.4.) If we think of B as the group Z_2, then $+$ is merely "mod 2" addition. It follows from Theorem 1 of Section 6.5 that $B^m = B \times B \times \cdots \times B$ (m factors) is a group under the operation $\oplus$ defined by

$$(x_1, x_2, \ldots, x_m) \oplus (y_1, y_2, \ldots, y_m) = (x_1 + y_1, x_2 + y_2, \ldots, x_m + y_m)$$

This group has been introduced in Example 2 of Section 6.5. Its identity is $\bar{0} = (0, 0, \ldots, 0)$ and every element is its own inverse. An element in B^m will be written as $(b_1, b_2, \ldots, b_m)$ or more simply as $b_1 b_2 \cdots b_m$. Observe that B^m has 2^m elements. That is, the order of the group B^m is 2^m.

Figure 1 shows the basic process of sending a word from one point to another point over a transmission channel. An element $x \in B^m$ is sent through the transmission channel and is received as an element $x_t \in B^m$. In actual practice, the transmission channel may suffer disturbances which are generally called **noise**, due to weather interference, electrical problems, and so on, that may cause a 0 to be received as a 1, or vice versa. This erroneous transmission of digits in a word being sent may give rise to the situation where the word received is different from the word that was sent; that is, $x \neq x_t$. If an error does occur, then actually x_t could be any element of B^m.

Figure 1

The basic task in the transmission of information is to reduce the likelihood of receiving a word that differs from the word that was sent. This is done as follows.

We first choose an integer $n > m$ and a one-to-one function $e : B^m \to B^n$. The function e is called an (m, n) **encoding function**, and we view it as a means of representing every word in B^m as a word in B^n. If $b \in B^m$, then $e(b)$ is called the **code word** representing b. The additional 0's and 1's can provide the means to detect or correct errors produced in the transmission channel.

We now transmit the code words by means of a transmission channel. Then each code word $x = e(b)$ is received as the word x_t in B^n. This situation is illustrated in Fig. 2.

Observe that we want an encoding function e to be one-to-one so that different words in B^m will be assigned different code words.

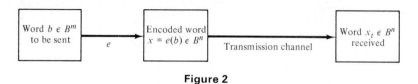

Figure 2

If the transmission channel is noiseless, then $x_t = x$ for all x in B^n. In this case $x = e(b)$ is received for each $b \in B^m$, and since e is a known function, b may be identified.

In general, errors in transmission do occur. We will say that the code word $x = e(b)$ has been transmitted with **k or fewer errors** if x and x_t differ in at least 1 but no more than k positions.

Let $e : B^m \to B^n$ be an (m, n) encoding function. We say that e **detects k or fewer errors** if whenever $x = e(b)$ is transmitted with k or fewer errors, then x_t is not a code word (thus x_t could not be x and therefore could not have been correctly transmitted). If $x \in B^n$, then the number of 1's in x is called the **weight** of x and is denoted by $|x|$.

Example 1   Find the weight each of the following words in B^5: (a) $x = 1000$; (b) $x = 11100$;   (c) $x = 00000$;   (d) $x = 11111$.

Solution.

(a) $|x| = 1$ (b)$|x| = 3$ (c)$|x| = 0$ (d)$|x| = 5$

Example 2   (**Parity check code**). The following encoding function $e : B^m \to B^{m+1}$ is called the **parity $(m, m + 1)$ check code**: If $b = b_1 b_2 \cdots b_m \in B^m$, define

$$e(b) = b_1 b_2 \cdots b_m b_{m+1},$$

where
$$b_{m+1} = \begin{cases} 0 & \text{if } |b| \text{ is even} \\ 1 & \text{if } |b| \text{ is odd} \end{cases}$$

Observe that b_{m+1} is zero if and only if the number of 1's in b is an even number. It then follows that every code word $e(b)$ has even weight. A single error in the transmission of a code word will change the received word to a word of odd weight, and therefore can be detected. In the same way we see that any odd number of errors can be detected.

For a concrete illustration of this encoding function, let $m = 3$. Then

$$e(100) = 1001,$$

$$e(011) = 0110,$$

$$e(111) = 1111$$

Suppose now that $b = 111$. Then $x = e(b) = 1111$. If the transmission channel transmits x as $x_t = 1101$, then $|x_t| = 3$, and we know that an odd number of errors (at least one) has occurred.

It should be noted that if the received word has even weight, then we cannot conclude that the code word was transmitted correctly, since this encoding function does not detect an even number of errors. Despite this limitation, the parity check code is widely used in contemporary technology.

Example 3   Consider the following $(m, 3m)$ encoding function $e : B^m \to B^{3m}$ if

$$b = b_1 b_2 \cdots b_m \in B^m$$

Define

$$e(b) = e(b_1 b_2 \cdots b_m) = b_1 b_2 \cdots b_m b_1 b_2 \cdots b_m b_1 b_2 \cdots b_m$$

That is, the encoding function e repeats each word of B^m three times. For a concrete example, let $m = 3$. Thus

$$e(100) = 100100100$$

$$e(011) = 011011011$$

$$e(001) = 001001001$$

Suppose now that $b = 011$. Then $e(011) = 011011011$. Assume now that the transmission channel makes an error in the underlined digit and that we receive the word 011111011. This is not a code word, so we have detected the error. It is not hard to see that any single error, and any two errors can be detected.

Let x and y be words in B^m. The **Hamming distance** $\delta(x, y)$ between x and y is the weight $|x \oplus y|$ of $x \oplus y$. Thus the distance between $x = x_1 x_2 \cdots x_m$ and $y = y_1 y_2 \cdots y_m$ is the number of values of i such that $x_i \neq y_i$, that is, the number of positions in which x and y differ.

Example 4   Find the distance between x and y:

(a) $x = 110110, y = 000101$
(b) $x = 001100, y = 010110$

Solution.

(a) $x \oplus y = 110011$, so $|x \oplus y| = 4$
(b) $x \oplus y = 011010$, so $|x \oplus y| = 3$

Theorem 1   **(Properties of the Distance Function).** Let x, y, and z be elements of B^m. Then

(a) $\delta(x, y) = \delta(y, x)$

(b) $\delta(x, y) \geq 0$

(c) $\delta(x, y) = 0$ if and only if $x = y$

(d) $\delta(x, y) \leq \delta(x, z) + \delta(z, y)$

Proof. Properties (a), (b), and (c) are simple to prove and are left as exercises.

(d) For a and b in B^m,

$$|a \oplus b| \leq |a| + |b|$$

Also, if $a \in B^m$, then $a \oplus a = \overline{0}$, the identity element in B^m. Then

$$\delta(x, y) = |x \oplus y| = |x \oplus \overline{0} \oplus y| = |x \oplus z \oplus z \oplus y|$$
$$\leq |x \oplus z| + |z \oplus y|$$
$$= \delta(x, z) + \delta(z, y)$$

The **minimum distance** of an encoding function $e : B^m \to B^n$ is the minimum of the distances between all distinct pairs of code words; that is,

$$\min \{\delta(e(x), e(y)) \mid x, y \in B^m\}$$

Example 5   Consider the following $(2, 5)$ encoding function e:

$$
\left.
\begin{aligned}
e(00) &= 00000 \\
e(10) &= 00111 \\
e(01) &= 01110 \\
e(11) &= 11111
\end{aligned}
\right\} \quad \text{code words}
$$

The minimum distance is 2 as can be checked by computing the minimum of the distances between all six distinct pairs of code words.

Theorem 2   An (m, n) encoding function $e : B^m \to B^n$ can detect k or fewer errors if and only if its minimum distance is at least $k + 1$.

Proof. Suppose that the minimum distance between any two code words is at least $k + 1$. Let $b \in B^m$, and let $x = e(b) \in B^n$ be the code word representing b. Then x is transmitted and is received as x_t. If x_t were a code word different from x, then

$\delta(x, x_t) \geq k + 1$, so x would be transmitted with $k + 1$ or more errors. Thus if x is transmitted with k or fewer errors, then x_t cannot be a code word. This means that e can detect k or fewer errors.

Conversely, suppose that the minimum distance between code words is $r \leq k$, and let x and y be code words with $\delta(x, y) = r$. If $x_t = y$, that is, if x is transmitted and is mistakenly received as y, then $r \leq k$ errors have been committed and have not been detected. Thus it is not true that e can detect k or fewer errors.

Example 6   Consider the (3, 8) encoding function $e : B^3 \rightarrow B^8$ defined by

$$e(000) = 00000000$$
$$e(001) = 10111000$$
$$e(010) = 00101101$$
$$e(011) = 10010101$$
$$e(100) = 10100100$$
$$e(101) = 10001001$$
$$e(110) = 00011100$$
$$e(111) = 00110001$$

How many errors will e detect?

Solution. The minimum distance of e is 3, as can be checked by computing the minimum of the distances between all 28 distinct pairs of code words. By Theorem 2, the code will detect k or fewer errors if and only if its minimum distance is at least $k + 1$. Since the minimum distance is 3, we have $3 \geq k + 1$ or $k \leq 2$. Thus the code will detect two or fewer errors.

Group Codes

So far, we have not made use of the fact that $(B^n, \oplus)$ is a group. We shall now consider an encoding function that makes use of this property of B^n.

An (m, n) encoding function $e : B^m \rightarrow B^n$ is called a **group code** if

$$e(B^m) = \{e(b) \,|\, b \in B^m\} = \text{Ran}\,(e)$$

is a subgroup of B^n.

Recall from Section 6.4 that N is a subgroup of B^n if (a) the identity of B^n is in N, (b) if x is in N then its inverse is in N, and (c) if x and y belong to N, then $x \oplus y \in N$. Property (b) need not be checked, since every element in B^n is its own inverse. Moreover, since B^n is abelian, every subgroup of B^n is a normal subgroup.

Example 7 Consider the $(3, 6)$ encoding function $e : B^3 \to B^6$ defined by

$$
\left.
\begin{aligned}
e(000) &= 000000 \\
e(001) &= 001100 \\
e(010) &= 010011 \\
e(011) &= 011111 \\
e(100) &= 100101 \\
e(101) &= 101001 \\
e(110) &= 110110 \\
e(111) &= 111010
\end{aligned}
\right\} \text{ code words}
$$

Show that this encoding function is a group code.

Solution. We must show that the set of all code words

$$N = \{000000, 001100, 010011, 011111, 100101, 101001, 110110, 111010\}$$

is a subgroup of B^6. This is done by first noting that the identity of B^6 belongs to N. Next we verify, by trying all possibilities, that if x and y are elements in N, then $x \oplus y$ is in N. Hence N is a subgroup of B^6 and the given encoding function is a group code.

Theorem 3 Let $e : B^m \to B^n$ be a group code. The minimum distance of e is the minimum weight of a nonzero code word.

Proof. Let δ be the minimum distance of the group code, and suppose that $\delta = \delta(x, y)$, where x and y are distinct code words. Also, let n be the minimum weight of a nonzero code word and suppose that $n = |z|$ for a code word z. Since e is a group code, $x \oplus y$ is a nonzero code word. Thus

$$\delta = \delta(x, y) = |x \oplus y| \geq n$$

On the other hand, since $\bar{0}$ and z are distinct code words

$$n = |z| = |z \oplus \bar{0}| = \delta(z, \bar{0}) \geq \delta$$

Hence $n = \delta$.

Example 8 The minimum distance of the group code in Example 7 is 2, since by Theorem 3 this distance is equal to the smallest number of 1's in any of the seven nonzero code words. To check this directly would require 28 different calculations.

We shall now take a brief look at a procedure for generating group codes. First, we need several results on Boolean matrices.

Consider the set $B = \{0, 1\}$ with the binary operation given in Table 2.

Table 2

·	0	1
0	0	0
1	0	1

This operation has been denoted in Section 1.8 by $\wedge$ and is just the GLB operation in the Boolean algebra B. Now let $\mathbf{D} = [d_{ij}]$ and $\mathbf{E} = [e_{ij}]$ be $m \times n$ Boolean matrices. We define the **mod 2 sum** $\mathbf{D} \oplus \mathbf{E}$ as the $m \times n$ Boolean matrix $\mathbf{F} = [f_{ij}]$, where

$$f_{ij} = d_{ij} + e_{ij}, \qquad 1 \le i \le m, \quad 1 \le j \le n \qquad \text{(here + is addition in } B\text{)}$$

Example 9 We have

$$
\begin{bmatrix} 1 & 0 & 1 & 1 \\ 0 & 1 & 1 & 0 \\ 1 & 0 & 0 & 1 \end{bmatrix} \oplus \begin{bmatrix} 1 & 1 & 0 & 1 \\ 1 & 1 & 0 & 1 \\ 0 & 1 & 1 & 1 \end{bmatrix}
$$

$$
= \begin{bmatrix} 1+1 & 0+1 & 1+0 & 1+1 \\ 0+1 & 1+1 & 1+0 & 0+1 \\ 1+0 & 0+1 & 0+1 & 1+1 \end{bmatrix} = \begin{bmatrix} 0 & 1 & 1 & 0 \\ 1 & 0 & 1 & 1 \\ 1 & 1 & 1 & 0 \end{bmatrix}
$$

Let $\mathbf{D} = [d_{ij}]$ be an $m \times p$ Boolean matrix and let $\mathbf{E} = [e_{ij}]$ be a $p \times n$ Boolean matrix. We define the **mod 2 Boolean product** $\mathbf{D} * \mathbf{E}$ as the $m \times n$ matrix $\mathbf{F} = [f_{ij}]$, where

$$f_{ij} = d_{i1} \cdot e_{1j} + d_{i2} \cdot e_{2j} + \cdots + d_{ip} \cdot e_{pj}, \qquad 1 \le i \le m, \quad 1 \le j \le n$$

Example 10 We have

$$
\begin{bmatrix} 1 & 1 & 0 \\ 0 & 1 & 1 \end{bmatrix} * \begin{bmatrix} 1 & 0 \\ 1 & 1 \\ 0 & 1 \end{bmatrix} = \begin{bmatrix} 1 \cdot 1 + 1 \cdot 1 + 0 \cdot 0 & 1 \cdot 0 + 1 \cdot 1 + 0 \cdot 1 \\ 0 \cdot 1 + 1 \cdot 1 + 1 \cdot 0 & 0 \cdot 0 + 1 \cdot 1 + 1 \cdot 1 \end{bmatrix} = \begin{bmatrix} 0 & 1 \\ 1 & 0 \end{bmatrix}
$$

The proof of the following theorem is left as an exercise.

Theorem 4   Let $\mathbf{D}$ and $\mathbf{E}$ be $m \times p$ Boolean matrices, and let $\mathbf{F}$ be a $p \times n$ Boolean matrix. Then

$$(\mathbf{D} \oplus \mathbf{E}) * \mathbf{F} = (\mathbf{D} * \mathbf{F}) \oplus (\mathbf{E} * \mathbf{F})$$

We shall now consider the element $x = x_1 x_2 \cdots x_n \in B^n$ as the $1 \times n$ matrix $[x_1 \ x_2 \ \cdots \ x_n]$.

Theorem 5   Let m and n be nonnegative integers with $m < n$, $r = n - m$, and let $\mathbf{H}$ be an $n \times r$ Boolean matrix. Then the function $f_H : B^n \to B^r$ defined by

$$f_H(x) = x * \mathbf{H}, \qquad x \in B^n$$

is a homomorphism from the group B^n to the group B^r.

Proof. If x and y are elements in B^n, then

$$f_H(x \oplus y) = (x \oplus y) * \mathbf{H}$$
$$= (x * \mathbf{H}) \oplus (y * \mathbf{H})$$
$$= f_H(x) \oplus f_H(y)$$

Hence f_H is a homomorphism from B^n to B^r.

Corollary 1   Let $m, n, r, \mathbf{H}$, and f_H be as in Theorem 5. Then

$$N = \{x \in B^n \,|\, x * \mathbf{H} = \bar{0}\}$$

is a normal subgroup of B^n.

Proof. It follows from the results in Section 6.5 that N is the kernel of the homomorphism f_H, so it is a normal subgroup of B^n.

Let $m < n$ and $r = n - m$. An $n \times r$ Boolean matrix

$$\mathbf{H} = \underbrace{\begin{bmatrix} h_{11} & h_{12} & \cdots & h_{1r} \\ h_{21} & h_{22} & \cdots & h_{2r} \\ \vdots & \vdots & & \vdots \\ h_{m1} & h_{m2} & \cdots & h_{mr} \\ 1 & 0 & \cdots & 0 \\ 0 & 1 & \cdots & 0 \\ \vdots & \vdots & & \vdots \\ 0 & 0 & \cdots & 1 \end{bmatrix}}_{}$$

$n - m = r$ rows $\Big\{$

whose last r rows form the $r \times r$ identity matrix, is called a **parity check matrix**. We use $\mathbf{H}$ to define an encoding function $e_H : B^m \to B^n$. If $b = b_1 b_2 \cdots b_m$, let

$x = e_H(b) = b_1 b_2 \cdots b_m x_1 x_2 \cdots x_r$, where

$$x_1 = b_1 \cdot h_{11} + b_2 \cdot h_{21} + \cdots + b_m \cdot h_{m1}$$
$$x_2 = b_1 \cdot h_{12} + b_2 \cdot h_{22} + \cdots + b_m \cdot h_{m2}$$
$$\vdots$$
$$x_r = b_1 \cdot h_{1r} + b_2 \cdot h_{2r} + \cdots + b_m \cdot h_{mr}$$

(1)

Theorem 6   Let $x = y_1 y_2 \cdots y_m x_1 \cdots x_r \in B^n$. Then $x * \mathbf{H} = \bar{0}$ if and only if $x = e_H(b)$ for some $b \in B^m$.

Proof. Suppose that $x * \mathbf{H} = \bar{0}$. Then

$$y_1 \cdot h_{11} + y_2 \cdot h_{21} + \cdots + y_m \cdot h_{m1} + x_1 = 0$$
$$y_1 \cdot h_{12} + y_2 \cdot h_{22} + \cdots + y_m \cdot h_{m2} + x_2 = 0$$
$$\vdots$$
$$y_1 \cdot h_{1r} + y_2 \cdot h_{2r} + \cdots + y_m \cdot h_{mr} + x_r = 0$$

The first equation is of the form

$$a + x_1 = 0, \text{ where } a = y_1 \cdot h_{11} + y_2 \cdot h_{21} + \cdots + y_m \cdot h_{m1}$$

Adding a to both sides, we obtain

$$a + (a + x_1) = a + 0 = a$$
$$(a + a) + x_1 = a$$
$$0 + x_1 = a \qquad (\text{since } a + a = 0)$$
$$x_1 = a$$

This can be done for each row, therefore

$$x_i = y_1 \cdot h_{1i} + y_2 \cdot h_{2i} + \cdots + y_m \cdot h_{mi}, \qquad 1 \leq i \leq r$$

Letting $b_1 = y_1$, $b_2 = y_2$, ..., $b_m = y_m$, we see that $x_1, x_2, \ldots, x_r$ satisfy the equations in (1). Thus, $b = b_1 b_2 \cdots b_m \in B^m$ and $x = e_H(b)$.

Conversely, if $x = e_H(b)$, the equations in (1) can be rewritten as

$$b_1 \cdot h_{11} + b_2 \cdot h_{21} + \cdots + b_m \cdot h_{m1} + x_1 = 0$$
$$b_1 \cdot h_{12} + b_2 \cdot h_{22} + \cdots + b_m \cdot h_{m2} + x_2 = 0$$
$$\vdots$$
$$b_1 \cdot h_{1r} + b_2 \cdot h_{2r} + \cdots + b_m \cdot h_{mr} + x_r = 0$$

which is merely $x * \mathbf{H} = \bar{0}$.

Corollary 2 $e_H(B^m) = \{e_H(b) \mid b \in B^m\}$ is a subgroup of B^n.

Proof. The result follows from the observation that

$$e_H(B^m) = \ker (f_H)$$

and from Corollary 1. Thus e_H is a group code.

Example 11   Let $m = 2$, $n = 5$, and

$$\mathbf{H} = \begin{bmatrix} 1 & 1 & 0 \\ 0 & 1 & 1 \\ 1 & 0 & 0 \\ 0 & 1 & 0 \\ 0 & 0 & 1 \end{bmatrix}$$

Determine the group code $e_H : B^2 \to B^5$.

Solution. We have $B^2 = \{00, 10, 01, 11\}$. Then

$$e(00) = 00x_1x_2x_3$$

where x_1, x_2, and x_3 are determined by the equations in (1). Thus

$$x_1 = x_2 = x_3 = 0$$

and

$$e(00) = 00000$$

Next,

$$e(10) = 10x_1x_2x_3$$

Using the equations in (1) with $b_1 = 1$ and $b_2 = 0$, we obtain

$$x_1 = 1 \cdot 1 + 0 \cdot 0 = 1$$
$$x_2 = 1 \cdot 1 + 0 \cdot 1 = 1$$
$$x_3 = 0 \cdot 1 + 1 \cdot 0 = 0$$

Thus $x_1 = 1$, $x_2 = 1$, and $x_3 = 0$, so

$$e(10) = 10110$$

Similarly,

$$e(01) = 01011$$

$$e(11) = 11101$$

EXERCISE SET 8.1

In Exercises 1 and 2, find the weights of the given words.

1. (a) 1011 (b) 0110 (c) 1110

2. (a) 011101 (b) 11111 (c) 010101

3. Consider the (3, 4) parity check code. For each of the received words, determine whether an error will be detected.

(a) 0100 (b) 1100 (c) 0010 (d) 1001

4. Consider the $(m, 3m)$ encoding function of example 3, where $m = 4$. For each of the received words, determine whether an error will be detected.

(a) 011010011111 (b) 110110010110
(c) 010010110010 (d) 001001111001

5. Find the distance between x and y.

(a) $x = 1100010$, $y = 1010001$
(b) $x = 0100110$, $y = 0110010$

6. Find the distance between x and y.

(a) $x = 00111001$, $y = 10101001$
(b) $x = 11010010$, $y = 00100111$

7. Prove (a), (b), and (c) of Theorem 1.

8. Find the minimum distance of the (2, 4) encoding function e.

$$e(00) = 0000$$

$$e(10) = 0110$$

$$e(01) = 1011$$

$$e(11) = 1100$$

9. Find the minimum distance of the (3, 8) encoding function e.

$$e(000) = 00000000$$

$$e(001) = 01110010$$

$$e(010) = 10011100$$

$$e(011) = 01110001$$

$$e(100) = 01100101$$

$$e(101) = 10110000$$

$$e(110) = 11110000$$

$$e(111) = 00001111$$

10. Consider the (2, 6) encoding function e.

$$e(00) = 000000$$

$$e(10) = 101010$$

$$e(01) = 011110$$

$$e(11) = 111000$$

(a) Find the minimum distance.
(b) How many errors will e detect?

11. Consider the (3, 9) encoding function e.

$$e(000) = 000000000$$

$$e(001) = 011100101$$

$$e(010) = 010101000$$

$$e(011) = 110010001$$

$$e(100) = 010011010$$

$$e(101) = 111101011$$

$$e(110) = 001011000$$

$$e(111) = 110000111$$

(a) Find the minimum distance.
(b) How many errors will e detect?

12. Show that the (2, 5) encoding function $e : B^2 \to B^5$ defined by

$$e(00) = 00000$$
$$e(01) = 01110$$
$$e(10) = 10101$$
$$e(11) = 11011$$

is a group code.

13. Show that the (3, 7) encoding function $e : B^3 \to B^7$ defined by

$$e(000) = 0000000$$
$$e(001) = 0010110$$
$$e(010) = 0101000$$
$$e(011) = 0111110$$
$$e(100) = 1000101$$
$$e(101) = 1010011$$
$$e(110) = 1101101$$
$$e(111) = 1111011$$

is a group code.

14. Find the minimum distance of the group code defined in Exercise 12.

15. Find the minimum distance of the group code defined in Exercise 13.

16. Compute

$$\begin{bmatrix} 1 & 1 & 0 \\ 0 & 1 & 1 \end{bmatrix} \oplus \begin{bmatrix} 1 & 1 & 1 \\ 0 & 1 & 1 \end{bmatrix}$$

17. Compute

$$\begin{bmatrix} 1 & 0 & 1 \\ 1 & 1 & 0 \\ 0 & 1 & 0 \\ 0 & 1 & 1 \end{bmatrix} \oplus \begin{bmatrix} 1 & 0 & 1 \\ 0 & 1 & 0 \\ 1 & 1 & 1 \\ 0 & 0 & 1 \end{bmatrix}$$

18. Compute

$$\begin{bmatrix} 1 & 0 \\ 1 & 1 \\ 0 & 1 \end{bmatrix} * \begin{bmatrix} 1 & 1 & 0 \\ 0 & 1 & 1 \end{bmatrix}$$

19. Compute

$$\begin{bmatrix} 1 & 0 & 1 \\ 0 & 1 & 1 \\ 1 & 0 & 1 \end{bmatrix} * \begin{bmatrix} 1 & 1 & 0 \\ 0 & 1 & 1 \\ 1 & 0 & 1 \end{bmatrix}$$

20. Let

$$\mathbf{H} = \begin{bmatrix} 0 & 1 & 1 \\ 0 & 1 & 1 \\ 1 & 0 & 0 \\ 0 & 1 & 0 \\ 0 & 0 & 1 \end{bmatrix}$$

be a parity check matrix. Determine the (2, 5) group code function $e_H : B^2 \to B^5$.

21. Let

$$\mathbf{H} = \begin{bmatrix} 1 & 0 & 0 \\ 0 & 1 & 1 \\ 1 & 1 & 1 \\ 1 & 0 & 0 \\ 0 & 1 & 0 \\ 0 & 0 & 1 \end{bmatrix}$$

be a parity check matrix. Determine the (3, 6) group code $e_H : B^3 \to B^6$.

22. Prove Theorem 4.

8.2 Decoding and Error Correction

Consider an (m, n) encoding function $e : B^m \to B^n$. Once the encoded word $x = e(b) \in B^n$, for $b \in B^m$, is received as the word x_t, we are faced with the problem of identifying the word b that was the original message.

An onto function $d : B^n \to B^m$ is called an (n, m) **decoding function** associated with e, if $d(x_t) = b' \in B^m$ is such that when the transmission channel has no noise, then $b' = b$, that is,

$$e \circ d = 1_{B^m}$$

where 1_{B^m} is the identity function on B^m. The decoding function d is required to be onto so that every received word can be decoded to give a word in B^m. It decodes properly received words correctly, but the decoding of improperly received words may or may not be correct.

Example 1 Consider the parity check code defined in Example 2 of Section 8.1. We now define the decoding function $d : B^{m+1} \to B^m$. If $y = y_1 y_2 \cdots y_m y_{m+1} \in B^{m+1}$, then

$$d(y) = y_1 y_2 \cdots y_m$$

Observe that if $b = b_1 b_2 \cdots b_m \in B^m$, then

$$(e \circ d)(b) = d(e(b)) = b$$

so $e \circ d = 1_{B^m}$.

For a concrete example, let $m = 4$. Then $d(10010) = 1001$ and $d(11001) = 1100$.

Let e be an (m, n) encoding function and let d be an (n, m) decoding function associated with e. We say that the pair (e, d) **corrects k or fewer errors** if whenever $x = e(b)$ is transmitted correctly or with k or fewer errors and x_t is received, then $d(x_t) = b$. Thus x_t is decoded as the correct message b.

Example 2 Consider the $(m, 3m)$ encoding function defined in Example 3 of Section 8.1. We now define the decoding function $d : B^{3m} \to B^m$. Let

$$y = y_1 y_2 \cdots y_m y_{m+1} \cdots y_{2m} y_{2m+1} \cdots y_{3m}$$

Then

$$d(y) = z_1 z_2 \cdots z_m$$

where

$$z_i = \begin{cases} 1 & \text{if} \quad \{y_i, y_{i+m}, y_{i+2m}\} \quad \text{has at least two 1's} \\ 0 & \text{if} \quad \{y_i, y_{i+m}, y_{i+2m}\} \quad \text{has less than two 1's} \end{cases}$$

That is, the decoding function d examines the ith digit in each of the three blocks transmitted. The digit that occurs at least twice in these three blocks is chosen as the decoded ith digit. For a concrete example, let $m = 3$. Then

$$e(100) = 100100100$$

$$e(011) = 011011011$$

$$e(001) = 001001001$$

Suppose now that $b = 011$. Then $e(011) = 011011011$. Assume now that the transmission channel makes an error in the underlined digit and that we receive the word $x_t = 011111011$. Then since the first digits in two out of the three blocks are 0, the first digit is decoded as 0. Similarly, the second digit is decoded as 1, since all three second digits in the three blocks are 1. Finally, the third digit is also decoded as 1, for the analogous reason. Hence $d(x_t) = 011$; that is, the decoded word is 011, which is the word that was sent. Therefore, the single error has been corrected. A similar analysis shows that, if e is this $(m, 3m)$ code for any value of m, and d is as above, then (e, d) corrects any single error.

Given an (m, n) encoding function $e : B^m \to B^n$, we often need to determine an (m, n) decoding function $d : B^n \to B^m$ associated with e. We now discuss a method, called the **maximum likelihood technique**, for determining a decoding function d for a given e.

Since B^m has m elements, there are 2^m code words in B^n. We first list the code words in a fixed order:

$$x^{(1)}, x^{(2)}, \ldots, x^{(2^m)}$$

If the received word is x_t, we compute $\delta(x^{(i)}, x_t)$ for $1 \le i \le 2^m$, and choose the first code word, say it is $x^{(s)}$, such that

$$\min_{1 \le i \le 2^m} \{\delta(x^{(i)}, x_t)\} = x^{(s)}$$

That is, $x^{(s)}$ is a code word that is closest to x_t, and the first in the list. If $x^{(s)} = e(b)$, we define the **maximum likelihood decoding function** d associated with e by

$$d(x_t) = b$$

Observe that d depends on the particular order in which the code words in $e(B^m)$ are listed. If the code words are listed in a different order, we may obtain a different maximum likelihood decoding function d associated with e.

Theorem 1 Suppose that e is an (m, n) encoding function and d is a maximum likelihood decoding function associated with e. Then (e, d) can correct k or fewer errors if and only if the minimum distance of e is at least $2k + 1$.

Proof. Assume that the minimum distance of e is at least $2k + 1$. Let $b \in B^m$, and $x = e(b) \in B^n$. Suppose that x is transmitted with k or fewer errors, and x_t is received. This means that $\delta(x, x_t) \leq k$. If z is any other code word, then

$$2k + 1 \leq \delta(x, z) \leq \delta(x, x_t) + \delta(x_t, z) \leq k + d(x_t, z)$$

Thus $\delta(x_t, z) \geq 2k + 1 - k = k + 1$. This means that x is the unique code word that is closest to x_t, so $d(x_t) = b$. Hence (e, d) corrects k or fewer errors.

Conversely, assume that the minimum distance between code words is $r \leq 2k$, and let $x = e(b)$ and $x' = e(b')$ be code words with $\delta(x, x') = r$. Suppose x' preceeds x in the list of code words used to define d. Write $x = b_1 b_2 \cdots b_n$, $x' = b'_1 b'_2 \cdots b'_n$. Then $b_i \neq b'_i$ for exactly r integers i between 1 and n. Assume, for simplicity, that $b_1 \neq b'_1$, $b_2 \neq b'_2$, ..., $b_r \neq b'_r$, but $b_i = b'_i$ when $i > r$. Any other case is handled in the same way.

(a) Suppose that $r \leq k$. If x is transmitted as $x_t = x'$, then $r \leq k$ errors have been committed, but $d(x_t) = b'$, so (e, d) has not corrected the r errors.

(b) Suppose that $k + 1 \leq r \leq 2k$, and let

$$y = b'_1 \, b'_2 \, \cdots \, b'_k \, b_{k+1} \, \cdots \, b_n$$

If x is transmitted as $x_t = y$, then $\delta(x_t, x') = r - k \leq k$ and $\delta(x_t, x) \geq k$. Thus x' is at least as close to x_t as x is, and x' preceeds x in the list of code words, so $d(x_t) \neq b$. Then we have committed k errors, which (e, d) has not corrected.

Example 3 Let e be the $(3, 8)$ encoding function defined in Example 6 of Section 8.1 and let d be an $(8, 3)$ maximum likelihood decoding function associated with e. How many errors can (e, d) correct?

Solution. Since the minimum distance of e is 3, we have $3 \geq 2k + 1$, so $k \leq 1$. Thus (e, d) can correct one error.

We now discuss a simple and effective technique for determining a maximum likelihood decoding function associated with a given group code. First, we prove the following result.

Theorem 2 If K is a subgroup of a group G, then every left coset of K in G has as many elements as K.

Proof. Let aK be a left coset of K in G, where $a \in G$. Consider the function $f: K \to aK$ defined by

$$f(k) = ak \qquad \text{for } k \in K$$

We show that f is one to one and onto.

To show that f is one to one, we assume that

$$f(k_1) = f(k_2), \qquad k_1, k_2 \in K$$

Then

$$ak_1 = ak_2$$

By Theorem 2 of section 6.4, $k_1 = k_2$. Hence f is one to one.

To show that f is onto, let b be an arbitrary element in aK. Then $b = ak$ for some $k \in K$. We now have

$$f(k) = ak = b$$

so f is onto. Since f is one to one and onto, K and aK have the same number of elements.

Let $e : B^m \rightarrow B^n$ be an (m, n) encoding function that is a group code. Thus the set N of code words in B^n is a subgroup of B^n whose order is 2^m, say $N = \{x^{(1)}, x^{(2)}, \ldots, x^{(2^m)}\}$.

Suppose that the code word $x = e(b)$ is transmitted and that the word x_t is received. The left coset of x_t is $x_t \oplus N = \{\varepsilon_1, \varepsilon_2, \ldots, \varepsilon_{2^m}\}$, where $\varepsilon_i = x_t \oplus x^{(i)}$. The distance from x_t to code word $x^{(i)}$ is just $|\varepsilon_i|$, the weight of ε_i. Thus if ε_j is a coset member with smallest weight, then $x^{(j)}$ must be a code word that is closest to x_t. In this case, $x^{(j)} = \bar{0} \oplus x^{(j)} = x_t \oplus x_t \oplus x^{(j)} = x_t \oplus \varepsilon_j$. An element ε_j, having smallest weight, is called a **coset leader**. Note that a coset leader need not be unique.

If $e : B^m \rightarrow B^n$ is a group code, we now state the following procedure for obtaining a maximum likelihood decoding function associated with e.

Step 1. Determine all the left cosets of $N = e(B^m)$ in B^n.

Step 2. For each coset, find a coset leader (a word of least weight). Steps 1 and 2 can be carried out in a systematic tabular manner which will be described below.

Step 3. If the word x_t is received, determine the coset of N to which x_t belongs. Since N is a normal subgroup of B^n, it follows from Theorems 3 and 4 in Section 6.5 that the cosets of N form a partition of B^n, so that each element of B^n belongs to one and only one coset of N in B^n. Moreover, there are $2^n/2^m = 2^r$ distinct cosets of N in B^n.

Step 4. Let ε be a coset leader for the coset determined in Step 3. Compute $x = x_t \oplus \varepsilon$. If $x = e(b)$, we let $d(x_t) = b$. That is, we decode x_t as b.

To implement the foregoing procedure, we must keep a complete list of all the cosets of N in B^n, usually in tabular form, with each row of the table containing one coset. We identify a coset leader in each row. Then, when a word x_t is received, we locate the row which contains it, find the coset leader for that row, and add it to x_t.

This gives us the code word closest to x_t. We can eliminate the need for these additions if we construct a more systematic table.

Before illustrating with an example, we make several observations. Let

$$N = \{x^{(1)}, x^{(2)}, \ldots, x^{(2^m)}\}$$

where $x^{(1)}$ is $\bar{0}$, the identity of B^n.

Steps 1 and 2 in the decoding algorithm above are carried out as follows. First, list all the elements of N in a row, starting with the identity $\bar{0}$ at the left. Thus we have

$$\bar{0} \qquad x^{(2)} \qquad x^{(3)} \qquad \cdots \qquad x^{(2^m)}$$

This row is the coset $[\bar{0}]$, and it has $\bar{0}$ as its coset leader. For this reason we will also refer to $\bar{0}$ as ε_1. Now choose any word y in B^n which has not been listed in the first row. List the elements of the coset $y \oplus N$ as the second row. This coset also has 2^m elements. Thus we have the two rows

$$\begin{array}{ccccc} \bar{0} & x^{(2)} & x^{(3)} & \cdots & x^{(2^m)} \\ y \oplus \bar{0} & y \oplus x^{(2)} & y \oplus x^{(3)} & \cdots & y \oplus x^{(2^m)} \end{array}$$

In the coset $y \oplus N$ pick an element of least weight, a coset leader, which we denote by $\varepsilon^{(2)}$. In case of ties, choose any element of least weight. Recall from Section 6.5 that since $\varepsilon^{(2)} \in y \oplus N$, we have $y \oplus N = \varepsilon^{(2)} \oplus N$. This means that every word in the second row can be written as $\varepsilon^{(2)} \oplus v$, $v \in N$. Now rewrite the second row as follows:

$$\varepsilon^{(2)} \qquad \varepsilon^{(2)} \oplus x^{(2)} \qquad \varepsilon^{(2)} \oplus x^{(3)} \qquad \cdots \qquad \varepsilon^{(2)} \oplus x^{(2^m)}$$

with $\varepsilon^{(2)}$ in the leftmost position.

Next, choose another element z in B^n that has not yet been listed in either of the first two rows, and form the third row $(z + x^{(j)})$, $1 \le j \le 2^m$ (another coset of N in B^n). This row can be rewritten in the form $\varepsilon^{(3)}$, $\varepsilon^{(3)} \oplus x^{(2)}, \ldots, \varepsilon^{(3)} \oplus x^{2^m}$, where $\varepsilon^{(3)}$ is a coset leader for the row.

Continue this process until all elements of B^n have been listed. The resulting table, shown in Table 1, is called a **decoding table**. Notice that it contains 2^r rows, one for each coset of N. If we receive the word x_t, we locate it in the table. If x is the

Table 1

$\bar{0}$	$x^{(2)}$	$x^{(3)}$	$\cdots$	$x^{(2^m-1)}$
$\varepsilon^{(2)}$	$\varepsilon^{(2)} \oplus x^{(2)}$	$\varepsilon^{(2)} \oplus x^{(3)}$	$\cdots$	$\varepsilon^{(2)} \oplus x^{(2^m-1)}$
$\vdots$	$\vdots$	$\vdots$		$\vdots$
$\varepsilon^{(2^r)}$	$\varepsilon^{(2^r)} \oplus x^{(2)}$	$\varepsilon^{(2^r)} \oplus x^{(3)}$	$\cdots$	$\varepsilon^{(2^r)} \oplus x^{(2^m-1)}$

element of N that is at the top of the column containing x_t, then x is the code word closest to x_t. Thus if $x = e(b)$, we let $d(x_t) = b$.

Example 4 Consider the (3, 6) group code defined in Example 7 of Section 8.1. Here

$$N = \{000000, 001100, 010011, 011111, 100101, 101001, 110110, 111010\}$$

$$= \{x^{(1)}, x^{(2)}, \ldots, x^{(8)}\}$$

defined in Example 1. We now implement the decoding procedure above for e as follows.

Steps 1 and 2. Determine all the left cosets of N in B^6, as rows of a table. For each row i, locate the coset leader ε_i, and rewrite the row in the order

$$\varepsilon_i, \quad \varepsilon_i \oplus 001100, \quad \varepsilon_i \oplus 010011, \quad \ldots, \quad \varepsilon_i \oplus 111010$$

The result is shown in Table 2.

Table 2

000000	001100	010011	011111	100101	101001	110110	111010
000001	001101	010010	011110	100100	101000	110111	111011
000010	001110	010001	011101	100111	101011	110100	111000
000100	001000	010111	011011	100001	101101	110010	111110
010000	011100	000011	001111	110101	111001	100110	101010
100000	101100	110011	111111	<u>000101</u>	001001	010110	011010
000110	001010	<u>010101</u>	011001	100011	101111	110000	111100
010100	011000	<u>000111</u>	001011	110001	111101	100010	101110

If we receive the word 000101, we decode it by first locating it in the decoding table: it appears in the fifth column, where it is underlined. The word at the top of the fifth column is 100101. Since $e(100) = 100101$, we decode 000101 as 100. Similarly, if we receive the word 010101, we first locate it in the third column of the decoding table, where it is underlined twice. The word at the top of the third column is 010011. Since $e(010) = 010011$, we decode 010101 as 010.

We make the following observations for this example. In determining the decoding table in Steps 1 and 2, there was more than one candidate for coset leader of the last two cosets. In row 7 we chose 00110 as coset leader. If we had chosen 001010 instead, row 7 would have appeared in the rearranged form

001010 001010 $\oplus$ 001100 $\cdots$ 001010 $\oplus$ 111010 =

001010 000110 011001 010101 101111 100011 111100 110000

The new decoding table is shown in Table 3.

Table 3

000000	001100	010011	011111	100101	101001	110110	111010
000001	001101	010010	011110	100100	101000	110111	111011
000010	001110	010001	011101	100111	101011	110100	111000
000100	001000	010111	011011	100001	101101	110010	111110
010000	011100	000011	001111	110101	111001	100110	101010
100000	101100	110011	111111	000101	001001	010110	011010
001010	000110	011001	<u>010101</u>	101111	100011	111100	110000
010100	011000	000111	001011	110001	111101	100010	101110

Now, if we receive the word 010101, we first locate it in the *fourth* column of Table 3. The word at the top of the fourth column is 011111. Since $e(011111) = 011$, we decode 010101 as 011.

Suppose now that the (m, n) group code is $e_H : B^m \to B^n$, where **H** is a given parity check matrix. In this case, the decoding technique above can be simplified. We now turn to a discussion of this situation.

Recall from Section 8.1 that $r = n - m$,

$$\mathbf{H} = \begin{bmatrix} h_{11} & h_{12} & \cdots & h_{1r} \\ h_{21} & h_{22} & \cdots & h_{2r} \\ \vdots & \vdots & & \vdots \\ h_{m1} & h_{m2} & \cdots & h_{mr} \\ 1 & 0 & \cdots & 0 \\ 0 & 1 & \cdots & 0 \\ \vdots & \vdots & & \vdots \\ 0 & 0 & \cdots & 1 \end{bmatrix}$$

and the function $f_H : B^n \to B^r$ defined by

$$f_H(x) = x * H$$

is a homomorphism from the group B^n to the group B^r.

Theorem 3   If $m, n, r, H,$ and f_H are as above, then f_H is onto.

Proof. Let $b = b_1 b_2 \cdots b_r$ be any element in B^r. Letting

$$x = \underbrace{00 \cdots 0}_{\substack{m \text{ zero} \\ \text{digits}}} b_1 b_2 \cdots b_r,$$

we obtain $x * H = b$. Thus $f_H(x) = b$, so f_H is onto.

It follows from Corollary 1 in Section 6.5 that B^r and B^n/N are isomorphic, where $N = \ker(f_H) = e_H(B^m)$, under the isomorphism $g : B^n/N \rightarrow B^r$ defined by

$$g(xN) = f_H(x) = x * H$$

The element $x * H$ is called the **syndrome** of x. We now have the following result.

Theorem 4   Let x and y be elements in B^n. Then x and y lie in the same left coset of N in B^n if and only if $f_H(x) = f_H(y)$, that is, if and only if they have the same syndrome.

Proof. It follows from Theorem 4 in Section 6.5 that x and y lie in the same left coset of N in B^n if and only if $x \oplus y = (-x) \oplus y \in N$. Since $N = \ker(f)$, $x \oplus y \in N$ if and only if

$$f_H(x \oplus y) = \bar{0}_{B^r}$$
$$f_H(x) \oplus f_H(y) = \bar{0}_{B^r}$$
$$f_H(x) = f_H(y)$$

In this case, the decoding procedure given previously can be modified as follows. Suppose that we compute the syndrome of each coset leader. If the word x_t is received, we also compute $f_H(x_t)$, the syndrome of x_t. By comparing $f_H(x_t)$ and the syndromes of the coset leaders, we find the coset in which x_t lies. Suppose that a coset leader of this coset is ε. We now compute $x = x_t \oplus \varepsilon$. If $x = e(b)$, we then decode x_t as b. Thus we need only the coset leaders and their syndromes in order to decode. We state the new procedure in detail.

Step 1. Determine all left cosets of $N = e_H(B^m)$ in B^n.

Step 2. For each coset, find a coset leader, and compute the syndrome of all leaders.

Step 3. If x_t is received, compute the syndrome of x_t, and find the coset leader ε having the same syndrome. Then $x_t \oplus \varepsilon = x$ is a code word $e_H(b)$, and $d(x_t) = b$.

For this procedure, we do not need to keep a table of cosets, and we can avoid the work of computing a decoding table. Simply list all cosets once, in any order, and select a coset leader from each coset. Then keep a table of these coset leaders and their syndromes. The above procedure is easily implemented with such a table.

Example 5   Consider the parity check matrix

$$\mathbf{H} = \begin{bmatrix} 1 & 1 & 0 \\ 1 & 0 & 1 \\ 0 & 1 & 1 \\ 1 & 0 & 0 \\ 0 & 1 & 0 \\ 0 & 0 & 1 \end{bmatrix}$$

and the (3, 6) group code $e_H : B^3 \to B^6$. Then

$$e(000) = 000000$$

$$e(001) = 001011$$

$$e(010) = 010101$$

$$e(011) = 011110$$

$$e(100) = 100110$$

$$e(101) = 101101$$

$$e(110) = 110011$$

$$e(111) = 111000$$

Thus

$$N = \{000000, 001011, 010101, 011110, 100110, 101101, 110011, 111000\}$$

We now implement the decoding procedure above as follows.

In Table 4 we give only the coset leaders together with their syndromes. Suppose now that we receive the word 001110. We compute the syndrome of $x_t = 001110$, obtaining $f_H(x_t) = x_t * H = 101$, which is the sixth entry in the first column of Table 4. This means that x_t lies in the coset whose leader is $\varepsilon = 010000$. We compute $x = x_t \oplus \varepsilon = 001110 \oplus 010000 = 011110$. Since $e(011) = 011110$, we decode 001110 as 011.

Table 4

Syndrome of coset leader	Coset leader
000	000000
001	000001
010	000010
011	001000
100	000100
101	010000
110	100000
111	001100

EXERCISE SET 8.2

1. Let d be the (4, 3) decoding function defined by letting m be 3 in Example 1. Determine $d(y)$ for the word y in B^4.
   (a) 0110   (b) 1011

2. Let d be the (6, 2) decoding function defined in Example 2. Determine $d(y)$ for the word y in B^6.
   (a) 111011   (b) 010100

In Exercises 3–8, let e be the indicated encoding function and let d be an associated maximum likelihood decoding function. Determine the number of errors that (e, d) will correct.

3. e is the encoding function in Exercise 8 of Section 8.1.

4. e is the encoding function in Exercise 9 of Section 8.1.

5. e is the encoding function in Exercise 10 of Section 8.1.

6. e is the encoding function in Exercise 11 of Section 8.1.

7. e is the encoding function in Exercise 12 of Section 8.1.

8. e is the encoding function in Exercise 13 of Section 8.1.

9. Consider the group code defined in Exercise 12 of Section 8.1. Decode the following words relative to a maximum likelihood decoding function.
    (a) 11110   (b) 10011   (c) 10100

10. Consider the (3, 5) group encoding function $e : B^3 \to B^5$ defined by

$$e(000) = 00000$$
$$e(001) = 00110$$
$$e(010) = 01001$$
$$e(011) = 01111$$
$$e(100) = 10011$$
$$e(101) = 10101$$
$$e(110) = 11010$$
$$e(111) = 11100$$

Decode the following words relative to a maximum likelihood decoding function.
    (a) 11001   (b) 01010   (c) 00111

11. Consider the (3, 6) group encoding function $e : B^3 \to B^6$ is defined by

$$e(000) = 000000$$
$$e(001) = 000110$$
$$e(010) = 010010$$
$$e(011) = 010100$$
$$e(100) = 100101$$
$$e(101) = 100011$$
$$e(110) = 110111$$
$$e(111) = 110001$$

Decode the following words relative to a maximum likelihood decoding function.
    (a) 011110   (b) 101011   (c) 110010

12. Let

$$H = \begin{bmatrix} 0 & 1 & 1 \\ 1 & 0 & 1 \\ 1 & 0 & 0 \\ 0 & 1 & 0 \\ 0 & 0 & 1 \end{bmatrix}$$

be a parity check matrix. Decode the following words relative to a maximum likelihood decoding function associated with e_H.
    (a) 10100   (b) 01101   (c) 11011

13. Let

$$H = \begin{bmatrix} 1 & 0 & 0 \\ 1 & 1 & 0 \\ 0 & 1 & 1 \\ 1 & 0 & 0 \\ 0 & 1 & 0 \\ 0 & 0 & 1 \end{bmatrix}$$

be a parity check matrix. Decode the following words relative to a maximum likelihood decoding function associated with e_H.
    (a) 011001   (b) 101011   (c) 111010

KEY IDEAS FOR REVIEW

☐ Message: finite sequence of characters from a finite alphabet.

☐ Word: sequence of 0's and 1's.

☐ (m, n) encoding function: one-to-one function $e : B^m \to B^n$, $m < n$.

☐ Code word: element in Ran (e).

☐ Weight of x, $|x|$: number of 1's in x.

☐ Parity check code: see page 354.

☐ Hamming distance between x and y, $\delta(x, y)$: $|x \oplus y|$.

☐ Theorem (Properties of the Distance Function): Let x, y, and z be elements of B^m. Then

(a) $\delta(x, y) = \delta(y, x)$
(b) $\delta(x, y) \geq 0$
(c) $\delta(x, y) = 0$ if and only if $x = y$
(d) $\delta(x, y) \leq \delta(x, z) + \delta(z, y)$

☐ Minimum distance of an (m, n) encoding function: minimum of the distances between all distinct pairs of code words.

☐ Theorem: An (m, n) encoding function $e : B^m \to B^n$ can detect k or fewer errors if and only if its minimum distance is at least $k + 1$.

☐ Group code: (m, n) encoding function $e : B^m \to B^n$ such that $e(B^m) = \{e(b) \mid b \in B^m\}$ is a subgroup of B^n.

☐ Theorem: The minimum distance of a group code is the minimum weight of a nonzero code word.

☐ Mod 2 sum of Boolean matrices **D** and **E**, $\mathbf{D} \oplus \mathbf{E}$: see page 359.

☐ Mod 2 Boolean product of Boolean matrices **D** and **E**, $\mathbf{D} * \mathbf{E}$: see page 359.

☐ Theorem: Let m and n be nonnegative integers with $m < n$, $r = n - m$, and let **H** be an $n \times r$ Boolean matrix. Then the function $f_H : B^n \to B^r$ defined by

$$f_H(x) = x * H, \qquad x \in B^n$$

is a homomorphism from the group B^n to the group B^r.

☐ Group code e_H corresponding to parity check matrix H: see page 360.

☐ (n, m) decoding function: see page 365.

☐ Maximum likelihood decoding function associated with e: see page 366.

☐ Theorem: Suppose that e is an (m, n) encoding function and d is a maximum likelihood decoding function associated with e. Then (e, d) can correct k or fewer errors if and only if the minimum distance is at least $2k + 1$.

☐ Decoding procedure for a group code: see page 368.

☐ Decoding procedure for a group code given by a parity check matrix: see page 372.

FURTHER READING

BERLEKAMP, ELWYN R., *Algebraic Coding Theory*, McGraw-Hill, New York, 1968.
FISHER, JAMES L., *Applications-Oriented Algebra*, Harper and Row, New York, 1977.
PETERSON, W. WESLEY, and E. J. WELDON, *Error-Correcting Codes*, 2nd ed., MIT Press, Cambridge, Mass., 1972.

Appendix: Logic

Logic is the discipline that deals with the methods of reasoning. Logic provides rules and techniques for determining whether a given argument is valid. Logical reasoning is used in mathematics to prove theorems, in computer science to verify the correctness of programs and to prove theorems, in the natural and physical sciences to draw conclusions from experiments, in the social sciences, and in our everyday lives to solve a multitude of problems. Indeed, we are constantly using logical reasoning. In this appendix we review a few of the basic ideas.

A **statement** or **proposition** is a declarative sentence that is either true or false, but not both.

Example 1   Which of the following are statements?

(a) $2 + 3 = 5$
(b) The earth is round.
(c) $3 - x = 5$
(d) Do you speak English?
(e) Take two aspirins.
(f) The temperature on the surface of the planet Venus is 800°F.
(g) The sun will come out tomorrow.

Solution.
(a) and (b) are statements that happen to be true.
(c) is a declarative sentence, but it is not a statement, since it is true or false depending on the value of x that is used.

(d) is a question, so it is not a statement.

(e) is not a statement, it is a command.

(f) is a declarative sentence whose truth or falsity we do not know at this time; however, we can in principle determine if it is true or false, so it is a statement.

(g) is a statement since it is either true or false, but not both, although we would have to wait until tomorrow to find out if it is true or false.

Logical Connectives and Compound Statements

In mathematics, the letters x, y, z, ... denote variables that can be replaced by real numbers, and these variables can be combined by the familiar operations $+$, $\times$, $-$, and $\div$. In logic, the letters p, q, r, ... denote **propositional variables**, that is, variables that can be replaced by statements. Thus we can write

> p: The sun is shining today.

> q: It is cold.

Statements or propositional variables can be combined by logical connectives to obtain **compound statements**. For example, we may combine the statements above by the connective *and* to form the compound statement

$$\text{The sun is shining today } \textit{and} \text{ it is cold}$$

or

$$p \quad and \quad q$$

The truth value of a compound statement depends on the truth values of the statements being combined and on the types of connectives being used. We shall now look at the most important connectives.

If p is a statement, the **negation** of p is the statement *not p*, denoted by $\sim p$. Thus $\sim p$ is the statement "it is not the case that p." From this definition, it follows that if p is true, then $\sim p$ is false, and if p is false, then $\sim p$ is true. The truth value of $\sim p$, relative to p, is given in Table 1. Such a table, giving the truth values of a compound statement in terms of the truth values of its component parts, is called a **truth table**. Strictly speaking, *not* is not a connective, since it does not join two statements, and $\sim p$ is not really a compound statement.

Table 1

p	$\sim p$
T	F
F	T

Example 2   Give the negations of the following statements.

(a) p: $2 + 3 > 1$

(b) q: It is cold.

   Solution.

   (a) $\sim p$: $2 + 3$ is not greater than 1. That is,

$$\sim p: \quad 2 + 3 \leq 1$$

Since p is true in this case, $\sim p$ is false.

   (b) $\sim q$:   It is not the case that it is cold, or more simply

$$\sim q: \quad \text{It is not cold.}$$

If p and q are statements, the **conjunction** of p and q is the compound statement "p and q", denoted by $p \wedge q$. The connective *and* is denoted by the symbol $\wedge$. The compound statement $p \wedge q$ is true when both p and q are true; otherwise, it is false. The truth values of $p \wedge q$ in terms of the truth values of p and of q are given in the truth table shown in Table 2.

   Observe that in giving the truth table of $p \wedge q$ we needed to look at four possible cases. This follows from the fact that p and q can each be true or false.

Table 2

p	q	$p \wedge q$
T	T	T
T	F	F
F	T	F
F	F	F

Example 3   Form the conjunction of p and q:

(a) p: It is snowing; q: I am cold.

(b) p: $2 < 3$; q: $-5 > -8$

(c) p: It is snowing; q: $3 < 5$

   Solution.

   (a) $p \wedge q$:   It is snowing and I am cold.

   (b) $p \wedge q$: $2 < 3$ and $-5 > -8$

   (c) $p \wedge q$:   It is snowing and $3 < 5$.

Part (c) of Example 3 shows that in logic, unlike in every day English, we may join two totally unrelated statements by the connective *and*.

If p and q are statements, the **disjunction** of p and q is the compound statement "p or q," denoted by $p \lor q$. The connective *or* is denoted by the symbol $\lor$. The compound statement $p \lor q$ is true if at least one of p or q is true; it is false when both p and q are false. The truth values of $p \lor q$ are given in the truth table shown in Table 3.

Table 3

p	q	$p \lor q$
T	T	T
T	F	T
F	T	T
F	F	F

Example 4   Form the disjunction of p and q.

(a) p: 2 is a positive integer q: $\sqrt{2}$ is a rational number.

(b) p: $2 + 3 \neq 5$ q: London is the capital of France.

Solution.

(a) $p \lor q$:   2 is a positive integer or $\sqrt{2}$ is a rational number. Since p is true, the disjunction $p \lor q$ is true, even though q is false.

(b) $p \lor q$: $2 + 3 \neq 5$ or London is the capital of France. Since p and q are both false, $p \lor q$ is false.

Example 4(b) shows that in logic, unlike in ordinary English, we may join two totally unrelated statements by the connective *or*.

The connective *or* is more complicated than the connective *and* because it is used in two different ways in English. Suppose that we say "I drove to work or I took the train to work." In this compound statement we have the disjunction of the statements p: "I drove to work" and q: "I took the train to work." Of course, exactly one of the two possibilities occurred. Both could not have occurred, so the connective or is being used in an *exclusive* sense. On the other hand, consider the disjunction "I passed mathematics or I failed French." In this case, at least one of the two possibilities occurred. However, both could have occurred, so the connective *or* is being used in an *inclusive* sense. In mathematics and computer science we agree to always use the connective *or* in the inclusive manner.

If p and q are statements, the compound statement if p then q, denoted by $p \to q$, is called a **conditional statement** or **implication**. Statement p is called the **antecedent**, and statement q is called the **consequent**. The connective *if $\cdots$ then* is denoted by the symbol $\to$.

Example 5 Write the implication $p \rightarrow q$.

    (a) p: I am hungry; q: I will eat.
    (b) p: It is snowing; q: $3 + 5 = 8$.

Solution.

    (a) If I am hungry, then I will eat.
    (b) If it is snowing, then $3 + 5 = 8$.

Example 5(b) shows that in logic we use conditional statements in a more general sense than is customary. Thus in English, when we say "if p then q," we are tacitly assuming that there is a cause-and-effect relationship between p and q. That is, we would never use the statement in Example 5(b) in ordinary English, since there is no way that statement p can have any effect on statement q.

In logic, implication is used in a much weaker sense. To say that the compound statement $p \rightarrow q$ is true simply asserts that if p is true, then q will also be found to be true. In other words, $p \rightarrow q$ says only that we will not have p true and q false at the same time. It does not say that p "causes" q in the usual sense. Table 4 describes the truth values of $p \rightarrow q$ in terms of the truth values of p and q. Notice that $p \rightarrow q$ is considered false only if p is true and q is false. In particular, if p is false, then $p \rightarrow q$ is true for any q. This fact is sometimes described by the misleading statement: "A false hypothesis implies any conclusion." Similarly, if q is true, then $p \rightarrow q$ will be true for any statement p. The implication

$$\text{"If } 2 + 2 = 5 \text{, then I am king of England"}$$

is true, simply because p ("$2 + 2 = 5$") is false, so it is not the case that p is true and q false simultaneously.

Table 4

p	q	$p \rightarrow q$
T	T	T
T	F	F
F	T	T
F	F	T

In the English language, and in mathematics, each of the following expressions is an equivalent form of the conditional statement $p \rightarrow q$:

$$p \text{ implies } q$$

$$q, \text{ if } p$$

p only if q

p is a sufficient condition for q

q is a necessary condition for p

If $p \to q$ is an implication, then the **converse** of $p \to q$ is the implication $q \to p$, and the **contrapositive** of $p \to q$ is the implication $\sim q \to \sim p$.

Example 6 Give the converse and contrapositive of the implication "If it is raining, then I get wet."

Solution. We have p: it is raining and q: I get wet. Then the converse is $q \to p$: If I get wet, then it is raining. The contrapositive is $\sim q \to \sim p$: If I do not get wet, then it is not raining.

If p and q are statements, the compound statement p if and only if q, denoted by $p \leftrightarrow q$, is called an **equivalence**. The connective *if and only if* is denoted by the symbol $\leftrightarrow$. The truth values of $p \leftrightarrow q$ are given in Table 5. Observe that $p \leftrightarrow q$ is true only when both p and q are true or when both are false. The equivalence $p \leftrightarrow q$ is also stated as p is a necessary and sufficient condition for q.

Table 5

p	q	$p \leftrightarrow q$
T	T	T
T	F	F
F	T	F
F	F	T

Example 7 Is the following equivalence true?

$$3 > 2 \quad \text{if and only if} \quad 0 < 3 - 2$$

Solution. Let p be the statement $3 > 2$ and let q be the statement $0 < 3 - 2$. Since p and q are both true, we conclude that $p \leftrightarrow q$ is true.

In general, a compound statement may have many component parts, each of which is itself a statement, represented by some propositional variable. The statement

$$s: \quad p \to (q \wedge (p \to r))$$

involves three propositions, p, q, and r, each of which may independently be true or false. There are altogether $2^3 = 8$ possible combinations of truth values for p, q, and

r, and the truth table for s must give the truth or falsity of s in all these cases. If a compound statement s contains n component statements, there will be 2^n entries needed in the truth table for s. Such a truth table may be systematically constructed in the following way.

Step 1. The first n columns of the table are labeled by the component propositional variables. Further columns are constructed for all intermediate combinations of statements, culminating in the given statement.

Step 2. Under each of the first n headings, we list the 2^n possible n-tuples of truth values of the component statement s. Each n-tuple is listed on a separate row.

Step 3. For each row we compute, in sequence, all remaining truth values.

Example 8   Compute the truth table of the statement $(p \to q) \leftrightarrow (\sim q \to \sim p)$

Solution. The following table is constructed using Steps 1, 2, and 3 above.

p	q	$p \to q$	$\sim q$	$\sim p$	$\sim q \to \sim p$	$(p \to q) \leftrightarrow (\sim q \to \sim p)$
T	T	T	F	F	T	T
T	F	F	T	F	F	T
F	T	T	F	T	T	T
F	F	T	T	T	T	T

A statement that is true for all possible values of its propositional variables is called a **tautology**. A statement that is always false is called a **contradiction** or an **absurdity**, and a statement that can be either true or false, depending on the truth values of its propositional variables, is called a **contingency**.

Example 9

(a)   The statement in Example 8 is a tautology.
(b)   The statement $p \wedge \sim p$ is an absurdity (verify).
(c)   The statement $(p \to q) \wedge (p \vee q)$ is a contingency.

In Table 6 we have listed a number of important tautologies involving equivalences, all of which can be proved by the use of a truth table. When an equivalence is shown to be a tautology, it means that its two component parts are always either both true or both false, for any values of the propositional variables. Thus the two sides are simply different ways of making the same statement and we say that they are **logically equivalent**.

Tautologies 6, 10, and 11 of Table 6 give equivalent statements for the negation of the compound statements $p \wedge q$, $p \vee q$, $p \to q$, and $p \leftrightarrow q$. Tautology 8 of Table

Table 6 *Tautologies*

1. $p \wedge p \leftrightarrow p$	$p \vee p \leftrightarrow p$	(Idempotent property)
2. $p \wedge q \leftrightarrow q \wedge p$	$p \vee q \leftrightarrow q \vee p$	(Commutative property)
3. $(p \wedge q) \wedge r \leftrightarrow p \wedge (q \wedge r)$	$(p \vee q) \vee r \leftrightarrow p \vee (q \vee r)$	(Associative property)
4. $p \wedge (q \vee r) \leftrightarrow (p \wedge q) \vee (p \wedge r)$	$p \vee (q \wedge r) \leftrightarrow (p \vee q) \wedge (p \vee r)$	(Distributive property)
5. $\sim \sim p \leftrightarrow p$		
6. $\sim(p \wedge q) \leftrightarrow (\sim p) \vee (\sim q)$	$\sim(p \vee q) \leftrightarrow (\sim p) \wedge (\sim q)$	(De Morgan's laws)
7. $(p \rightarrow q) \leftrightarrow ((\sim p) \vee q)$		
8. $(p \rightarrow q) \leftrightarrow (\sim q \rightarrow \sim p)$		
9. $(p \leftrightarrow q) \leftrightarrow [(p \rightarrow q) \wedge (q \rightarrow p)]$		
10. $\sim(p \rightarrow q) \leftrightarrow p \wedge \sim q$		
11. $\sim(p \leftrightarrow q) \leftrightarrow [(p \wedge \sim q) \vee (q \wedge \sim p)]$		

6 shows us that any implication is logically equivalent with its contrapositive. We will use this property in proofs. When attempting to prove an implication, it is equivalent, and sometimes simpler, to prove the contrapositive.

Table 7 lists several important tautologies that are implications. These will be used extensively in proving results in mathematics and computer science and we will illustrate them below.

Table 7

1. $(p \wedge q) \rightarrow p$	$(p \wedge q) \rightarrow q$
2. $p \rightarrow (p \vee q)$	$q \rightarrow (p \vee q)$
3. $\sim p \rightarrow (p \rightarrow q)$	
4. $\sim(p \rightarrow q) \rightarrow p$	
5. $[p \wedge (p \rightarrow q)] \rightarrow q$	
6. $[\sim p \wedge (p \vee q)] \rightarrow q$	
7. $[\sim q \wedge (p \rightarrow q)] \rightarrow \sim p$	
8. $[(p \rightarrow q) \wedge (q \rightarrow r)] \rightarrow [p \rightarrow r]$	

Methods of Proof

If an implication $p \rightarrow q$ is true, where p and q may be compound statements involving any number of logical variables, we say that q **logically follows** from p. If an implication of the form $(p_1 \wedge p_2 \wedge \cdots \wedge p_n) \rightarrow q$ is true, we say that q **logically follows** from $p_1, p_2, \ldots, p_n$. When q logically follows from $p_1, p_2, \ldots, p_n$, we write

$$p_1$$
$$p_2$$
$$\vdots$$
$$\underline{p_n}$$
$$\therefore q$$

(where the symbol $\therefore$ means "therefore"). This means that if we know that p_1 is true, p_2 is true, $\ldots$, and p_n is true, then we know q is true. A statement q may or may not logically follow from statements $p_1, p_2, \ldots, p_n$, depending on the truth values of the propositional variables involved in these statements. This is because the truth value of the implication $(p_1 \wedge p_2 \wedge \cdots \wedge p_n) \rightarrow q$ depends on the truth values of these component variables.

Virtually all mathematical theorems are composed of implications of the type

$$(p_1 \wedge p_2 \wedge \cdots \wedge p_n) \rightarrow q.$$

The p_i's are called the **hypotheses**, and q is called the **conclusion**. To "prove the theorem" means to show that the *implication* is true. Note that we are not trying to show that q (the conclusion) is true, but only that q will be true if the p_i are all true. For this reason, mathematical proofs often begin with the statement "suppose that $p_1, p_2, \ldots,$ and p_n are true" and conclude with the statement "therefore, q is true." The proof does not show that q is true but simply shows that q has to be true if the p_i are all true; that is, the implication is true.

When an implication $p \rightarrow q$ or $(p_1 \wedge p_2 \wedge \cdots \wedge p_n) \rightarrow q$ is a tautology, then it is always true, regardless of the truth values of any component statements of q or the p_i's. In this case, the argument

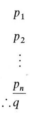

$$p_1$$
$$p_2$$
$$\vdots$$
$$p_n$$
$$\therefore q$$

is universally valid, no matter what actual statements are substituted for the variables in q and the p_i's. The validity depends on the form of the statements and not on their truth values. Such universally correct arguments therefore represent general methods of correct reasoning. They are called **rules of inference**. The various steps in the mathematical proof of a theorem must follow from the use of various rules of inference, and a mathematical proof of a theorem must begin with the hypotheses, proceed through various steps, each justified by some rule of inference, and arrive at the conclusion.

Example 10   According to item 8 of Table 7, $[(p \rightarrow q) \wedge (q \rightarrow r)] \rightarrow (p \rightarrow r)$ is a tautology. Thus the argument

$$p \rightarrow q$$
$$q \rightarrow r$$
$$\therefore p \rightarrow r$$

is universally valid, and so is a rule of inference.

Example 11   Is the following argument valid?

If you invest in the stock market, then you will get rich.
If you get rich, then you will be happy.
∴If you invest in the stock market, then you will be happy.

Solution. The argument is of the form

$$p \rightarrow q$$

$$q \rightarrow r$$
$$\therefore p \rightarrow r$$

Hence the argument is valid, although the conclusion may be false.

Example 12   The tautology $(p \leftrightarrow q) \leftrightarrow [(p \rightarrow q) \wedge (q \rightarrow p)]$ is given in Table 6, number 9. Thus both of the following arguments are valid:

$$p \leftrightarrow q$$
$$\therefore (p \rightarrow q) \wedge (q \rightarrow p)$$

$$p \rightarrow q$$
$$q \rightarrow p$$
$$\therefore p \leftrightarrow q$$

Some mathematical theorems are equivalences, that is, they are of the form $p \leftrightarrow q$. They are usually stated "p if and only if q." By Example 12, the proof of such a theorem is logically equivalent with proving both $p \rightarrow q$ and $q \rightarrow p$, and this is almost always the way in which equivalences are proved. We first assume that p is true, and show that q must then be true; next we assume that q is true and show that p must then be true.

A very important rule of inference is

$$p$$

$$p \rightarrow q$$
$$\therefore q$$

That is, "p is true, and $p \rightarrow q$ is true, so q is true." This follows from tautology number 5 in Table 7.

Some rules of inference were given Latin names by classical scholars. Tautology 5 is referred to as **modus ponens**, or loosely, the method of asserting.

Example 13   Is the following argument valid?

Smoking is healthy.
If smoking is healthy, then cigarettes are prescribed by physicians.
∴Cigarettes are prescribed by physicians.

Solution. The argument is valid since it is of the form modus ponens, discussed above. However, the conclusion is false. Observe that the first premise p: "smoking is healthy" is false. The second premise $p \to q$ is then true and $[p \land (p \to q)]$, the conjunction of the two premises is false.

Example 14   Is the following argument valid?

> If taxes were lowered, then my income rises.
> My income rises.
> ∴ Taxes were lowered.

Solution. Let p and q be the statements

$$p: \quad \text{taxes were lowered} \qquad q: \quad \text{my income rises}$$

Then the argument is of the form

$$p \to q$$
$$\dfrac{q}{\therefore p}$$

Assume that $p \to q$ and q are both true. Now $p \to q$ may be true with p being false. Then the conclusion p is false. Hence the argument is not valid. Another approach is to verify whether the statement $[(p \to q) \land q]$ logically implies the statement p. A truth table shows that this is not the case.

An important technique of proof called the **indirect method** of proof follows from the tautology

$$(p \to q) \leftrightarrow [(\sim q) \to (\sim p)]$$

This states, as we have mentioned previously, that an implication is equivalent with its contrapositive. Thus to prove $p \to q$ indirectly, we assume that q is false (the statement $\sim q$) and show that p is then false (the statement $\sim p$).

Example 15   Let n be an integer. Prove that if n^2 is odd, then n is odd.

Solution. Let p and q be the statements "n^2 is odd" and "n is odd," respectively. We have to prove that $p \to q$ is true. Instead, we prove the contrapositive $\sim q \to \sim p$. Thus suppose that n is not odd, so that n is even. Then $n = 2k$, where k is an integer. We have $n^2 = (2k)^2 = 4k^2 = 2(2k^2)$, so n^2 is even. We thus show that if n is even, then n^2 is even, which is the contrapositive of the given statement. Hence the given statement has been proved.

Another important technique of proof is **proof by contradiction**. This method is based on the tautology

$$[(p \to q) \land (\sim q)] \to (\sim p)$$

Thus the rule of inference

$$p \to q$$

$$\frac{\sim q}{\therefore \sim p}$$

is valid. Informally, this states that if a statement p implies a false statement q, then p must be false. This is often applied to the case where q is an absurdity or contradiction, that is, a statement which is always false, no matter what truth values its propositional variables have. An example is given by taking q as the contradiction $r \land (\sim r)$. Thus any statement that implies a contradiction must be false. In order to use proof by contradiction, suppose we wish to show that a statement q logically follows from statements $p_1, p_2, \ldots, p_n$. Assume that $\sim q$ is true (that is, q is false) as an extra hypothesis, and that $p_1, p_2, \ldots, p_n$ are also true. If this enlarged hypothesis $p_1 \land p_2 \land \cdots \land P_n \land (\sim q)$ implies a contradiction, then at least one of the statements $p_1, p_2, \ldots, p_n, (\sim q)$ must be false. This means that if all p_i's are true, then $\sim q$ must be false, so q must be true. Thus q follows from $p_1, p_2, \ldots, p_n$. This is proof by contradiction.

Example 16 Let a be an integer. Prove that if 2 divides a, then a is even. This fundamental result was used in Example 15.

Solution. Let p be the statement "2 divides a," and let q be the statement "a is even." We present a proof by contradiction. Assume that a is odd (not even). Then $a = 2k + 1$, where k is an integer. If 2 divides a, then $a = 2t$, where t is an integer. Hence $2t = 2k + 1$. Subtracting, we have $2t - 2k = 1$, or $2(t - k) = 1$. The left side of this equation is either negative or is ≥ 2. In either case, it cannot equal 1. Thus we have reached a contradiction, so the assumption that a is odd is incorrect. Hence a is even.

We have presented several rules of inference and logical equivalences which correspond to valid techniques of proof. In order to prove a theorem of the (typical) form $(p_1 \land p_2 \land \cdots \land p_n) \to q$, we begin with the hypotheses $p_1, p_2, \ldots, p_n$ and show that some result r_1, logically follows. Then, using $p_1, p_2, \ldots, p_n, r_1$, we show that some other statement r_2 logically follows. We continue this process, producing intermediate statements $r_1, r_2, \ldots, r_k$, called **steps in the proof**, until we can finally show that the conclusion q logically follows from $p_1, p_2, \ldots, p_n, r_1, \ldots, r_k$. Each logical step must be justified by some valid technique of proof, based on the rules of inference we have developed, or on other rules which come from tautological implications we have not discussed. At any stage, we may replace a statement that needs to be derived by its contrapositive statement, or by any other equivalent form.

In practice, the construction of proofs is an art, and must be learned in part from observation and experience. The choice of intermediate steps and methods of deriving them is a creative activity which cannot be precisely described.

Example 17   Let n and m be integers. Prove that $n^2 = m^2$ if and only if $n = m$ or $n = -m$.

Let us analyze this proof as we present it. Suppose that p is the statement "$n^2 = m^2$," q is the statement "$n = m$," and r is the statement "$n = -m$." Then we are asked to prove the theorem

$$p \leftrightarrow (q \lor r)$$

We know from previous discussion that we may instead prove that the two statements $s : p \to (q \lor r)$ and $t : (q \lor r) \to p$ are true. Let us begin by showing t to be true. Thus we assume that either $q : n = m$ or $r : n = -m$ is true. If q is true, then $n^2 = m^2$, and if r is true, then $n^2 = (-m)^2 = m^2$, so in either case p is true. We have therefore shown that the implication $t : q \lor r \to p$ is true.

Now we must prove that $s : p \to q \lor r$ is true; that is, we assume p and try to prove q or r. If p is true, then $n^2 = m^2$, so $n^2 - m^2 = (n - m)(n + m) = 0$. If r_1 is the intermediate statement $(n - m)(n + m) = 0$, we have shown that $p \to r_1$ is true. We now show that $r_1 \to (q \lor r)$ is true, by showing that the contrapositive statement $\sim(q \lor r) \to (\sim r_1)$ is true. Now $\sim(q \lor r)$ is equivalent with $(\sim q) \land (\sim r)$, so we show that $(\sim q) \land (\sim r) \to (\sim r_1)$. Thus, if $(\sim q) : n \neq m$ and $(\sim r) : n \neq -m$ are true, then $(n - m) \neq 0$ and $(n + m) \neq 0$, so $(n - m)(n + m) \neq 0$ and r_1 is false. We have therefore shown that $r_1 \to (q \lor r)$ is true. Finally, from the truth of $p \to r_1$ and $r_1 \to (q \lor r)$, we can conclude that $p \to (q \lor r)$ is true, and we are done.

We do not usually analyze proofs in this detailed manner. We have done so only to illustrate that proofs are devised by piecing together equivalences and valid steps resulting from rules of inference.

As a final remark, we caution the reader that many mathematical theorems actually state that some statement is true for all objects of a certain type. Sometimes this is not evident. Thus the theorem given in Example 16 really states that for all integers n and m, $n^2 = m^2$ if and only if $n = m$ or $n = -m$.

Similarly, the statement "If x and y are real numbers, and $x \neq y$, then $x < y$ or $y < x$" is a statement about all real numbers x and y. To prove such a theorem, we must make sure that the steps in the proof are each valid for every number. We could not assume, for example, that x was 2, or that y was π, or $\sqrt{3}$.

On the other hand, in order to prove that such a theorem is false, we need only find a single example where the property fails.

Example 18   Prove or disprove the statement that if x and y are real numbers,

$$(x^2 = y^2) \rightarrow (x = y).$$

To prove this result, we would need to provide steps, each of which would be true for all x and y. To disprove the result, we only need to find one example for which the implication is false. This is because the theorem claims to be true for all x and y.

Since $(-3)^2 = 3^2$, but $-3 \neq 3$, the theorem is false. This is called a **counterexample**, and any other counterexample would do just as well.

In summary, if a theorem claims that a property holds for all objects of a certain type, then to prove it, we must use steps that are valid for all objects of that type, and that do not make reference to any particular object. To disprove the theorem, we need only show one counterexample, that is, one particular object or set of objects for which the claim fails.

EXERCISE SET

1. Which of the following are statements?
   (a) Is 2 a positive number?
   (b) $x^2 + x + 1 = 0$
   (c) Study logic.
   (d) There will be snow in January.
   (e) If stocks fall, then I will lose money.

2. Give the negations of the following statements.
   (a) $2 + 7 \leq 11$
   (b) 2 is an even integer and 8 is an odd integer.
   (c) It will rain tomorrow or it will not snow tomorrow.
   (d) If you drive, then I will walk.

3. In each of the following, form the conjunction and the disjunction of p and q.
   (a) p: $3 + 1 < 5$; q: $7 = 3 \times 6$
   (b) p: I am rich; q: I am happy.
   (c) p: I will drive my car; q: I will be late.

4. Determine the truth or falsity of each of the following statements.
   (a) $2 < 3$ and 3 is a positive integer.
   (b) $2 \geq 3$ and 3 is a positive integer.
   (c) $2 < 3$ and 3 is not a positive integer.
   (d) $2 \geq 3$ and 3 is not a positive integer.

5. Determine the truth or falsity of each of the following statements.
   (a) $2 < 3$ or 3 is a positive integer.
   (b) $2 \geq 3$ or 3 is a positive integer.
   (c) $2 < 3$ or 3 is not a positive integer.
   (d) $2 \geq 3$ or 3 is not a positive integer.

6. Which of the following statements is the negation of the statements "2 is even and -3 is negative"?
   (a) 2 is even and -3 is not negative.
   (b) 2 is odd and -3 is not negative.
   (c) 2 is even or -3 is not negative.
   (d) 2 is odd or -3 is not negative.

7. Which of the following statements is the negation of the statement "2 is even or -3 is negative"?
   (a) 2 is even or -3 is not negative.
   (b) 2 is odd or -3 is not negative.
   (c) 2 is even and -3 is not negative.
   (d) 2 is odd and -3 is not negative.

8. State the converse of each of the following implications.
   (a) If $2 + 2 = 4$, then I am not King of England.
   (b) If I am not President of the United States, then I will walk to work.
   (c) If I am late, then I did not take the train to work.
   (d) If I have time and I am not too tired, then I will go to the store.
   (e) If I have enough money, then I will buy a car and I will buy a house.

9. State the contrapositive of each implication in Exercise 8.

10. Determine the truth or falsity of each of the following statements.
    (a) If 2 is even, then New York has a large population.
    (b) If 2 is even, then New York has a small population.
    (c) If 2 is odd, then New York has a large population.
    (d) If 2 is odd, then New York has a small population.

In Exercises 11 and 12, let p, q, and r be the following statements:

    p: I will study discrete structures.

    q: I will go to a movie.

    r: I am in a good mood.

11. Write the following statements in terms of p, q, and r and logical connectives.
    (a) If I am not in a good mood, then I will go to a movie.
    (b) I will not go to a movie, and I will study discrete structures.
    (c) I will go to a movie only if I will not study discrete structures.
    (d) If I will not study discrete structures, then I am not in a good mood.

12. Write the English sentences corresponding to the following statements.
    (a) $(\sim p) \wedge q$
    (b) $r \to (p \vee q)$
    (c) $(\sim r) \to ((\sim q) \vee p)$
    (d) $(q \wedge (\sim p)) \leftrightarrow r$

13. By examining truth tables, determine whether each of the following is a tautology, a contingency, or an absurdity.
    (a) $p \wedge \sim p$
    (b) $p \to (q \to p)$
    (c) $q \to (q \to p)$

    (d) $q \vee (\sim q \wedge p)$
    (e) $(q \wedge p) \vee (q \wedge \sim p)$
    (f) $(p \wedge q) \to p$
    (g) $p \to (q \wedge p)$

In Exercises 14–20, state whether the arguments given are valid or are not valid. If they are valid, identify the tautology or tautologies used.

14. If I drive to work, then I will arrive tired.
I am not tired when I arrive at work.

∴ I do not drive to work.

15. If I drove to work, then I arrive at work tired.
I arrive at work tired.

∴ I drove to work.

16. If I drive to work, then I will arrive tired.
I do not drive to work.

∴ I will not arrive tired.

17. If I drive to work, then I will arrive tired.
I drive to work.

∴ I will arrive tired.

18. I will become famous or I will not become a writer.
I will become a writer.

∴ I will become famous.

19. I will become famous or I will be a writer.
I will not be a writer.

∴ I will become famous.

20. If I try hard and I have talent, then I will become a musician.
If I become a musician, then I will be happy.

∴ If I will not be happy, then I did not try hard, or I do not have talent.

KEY IDEAS FOR REVIEW

☐ Statement: declarative sentence that is either true or false, but not both.
☐ Propositional variable: letter denoting a statement.
☐ Compound statement: statement obtained by combining two or more statements by a logical connective.

- [ ] Logical connectives: not ($\sim$), and ($\wedge$), or ($\vee$), if then ($\rightarrow$), if and only if ($\leftrightarrow$).
- [ ] Conjunction: $p \wedge q$ (p and q).
- [ ] Disjunction: $p \vee q$ (p or q).
- [ ] Conditional statement or implication: $p \rightarrow q$ (if p then q); p is the antecedent or hypothesis, q is the consequent or conclusion.
- [ ] Converse of $p \rightarrow q$: $q \rightarrow p$.
- [ ] Contrapositive of $p \rightarrow q$: $\sim q \rightarrow \sim p$.
- [ ] Equivalence: $p \leftrightarrow q$.
- [ ] Tautology: statement that is true for all possible values of its propositional variables.
- [ ] Absurdity: statement that is false for all possible values of its propositional variables.
- [ ] Contingency: statement that may be true or false, depending on the truth values of its propositional variables.
- [ ] Logically equivalent statements p and q: $p \leftrightarrow q$.
- [ ] Methods of proof:
    q logically follows from p: see page 383.
    Rules of inference: see page 384.
    Modus ponens: see page 385.
    Indirect method: see page 386.
    Proof by contradiction: see page 387.
- [ ] Counterexample: single instance that disproves a theorem or proposition.

FURTHER READING

MENDELSON, ELLIOTT, *Introduction to Mathematical Logic*, 2nd ed., D. Van Nostrand, Princeton, N.J., 1979.

SOLOW, DANIEL, *How to Read and Do Proofs*, Wiley, New York, 1982.

Answers
to Odd-numbered
Exercises

Section 1.1

1. (a) T   (b) F   (c) T   (d) F   (e) T   (f) F
3. (a) {A, R, D, V, K}   (b) {B, O, K}   (c) {M, I, S, P}
5. (a) $\{x \mid x$ is a positive even integer and $x \le 10\}$
   (b) $\{x \mid x$ is a vowel of the English alphabet$\}$
   (c) $\{x \mid x$ is a square of a positive integer $y \le 6\}$
   (d) $\{x \mid x \in Z$ and $|x| \le 2\}$.
7. (b) (c) and (e).
9. $\varnothing$, {ALGOL}, {BASIC}, {FORTRAN}, {ALGOL, BASIC}, {ALGOL, FORTRAN}, {BASIC, FORTRAN}, {ALGOL, BASIC, FORTRAN}
11. (a) T   (b) F   (c) T   (d) F   (e) T   (f) T   (g) T   (h) T
13. (a) $\subseteq$ (b) $\subseteq$ (c) $\nsubseteq$ (d) $\subseteq$ (e) $\nsubseteq$ (f) $\subseteq$
15. (a) F   (b) T   (c) F   (d) T   (e) T   (f) F

Section 1.2

1. {2, 1}   3. $\{x \mid x \in Z^{+}$ and x is odd$\}$ 5. {a, b, c, d}   7. 2, 4, 8, 16
9. $-1, 10, 27, 50$ 11. $x, y, z; x, x, y, y, z, z; x, y, z, x, y, z, x, \ldots$
13. $2n - 1$ 15. $(-1)^{n+1}$ 17. $3n - 2$

Section 1.3

1. (a) $\{a, b, c, d, e, f, g\}$ (b) $\{a, c, d, e, f, g\}$ (c) $\{a, c\}$ (d) $\{f\}$
   (e) $\{a, b, c\}$ (f) $\{d, e, f, h, k\}$ (g) $\{a, b, c, d, e, f\}$ (h) $\{b, f, g\}$
3. (a) $\{a, b, c, d, e, f, g\}$ (b) $\varnothing$ (c) $\{a, c, g\}$
   (d) $\{a, c, f\}$ (e) $\{h, k\}$ (f) $\{a, b, c, d, e, f, h, k\}$
5. (a) $\{1, 2, 4, 5, 6, 8, 9\}$ (b) $\{1, 2, 3, 4, 6, 8\}$ (c) $\{1, 2, 4\}$ (d) $\varnothing$
   (e) $\{1, 6, 8\}$ (f) $\{5, 9\}$ (g) $\{3, 5, 7, 9\}$ (h) $\{1, 5, 6, 8, 9\}$

7. (a) $\{1, 2, 3, 4, 5, 6, 8, 9\}$ (b) $\{1, 2, 3, 4, 5, 7, 8, 9\}$ (c) $\{2, 4\}$
(d) $\varnothing$ (e) $\{1, 2, 4, 6, 8\}$ (f) $\{\varnothing\}$

9. (a) $\{b, d, e, h\}$ (b) $\{b, c, d, f, g, h\}$ (c) $\{b, d, h\}$
(d) $\{b, c, d, e, f, g, h\}$ (e) $\varnothing$ (f) $\{c, f, g\}$

11. (a) $\{x \,|\, x \text{ is a real number and } x \neq \pm 1\}$. (b) $\{x \,|\, x \text{ is a real number and } x \neq -1, 4\}$.
(c) $\{x \,|\, x \text{ is a real number and } x \neq \pm 1, 4\}$. (d) $\{x \,|\, x \text{ is a real number and } x \neq -1\}$.

13. (a) T (b) T (c) F (d) F   **19.** (a) 20 (b) 325

21. (a) 10101011 (b) 00001010 (c) 10101010 (d) 11011111 (e) 00110000

Section 1.4

1. 67,600   **3.** 16   **5.** 1296   **7.** (a) 24 (b) 120 (c) 720

9. (a) $12! = 479{,}001{,}600$ (b) $2 \cdot (6!)^2 = 1{,}036{,}800$ **11.** 120   **13.** 120

15. (a) 1 (b) 35 (c) 6 (d) n (e) $\dfrac{n(n-1)}{2}$ (f) $\dfrac{n(n+1)}{2}$

17. 980   **19.** $C_7^{12} \cdot C_7^{15} = 5{,}096{,}520$ **21.** $C_5^{26} \cdot C_3^{26} = 171{,}028{,}000$

23. (a) $C_5^8 = 56$ (b) $C_5^7 = 21$ (c) $C_2^8 \cdot C_3^7 = 980$ (d) $C_3^8 \cdot C_2^7 = 1176$

Section 1.5

1. FUNCTION TAX (INCOME)
  1. IF (INCOME $\geq$ 30,000) THEN
    a. TAXDUE $\leftarrow$ 6000
  2. ELSE
    a. IF (INCOME $\geq$ 20,000) THEN
      1. TAXDUE $\leftarrow$ 2500
    b. ELSE
      1. TAXDUE $\leftarrow$ INCOME $\times$ 0.1
  3. RETURN (TAXDUE)
  END OF FUNCTION TAX

3. 1. SUM $\leftarrow$ 0
  2. FOR $I = 1$ THRU N
    a. SUM $\leftarrow$ SUM $+ X[I]$
  3. AVERAGE $\leftarrow$ SUM$/N$

5. 1. DOTPROD $\leftarrow$ 0
  2. FOR $I = 1$ THRU 3
    a. DOTPROD $\leftarrow$ DOTPROD $+ (X[I])(Y[I])$

7. 1. RAD $\leftarrow (A[2])^2 - 4(A[1])(A[3])$
  2. IF (RAD < 0) THEN
    a. PRINT ('ROOTS ARE IMAGINARY')
  3. ELSE
    a. IF (RAD $= 0$) THEN
      1. $R1 \leftarrow -A[2]/(2A[1])$
      2. PRINT ('ROOTS ARE REAL AND EQUAL')
    b. ELSE
      1. $R1 \leftarrow (-A[2] + \text{SQ(RAD)})/(2A[1])$
      2. $R2 \leftarrow (-A[2] - \text{SQ(RAD)})/(2A[1])$

9. 1. FOR $I = 1$ THRU N
    a. IF $(A[I] \neq B[I])$ THEN
        1. $C[I] \leftarrow 1$
    b. ELSE
        1. $C[I] \leftarrow 0$

11. 1. FOR $I = 1$ THRU N
    a. IF $(A[I] = 0$ AND $B[I] = 0)$ THEN
        1. $C[I] \leftarrow 1$
    b. ELSE
        1. $C[I] \leftarrow 0$

13. 1. SUM $\leftarrow 0$
    2. FOR $I = 0$ THRU $2(N - 1)$ BY 2
    a. SUM $\leftarrow$ SUM $+ I$

15. 1. PROD $\leftarrow 1$
    2. FOR $I = 2$ THRU $2N$ BY 2
    a. PROD $\leftarrow$ (PROD) $\times I$

17. 1. SUM $\leftarrow 0$
    2. FOR $I = 1$ THRU 77
    a. SUM $\leftarrow$ SUM $+ I^2$

19. 1. SUM $\leftarrow 0$
    2. FOR $I = 1$ THRU 10
    a. SUM $\leftarrow$ SUM $+ (1/(3I + 1))$

21. Finds the larger of the numbers X and Y.   **23.** Finds $|X|$.
25. If $M = kN$, $k \geq 1$, then assigns R the value 1; otherwise, assigns R the value 0.
27. $I = N + 1; X = \sum_{j=1}^{N} j$ **29.** $X = 25; I = 49$

Section 1.6

27. $\{x \mid x = 3k, k \in Z, k \geq 0\}$ **29.** 0, 1, 1, 3, 5, 11
33. (a) Belongs   (b) Belongs   (c) Belongs   (d) Does not belong   (e) Does not belong

Section 1.7

1. $20 = 6 \cdot 3 + 2$ **3.** $3 = 0 \cdot 22 + 3$ **5.** $8 = (-2)(-3) + 2$ **7.** (b), (c), (h)
9. $1; 1 = 11(32) - 13(27)$ **11.** $8; 8 = 1(88) - 2(40)$ **13.** $2; 2 = 12(34) - 7(58)$
15. 216   **17.** 1225

Section 1.8

1. (a) $a_{12} = -2, a_{22} = 1, a_{23} = 2$ (b) $b_{11} = 3, b_{31} = 4$
    (c) $c_{13} = 4, c_{21} = 5, c_{33} = 8$ (d) 2, 6, 8
3. $a = 3, b = 1, c = 8, d = -2$
5. (a) $\begin{bmatrix} 4 & 0 & 2 \\ 9 & 6 & 2 \\ 3 & 2 & 4 \end{bmatrix}$ (b) $\mathbf{AB} = \begin{bmatrix} 7 & 13 \\ -3 & 0 \end{bmatrix}$ $\mathbf{BA} = \begin{bmatrix} 4 & 1 & -2 \\ 10 & 3 & -1 \\ 16 & 5 & 0 \end{bmatrix}$ (c) $\begin{bmatrix} -5 & 0 \\ 4 & -7 \end{bmatrix}$

    (d) not possible   (e) $\begin{bmatrix} 21 & 14 \\ -7 & 17 \end{bmatrix}$ (f) $\begin{bmatrix} 0 & 8 \\ 8 & 16 \\ 16 & 24 \end{bmatrix}$

7. (a) Not possible   (b) Not possible   (c) Not possible

(d) $\begin{bmatrix} 10 & 0 & -25 \\ 40 & 14 & 12 \end{bmatrix}$ (e) $\begin{bmatrix} -16 & 24 \\ 32 & 40 \end{bmatrix}$ (f) Not possible

9. (a) $\begin{bmatrix} 5 & -3 \\ -10 & -29 \\ 15 & 41 \end{bmatrix}$ (b) $\begin{bmatrix} 22 & 34 \\ 3 & 11 \\ -31 & 3 \end{bmatrix}$ (c) $\begin{bmatrix} 15 & 48 \\ 15 & 90 \\ 21 & -9 \end{bmatrix}$

(d) Not possible   (e) $\begin{bmatrix} 25 & 5 & 26 \\ 20 & -3 & 32 \end{bmatrix}$ (f) Not possible

13. (a) $\begin{bmatrix} 27 & 0 & 0 \\ 0 & -8 & 0 \\ 0 & 0 & 64 \end{bmatrix}$ (b) $\begin{bmatrix} 3^k & 0 & 0 \\ 0 & (-2)^k & 0 \\ 0 & 0 & 4^k \end{bmatrix}$

19. $\mathbf{A} = \begin{bmatrix} 1 & 0 \\ 0 & 0 \end{bmatrix}$ $\mathbf{B} = \begin{bmatrix} 0 & 0 \\ 0 & 1 \end{bmatrix}$

29. (a) $\mathbf{A} \vee \mathbf{B} = \begin{bmatrix} 1 & 1 \\ 1 & 1 \end{bmatrix}$; $\mathbf{A} \wedge \mathbf{B} = \begin{bmatrix} 0 & 0 \\ 0 & 1 \end{bmatrix}$; $\mathbf{A} \odot \mathbf{B} = \begin{bmatrix} 1 & 1 \\ 1 & 1 \end{bmatrix}$

(b) $\mathbf{A} \vee \mathbf{B} = \begin{bmatrix} 0 & 1 & 1 \\ 1 & 1 & 0 \\ 1 & 0 & 1 \end{bmatrix}$; $\mathbf{A} \wedge \mathbf{B} = \begin{bmatrix} 0 & 0 & 1 \\ 1 & 1 & 0 \\ 1 & 0 & 0 \end{bmatrix}$; $\mathbf{A} \odot \mathbf{B} = \begin{bmatrix} 1 & 0 & 1 \\ 1 & 1 & 1 \\ 0 & 1 & 1 \end{bmatrix}$

Section 2.1

1. (a) $x = 4$ (b) $y = 3$ (c) $x = 2$ (d) $y = $ ALGOL

3. (a) $\{(a, 4), (a, 5), (a, 6), (b, 4), (b, 5), (b, 6)\}$

(b) $\{(4, a), (5, a), (6, a), (4, b), (5, b), (6, b)\}$

(c) $\{(a, a), (a, b), (b, a), (b, b)\}$

(d) $\{(4, 4), (4, 5), (4, 6), (5, 4), (5, 5), (5, 6), (6, 4), (6, 5), (6, 6)\}$

5. (a) 6   (b) $\{(m, s), (m, m), (m, l), (f, s), (f, m), (f, l)\}$ **7.** 676

9. $\{(a, 1, r), (a, 1, s), (a, 2, r), (a, 2, s), (b, 1, r), (b, 1, s), (b, 2, r), (b, 2, s), (c, 1, r), (c, 1, s), (c, 2, r), (c, 2, s)\}$

11.   **15.** No

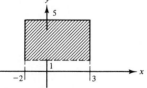

17. $\{\{a\}, \{b\}, \{c\}, \{d\}\}, \{\{a, b\}, \{c, d\}\}, \{\{a, c\}, \{b, d\}\}, \{\{a, d\}, \{b, c\}\}, \{\{a, b\}, \{c\}, \{d\}\},$
$\{\{a, c\}, \{b\}, \{d\}\}, \{\{a, d\}, \{b\}, \{c\}\}, \{\{b, c\}, \{a\}, \{d\}\}, \{\{b, d\}, \{a\}, \{c\}\},$
$\{\{c, d\}, \{a\}, \{b\}\}, \{\{a, b, c\}, \{d\}\} \{\{a, b, d\}, \{c\}\}, \{\{a, c, d\}, \{b\}\}, \{\{b, c, d\}, \{a\}\}, \{\{a, b, c, d\}\}$

Section 2.2

1. (b), (d), (e), (f)

3. Dom $(R) = A$, Ran $R = \{1, 2\}$

$$\mathbf{M}_R = \begin{bmatrix} 1 & 1 & 0 \\ 1 & 0 & 0 \\ 0 & 1 & 0 \\ 1 & 0 & 0 \end{bmatrix}$$

5. Dom (R) = {IBM, Univac, Commodore, Atari}
Ran (R) = {370, 1110, 4000, 800}

$$\mathbf{M}_R = \begin{bmatrix} 1 & 0 & 0 & 0 & 0 & 0 & 0 \\ 0 & 1 & 0 & 0 & 0 & 0 & 0 \\ 0 & 0 & 1 & 0 & 0 & 0 & 0 \\ 0 & 0 & 0 & 0 & 0 & 1 & 0 \\ 0 & 0 & 0 & 0 & 0 & 0 & 0 \end{bmatrix}$$

7. Dom (R) = A = Ran (R)

$$\mathbf{M}_R = \begin{bmatrix} 1 & 0 & 0 & 0 & 0 \\ 0 & 1 & 0 & 0 & 0 \\ 0 & 0 & 1 & 0 & 0 \\ 0 & 0 & 0 & 1 & 0 \\ 0 & 0 & 0 & 0 & 1 \end{bmatrix}$$

9. Dom (R) = A = Ran (R)

$$\mathbf{M}_R = \begin{bmatrix} 1 & 0 & 0 & 0 & 0 \\ 1 & 1 & 0 & 0 & 0 \\ 1 & 0 & 1 & 0 & 0 \\ 1 & 1 & 0 & 1 & 0 \\ 1 & 1 & 1 & 0 & 1 \end{bmatrix}$$

11. Dom (R) = {3, 5, 7, 9}, Ran (R) = {2, 4, 6, 8}

$$\mathbf{M}_R = \begin{bmatrix} 0 & 0 & 0 & 0 \\ 1 & 0 & 0 & 0 \\ 1 & 1 & 0 & 0 \\ 1 & 1 & 1 & 0 \\ 1 & 1 & 1 & 1 \end{bmatrix}$$

13. Dom (R) = {1, 2, 3, 4}, Ran (R) = {2, 3, 4, 5}

$$\mathbf{M}_R = \begin{bmatrix} 0 & 1 & 0 & 0 & 0 & 0 \\ 0 & 0 & 1 & 0 & 0 & 0 \\ 0 & 0 & 0 & 1 & 0 & 0 \\ 0 & 0 & 0 & 0 & 1 & 0 \\ 0 & 0 & 0 & 0 & 0 & 0 \\ 0 & 0 & 0 & 0 & 0 & 0 \end{bmatrix}$$

15. Dom (R) = $[-5, 5]$ = Ran (R)
17. $R = \{(a_1, b_1), (a_1, b_2), (a_1, b_5), (a_2, b_1), (a_2, b_3), (a_3, b_1), (a_3, b_3), (a_4, b_5), (a_5, b_1), (a_5, b_5), (a_5, b_6)\}$
19. $R = \{(1, 1), (1, 3), (2, 3), (3, 1), (4, 1), (4, 2), (4, 4)\}$
21. $R = \{(1, 2), (1, 3), (1, 4), (2, 2), (2, 3), (4, 1), (4, 4), (4, 5)\}$
23.

	Vertex 1	Vertex 2	Vertex 3	Vertex 4	Vertex 5
In-degree	1	2	1	3	0
Out-degree	1	2	1	1	2

25. $\{(1, 2), (1, 4), (2, 5)\}$

Section 2.3

1. (1, 2), (1, 6), (2, 3), (3, 3), (3, 4), (4, 1), (4, 3), (4, 5), (6, 4)

3. (3, 3, 4, 1), (3, 3, 3, 3), (3, 3, 4, 5), (3, 4, 3, 4), (3, 4, 1, 2), (3, 3, 3, 4), (3, 4, 1, 6), (3, 4, 3, 3), (3, 3, 4, 3)

5. (6, 4, 1, 6) or (6, 4, 3, 4, 1, 6)

7. (1, 2, 3, 3), (1, 2, 3, 4), (1, 6, 4, 1), (1, 6, 4, 3), (1, 6, 4, 5), (2, 3, 3, 3),
(2, 3, 3, 4), (2, 3, 4, 3), (2, 3, 4, 1), (2, 3, 4, 5), (3, 3, 3, 3), (3, 3, 3, 4),
(3, 3, 4, 3), (3, 3, 4, 1), (3, 3, 4, 5), (3, 4, 1, 2), (3, 4, 1, 6), (3, 4, 3, 4),
(3, 4, 3, 3), (4, 3, 3, 3), (4, 3, 4, 3), (4, 3, 4, 1), (4, 1, 6, 4), (4, 3, 4, 5),
(4, 1, 2, 3), (6, 4, 3, 4), (6, 4, 3, 3), (6, 4, 1, 6), (6, 4, 1, 2)

9. $\begin{bmatrix} 0 & 0 & 1 & 1 & 0 & 0 \\ 0 & 0 & 1 & 1 & 0 & 0 \\ 1 & 0 & 1 & 1 & 1 & 0 \\ 0 & 1 & 1 & 1 & 0 & 1 \\ 0 & 0 & 0 & 0 & 0 & 0 \\ 1 & 0 & 1 & 0 & 1 & 0 \end{bmatrix}$
11. $\begin{bmatrix} 1 & 1 & 1 & 1 & 1 & 1 \\ 1 & 1 & 1 & 1 & 1 & 1 \\ 1 & 1 & 1 & 1 & 1 & 1 \\ 1 & 1 & 1 & 1 & 1 & 1 \\ 0 & 0 & 0 & 0 & 0 & 0 \\ 1 & 1 & 1 & 1 & 1 & 1 \end{bmatrix}$

13. $(c, d, c), (c, e, f), (c, d, b)$ **15.** (c, d, c) or (c, d, b, f, d, c)

17. $(a, b, b), (a, b, f), (a, c, d), (a, c, e), (b, b, b), (b, b, f), (b, f, d), (c, d, c), (c, d, b), (c, e, f), (d, c, d), (d, c, e), (d, b, b),$
$(d, b, f), (e, f, d), (f, d, b), (f, d, c)$

19.

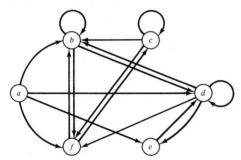

21. $\{(a, b), (a, c), (a, d), (a, e), (a, f), (b, b), (b, c), (b, d), (b, e), (b, f), (c, b), (c, c), (c, d), (c, e), (c, f), (d, b), (d, c), (d, e),$
$(d, d), (d, f), (e, b), (e, c), (e, d), (e, e), (e, f), (f, b), (f, c), (f, d), (f, e), (f, f)\}$

25. 1, 2, 4, 3, 5, 6, 4   **27.** 3, 4, 7, 6, 5

Section 2.4

	Reflexive	Irreflexive	Symmetric	Asymmetric	Antisymmetric	Transitive
1.	Y	N	Y	N	N	Y
3.	N	N	N	N	N	N
5.	N	Y	Y	Y	Y	Y
7.	N	N	N	N	N	Y
9.	N	N	N	N	Y	Y
11.	N	Y	Y	N	N	N
13.	Y	N	N	N	N	N
15.	Y	N	N	N	Y	Y
17.	N	Y	Y	N	N	N
19.	N	N	Y	N	N	N

21.

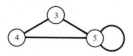

23. $\{(1, 2), (2, 1), (1, 5), (5, 1), (1, 6), (6, 1), (2, 3), (3, 2), (2, 7), (7, 2), (6, 5), (5, 6)\}$
25. Yes **27.** Yes **29.** No **31.** No **33.** Yes **35.** Yes
37. (b) $\{\{(1, 1), (2, 2), (3, 3), (4, 4), (5, 5)\}, \{(1, 2), (2, 4)\}, \{(2, 1), (4, 2)\}, \{(1, 3)\}, \{(3, 1)\}, \{(1, 4)\}, \{(4, 1)\}, \{(1, 5)\},$
$\{(5, 1)\}, \{(2, 3)\}, \{(3, 2)\}, \{(2, 5)\}, \{(5, 2)\}, \{(3, 4)\}, \{(4, 3)\}, \{(3, 5)\}, \{(5, 3)\}, \{(4, 5)\}, \{(5, 4)\}\}$
39. $\{(1, 1), (1, 3), (1, 5), (3, 1), (3, 3), (3, 5), (5, 1), (5, 3), (5, 5), (2, 2), (2, 4), (4, 2), (4, 4)\}$

Section 2.5

5.
$$\mathbf{M}_R = \begin{bmatrix} 1 & 1 & 1 & 0 \\ 0 & 0 & 1 & 1 \\ 1 & 0 & 0 & 1 \\ 0 & 1 & 0 & 0 \end{bmatrix}$$

VERT	TAIL	HEAD	NEXT
1	1	1	2
4	1	2	3
6	1	3	0
8	2	3	5
	2	4	0
	3	1	7
	3	4	0
	4	2	0

7.
$$\mathbf{M}_R = \begin{bmatrix} 0 & 1 & 1 & 0 \\ 0 & 1 & 1 & 0 \\ 0 & 0 & 0 & 1 \\ 1 & 0 & 1 & 1 \end{bmatrix}$$

Section 2.6

1. (a) $\{(1, 3), (2, 2), (3, 1), (3, 3), (4, 1), (4, 2)\}$
  (b) $\{(1, 1), (1, 2), (1, 3), (2, 1), (2, 3), (3, 2), (3, 3), (4, 3)\}$
  (c) $\{(2, 3), (3, 2)\}$ (d) $\{(3, 1), (3, 2), (2, 3), (3, 3)\}$
3. (a) $\{(1, 3), (2, 1), (2, 2), (3, 2), (3, 3)\}$
  (b) $\{(1, 1), (1, 2), (2, 1), (2, 3), (3, 1), (3, 2), (3, 3)\}$
  (c) $\{(3, 1)\}$ (d) $\{(1, 2), (1, 3), (2, 3), (3, 3)\}$
5. (a) $\{(1, 4), (2, 1), (3, 1), (3, 2), (3, 3), (4, 2), (4, 3), (4, 4)\}$
  (b) $\{(1, 1), (1, 2), (2, 2), (2, 3), (2, 4), (4, 1)\}$

(c) $\{(1, 1), (1, 2), (1, 3), (1, 4), (2, 2), (2, 3), (2, 4), (3, 1), (3, 2), (3, 4), (4, 1), (4, 3), (4, 4)\}$

(d) $\{(1, 1), (2, 1), (4, 1), (2, 2), (3, 2), (4, 2), (2, 3), (1, 3), (4, 4), (1, 4), (3, 4)\}$.

7. (a) $\{(1, 1), (1, 4), (2, 2), (2, 3), (3, 3), (3, 4)\}$ (b) $\{(1, 2), (2, 4), (3, 1), (3, 2)\}$

    (c) $\{(1, 1), (1, 2), (1, 3), (1, 4), (2, 1), (2, 4), (3, 1), (3, 2), (3, 3)\}$

    (d) $\{(1, 1), (1, 3), (2, 1), (2, 3), (3, 3), (4, 1), (4, 2)\}$

9. (a) $\{(3, 1), (1, 2), (3, 2), (1, 3), (2, 3), (4, 3), (2, 4)\}$

(b)

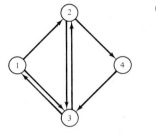

(c)
$$\mathbf{M}_{\bar{R}} = \begin{bmatrix} 1 & 1 & 0 & 1 \\ 0 & 1 & 0 & 1 \\ 0 & 0 & 1 & 0 \\ 1 & 0 & 1 & 1 \end{bmatrix}$$

11. (a) $\begin{bmatrix} 0 & 0 & 0 \\ 0 & 0 & 1 \\ 0 & 0 & 0 \\ 0 & 1 & 0 \end{bmatrix}$ (b) $\begin{bmatrix} 1 & 1 & 1 \\ 1 & 1 & 1 \\ 0 & 1 & 1 \\ 1 & 1 & 1 \end{bmatrix}$ (c) $\begin{bmatrix} 0 & 0 & 0 & 1 \\ 1 & 1 & 0 & 1 \\ 0 & 1 & 1 & 1 \end{bmatrix}$ (d) $\begin{bmatrix} 0 & 1 & 0 \\ 0 & 1 & 0 \\ 1 & 0 & 1 \\ 1 & 0 & 1 \end{bmatrix}$

17. Partition corresponding to $R \cap S$ is

    $\{\{1, 2\}, \{3\}, \{4\}, \{5\}, \{6\}\}$.

19. (a) Yes   (b) No   (c) $\{(a, \square), (a, \diamondsuit), (a, \triangle), (c, \square), (b, \triangle), (c, \diamondsuit), (d, \triangle)\}$.

21. All pairs of people in A who are either brothers or sisters.

23. All pairs of people in A who have the same last name and receive the same benefits.

25. $R = \{(2, 6), (2, 12), (6, 2), (12, 2), (2, 2), (6, 6), (3, 3), (12, 12), (6, 12), (12, 6)\}$

    $S = \{(3, 6), (6, 3), (3, 12), (12, 3), (3, 3), (6, 6), (12, 12), (12, 6), (6, 12), (2, 2)\}$

    (a) $\{(2, 3), (3, 2), (3, 6), (3, 12), (12, 3), (6, 3)\}$

    (b) $\{(3, 3), (6, 6), (12, 12), (6, 12), (12, 6), (2, 2)\}$

    (c) $\{(2, 2), (2, 6), (2, 12), (3, 3), (3, 6), (3, 12), (6, 2), (6, 3), (6, 6), (6, 12), (12, 2), (12, 3), (12, 6), (12, 12)\}$.

    (d) S

27. (a) Yes   (b) Yes   (c) Yes

    (d) $\{(1, 1), (1, 2), (1, 3), (1, 4), (2, 4), (2, 1), (2, 2), (3, 1), (3, 2), (4, 1), (4, 2), (4, 3), (4, 4)\}$

    (e) $\{(1, 1), (1, 4), (1, 3), (2, 1), (2, 4), (3, 4), (4, 1), (4, 3), (4, 4)\}$

    (f) $\{(3, 1), (3, 2), (4, 1), (4, 2), (2, 4), (2, 1), (2, 2), (1, 1), (1, 2)\}$

    (g) $\{(3, 1), (3, 4), (4, 4), (2, 1), (2, 4), (1, 1), (1, 4)\}$

31. (a) $\begin{bmatrix} 1 & 1 & 0 & 1 & 1 \\ 0 & 1 & 0 & 1 & 1 \\ 1 & 1 & 0 & 1 & 1 \\ 1 & 1 & 0 & 1 & 1 \\ 1 & 1 & 0 & 0 & 1 \end{bmatrix}$ (b) $\begin{bmatrix} 1 & 0 & 1 & 1 & 1 \\ 1 & 1 & 0 & 1 & 1 \\ 1 & 0 & 1 & 1 & 1 \\ 1 & 1 & 1 & 1 & 1 \\ 0 & 0 & 0 & 1 & 1 \end{bmatrix}$

    (c) $\begin{bmatrix} 1 & 1 & 0 & 1 & 1 \\ 1 & 1 & 0 & 0 & 1 \\ 0 & 1 & 0 & 1 & 1 \\ 1 & 1 & 0 & 1 & 1 \\ 1 & 1 & 0 & 0 & 1 \end{bmatrix}$ (d) $\begin{bmatrix} 1 & 1 & 1 & 1 & 1 \\ 1 & 0 & 1 & 1 & 1 \\ 1 & 1 & 0 & 1 & 1 \\ 1 & 1 & 1 & 1 & 1 \\ 0 & 1 & 0 & 1 & 1 \end{bmatrix}$

Section 2.7

1. $\begin{bmatrix} 1 & 1 & 1 \\ 1 & 1 & 1 \\ 1 & 1 & 1 \end{bmatrix}$ **3.** $\begin{bmatrix} 1 & 1 & 1 \\ 1 & 1 & 1 \\ 1 & 1 & 1 \end{bmatrix}$

7. $\begin{bmatrix} 1 & 0 & 0 & 1 \\ 1 & 1 & 0 & 1 \\ 0 & 0 & 1 & 0 \\ 0 & 0 & 0 & 1 \end{bmatrix}$ **9.** $\begin{bmatrix} 1 & 0 & 0 & 1 \\ 0 & 1 & 1 & 0 \\ 0 & 1 & 1 & 0 \\ 1 & 0 & 0 & 1 \end{bmatrix}$ **11.** $\begin{bmatrix} 1 & 1 & 1 & 0 & 0 \\ 1 & 1 & 1 & 0 & 0 \\ 1 & 1 & 1 & 0 & 0 \\ 0 & 0 & 0 & 1 & 1 \\ 0 & 0 & 0 & 1 & 1 \end{bmatrix}$

Section 3.1

1. Function, range $= \{1, 2\}$ **3.** Not a function   **5.** Function
11. (a) 3   (b) 1   (c) $(x-1)^2$ (d) x^2-1 (e) $x-2$ (f) x^4
15. One-to-one and onto   **17.** Not one-to-one, onto
19. Not one-to-one, not onto   **21.** Not one-to-one, onto
27. $f^{-1}(b) = \sqrt[3]{b-1}$ **29.** $f^{-1} = \{(1, 5), (2, 2), (3, 1), (4, 3), (5, 4)\}$

31. Yes   **33.** n^n **35.** n^m

Section 3.2

1. (a), (c)

3. (a) $\begin{pmatrix} 1 & 2 & 3 & 4 & 5 & 6 \\ 3 & 4 & 1 & 2 & 6 & 5 \end{pmatrix}$ (b) $\begin{pmatrix} 1 & 2 & 3 & 4 & 5 & 6 \\ 2 & 5 & 6 & 3 & 1 & 4 \end{pmatrix}$

(c) $\begin{pmatrix} 1 & 2 & 3 & 4 & 5 & 6 \\ 5 & 2 & 1 & 6 & 3 & 4 \end{pmatrix}$ (d) $\begin{pmatrix} 1 & 2 & 3 & 4 & 5 & 6 \\ 4 & 5 & 1 & 2 & 6 & 3 \end{pmatrix}$

5. $\begin{pmatrix} 1 & 2 & 3 & 4 & 5 & 6 & 7 & 8 \\ 5 & 1 & 2 & 4 & 7 & 6 & 8 & 3 \end{pmatrix}$ **7.** $\begin{pmatrix} 1 & 2 & 3 & 4 & 5 & 6 & 7 & 8 \\ 5 & 1 & 3 & 2 & 6 & 7 & 4 & 8 \end{pmatrix}$

9. $\begin{pmatrix} a & b & c & d & e & f & g \\ f & c & d & e & b & g & a \end{pmatrix}$ **11.** (1, 4, 5) (2, 3) (6, 8)

13. (1, 6, 3, 7, 2, 5, 4, 8)   **15.** (a, g, e, c, b, d)
17. (2, 1) (2, 4) (2, 5) (2, 8) (2, 6)   **19.** Even   **21.** Even

Section 4.1

1. No   **3.** Yes   **5.** Yes   **7.** No
9. $\{(a, a), (b, b), (c, c), (a, b)\}$
$\{(a, a), (b, b), (c, c), (a, b), (a, c)\}$
$\{(a, a), (b, b), (c, c), (a, b), (c, b)\}$
$\{(a, a), (b, b), (c, c), (a, b), (b, c), (a, c)\}$
$\{(a, a), (b, b), (c, c), (a, b), (c, b), (c, a)\}$

11. 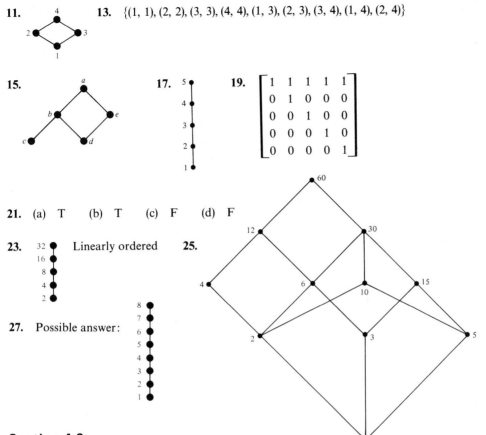 **13.** $\{(1, 1), (2, 2), (3, 3), (4, 4), (1, 3), (2, 3), (3, 4), (1, 4), (2, 4)\}$

15.   **17.**   **19.**
$$\begin{bmatrix} 1 & 1 & 1 & 1 & 1 \\ 0 & 1 & 0 & 0 & 0 \\ 0 & 0 & 1 & 0 & 0 \\ 0 & 0 & 0 & 1 & 0 \\ 0 & 0 & 0 & 0 & 1 \end{bmatrix}$$

21. (a) T   (b) T   (c) F   (d) F

23. Linearly ordered   **25.**

27. Possible answer:

Section 4.2

1. Maximal elements: 3, 5; minimal elements: 1, 6
3. Maximal elements: e, f; minimal elements: a **5.** No maximal or minimal elements
7. Maximal elements: 1; minimal elements: none   **9.** Greatest element: f; least element: a
11. No greatest element; no least element   **13.** No greatest element; no least element
15. Greatest element: 72; least element: 2
17. (a) Upper bounds: f, g, h (b) Lower bounds: c, a, b (c) LUB: f (d) GLB: c
19. (a) Upper bounds: d, e, f (b) Lower bounds: b, a (c) LUB: d (d) GLB: b
21. (a) Upper bounds: none   (b) Lower bounds: b (c) LUB: none   (d) GLB: b
23. (a) Upper bounds: none   (b) Lower bounds: 1, 2, 3   (c) LUB: none   (d) GLB: 3
25. (a) Upper bounds: $\{x \mid x \geq 2\}$ (b) Lower bounds: $\{x \mid x \leq 1\}$ (c) LUB: 2   (d) GLB: 1
27. (a) Upper bounds: 12, 24, 48   (b) Lower bounds: 2   (c) LUB: 12   (d) GLB: 2

29. Possible answer:

Section 4.3

1. Yes **3.** No **5.** Yes **7.** No

13. One element: •; two elements: •; three elements: •; four elements: • , •;

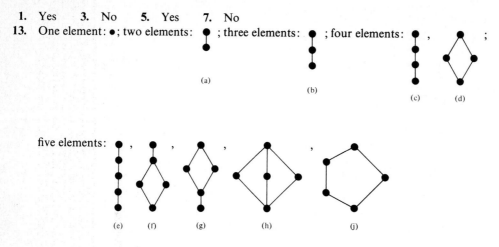

(a)   (b)   (c)   (d)

five elements: • , • , • , • , •

(e)   (f)   (g)   (h)   (j)

21. Yes **27.** $1' = 105; 3' = 35; 5' = 21; 7' = 15; 15' = 7; 35' = 3; 21' = 5; 105' = 1$

29. Nondistributive, noncomplemented **31.** Distributive, noncomplemented

33. $a' = e, b' = c, c' = b$ or $d, d' = c, e' = a$

Section 4.4

1. No **3.** No **5.** Yes **7.** Yes **9.** Yes

23.

x	y	z	$f(x, y, z) = x \wedge (y \vee z')$
0	0	0	0
0	0	1	0
0	1	0	0
0	1	1	0
1	0	0	1
1	0	1	0
1	1	0	1
1	1	1	1

25.

x	y	z	$f(x, y, z) = (x \wedge y') \vee (y \wedge (x' \vee y))$
0	0	0	0
0	0	1	0
0	1	0	1
0	1	1	1
1	0	0	1
1	0	1	1
1	1	0	1
1	1	1	1

27.

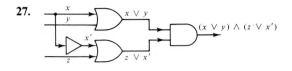
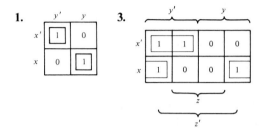

Section 4.5

1.

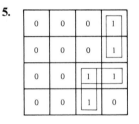

3.

5.

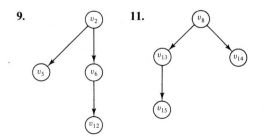

7. $(x' \land y') \lor (x \land y)$

9. $(x' \land z) \lor z'$

11. $(x' \land z') \lor (x' \land z \land w') \lor (x \land y' \land w')$

13. y' **15.** $(x' \land z) \lor (y \land z) \lor (x' \land y)$

17. $(x' \land y' \land w') \lor (x' \land y \land z' \land w) \lor (x' \land y \land z \land w') \lor (x \land y \land w')$

Section 5.1

1. Tree; root is b.   **3.** Tree; root is f.   **5.** Not a tree.   **7.** Tree; root is t.

9.

11.

Section 5.2

1.

3.

5.

7.

9.

11.

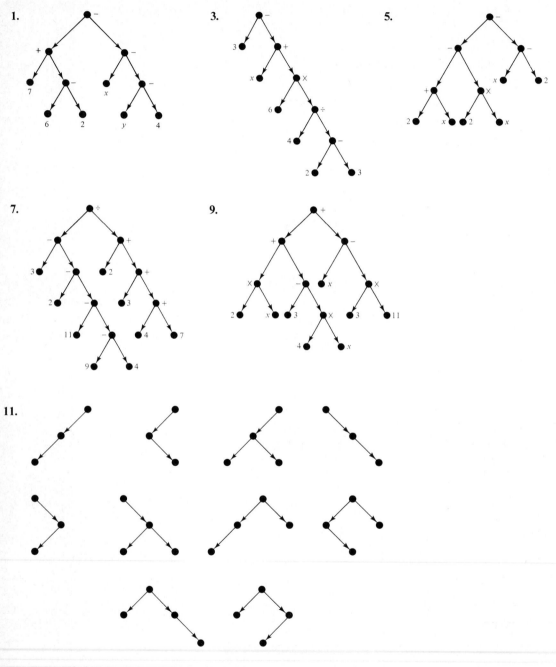

13. 721

Section 5.3

1. $\{x^n y^m z \mid n \geq 0, m \geq 1\}$ **3.** $\{a^{2n}b \mid n \geq 0\} \cup \{a^{2n+1} \mid n \geq 0\}$

5. $\{(\!(\cdots \underbrace{(a + a + \cdots + a))}_{n} \underbrace{\cdots)}_{k} \mid k \geq 0, n \geq 3\}$
$\underbrace{}_{k}$

7. $\{x^n\, y\, z^m \mid n \geq 1, m \geq 0\}$ **9.** (a), (c), (e), (h), (i)

11.   **13.**   **15.** (1) v_0 (2) v_0

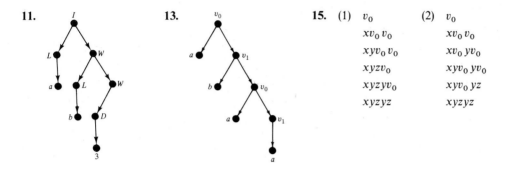

$xv_0 v_0$	$xv_0 v_0$
$xyv_0 v_0$	$xv_0 yv_0$
$xyzv_0$	$xyv_0 yv_0$
$xyzyv_0$	$xyv_0 yz$
$xyzyz$	$xyzyz$

17. (1) I   (2) I
    LW   LW
    aW   LDW
    aDW   LDDW
    a1W   LDDD
    a1DW   L1DD
    a1DD   L10D
    a10D   L100
    a100   a100

19. $G = (V, S, v_0, \mapsto)$
$V = \{v_0, 0, 1\}, S = \{0, 1\}$
$\mapsto : v_0 \mapsto 0v_0 1$
$\quad\quad v_0 \mapsto 1v_0 0$
$\quad\quad 0v_0 \mapsto v_0 0$
$\quad\quad v_0 0 \mapsto 0v_0$
$\quad\quad 1v_0 \mapsto v_0 1$
$\quad\quad v_0 1 \mapsto 1v_0$
$\quad\quad v_0 \mapsto \Lambda$

21. $G = (V, S, v_0, \mapsto)$
$V = \{v_0, a, b\}, S = \{a, b\}$
$\mapsto : v_0 \mapsto av_0 b$
$\quad\quad v_0 \mapsto ab$

23. $G = (V, S, v_0, \mapsto)$
$V = \{v_0, x, y\}, S = \{x, y\}$
$\mapsto : v_0 \mapsto v_0 yy$
$\quad\quad v_0 \mapsto xv_0$
$\quad\quad v_0 \mapsto xx$

25. (a), (d)

Section 5.4

1. *BNF*

$\langle v_0 \rangle ::= x\langle v_0 \rangle \mid y\langle v_1 \rangle$

$\langle v_1 \rangle ::= y\langle v_1 \rangle \mid z$

Syntax Diagram

3. *BNF*

$$\langle v_0 \rangle ::= a \langle v_1 \rangle$$

$$\langle v_1 \rangle ::= b \langle v_0 \rangle \mid a$$

Syntax Diagram

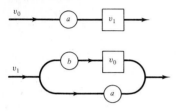

5. *BNF*

$$\langle v_0 \rangle ::= aa \langle v_0 \rangle \mid b \langle v_1 \rangle$$

$$\langle v_1 \rangle ::= c \langle v_2 \rangle b \mid cb$$

$$\langle v_2 \rangle ::= bb \langle v_2 \rangle \mid bb$$

Syntax Diagram

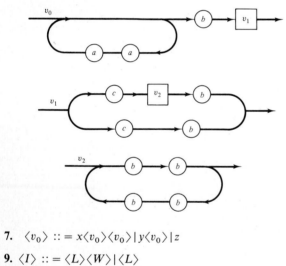

7. $\langle v_0 \rangle ::= x \langle v_0 \rangle \langle v_0 \rangle \mid y \langle v_0 \rangle \mid z$

9. $\langle I \rangle ::= \langle L \rangle \langle W \rangle \mid \langle L \rangle$

$\langle W \rangle ::= \langle L \rangle \langle W \rangle \mid \langle D \rangle \langle W \rangle \mid \langle L \rangle \mid \langle D \rangle$

$\langle L \rangle ::= a \mid b \mid c$

$\langle D \rangle ::= 0 \mid 1 \mid 2 \mid 3 \mid 4 \mid 5 \mid 6 \mid 7 \mid 8 \mid 9$

11. $\langle v_0 \rangle ::= ab \mid ab \langle v_1 \rangle$

$\langle v_1 \rangle ::= d \mid d \langle v_1 \rangle \mid d \langle v_2 \rangle \langle v_1 \rangle$

$\langle v_2 \rangle ::= c \mid d$

13. $(aa)^*(b \vee a)$ **15.** $(ab)^*aa$ **17.** $abc^*d(dd)^*$

Section 5.5

1. *x y s z t u v*   **3.** *a b c g h i d k e j f*   **5.** *y s x z v u t*
7. *g c h b i a k d j e f*   **9.** *s y v u t z x*   **11.** *g h c i b k j f e d a*
13. I NEVER SAW A PURPLE COW I HOPE I NEVER SEE ONE.

15.

INDEX	LEFT	DATA	RIGHT
1	2	⊠	0
2	3	×	5
3	4	−	7
4	6	+	8
5	0	2	0
6	0	2	0
7	0	1	0
8	0	3	0

17.

INDEX	LEFT	DATA	RIGHT
1	2	⊠	0
2	3	1	4
3	5	2	0
4	0	3	6
5	7	4	0
6	0	5	8
7	0	6	0
8	0	7	0

19.   **21.**

Section 5.6 (Solutions 1–11 represent possible answers. Others are also correct.)

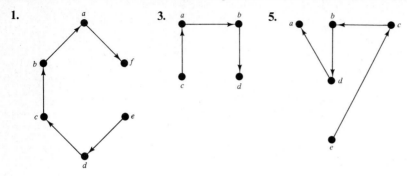

1.

3.

5.

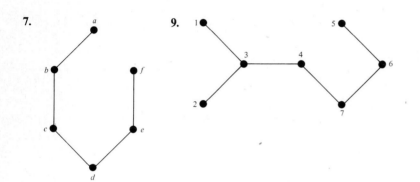

7.

9.

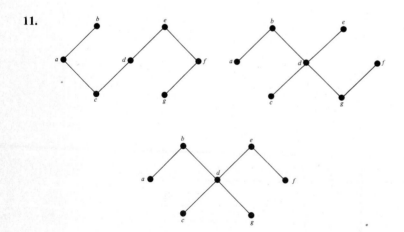

11.

13.

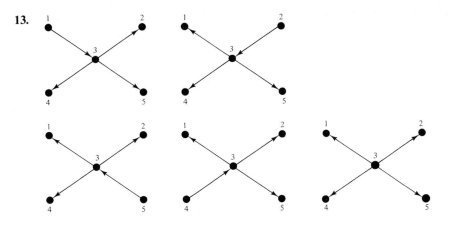

There are five.

Section 6.1

1. Yes **3.** No **5.** No **7.** No
9. Commutative, associative **11.** Not commutative, associative
13. Commutative, associative **15.** Commutative, associative **17.** Commutative, associative

19.

*	a	b	c
a	b	c	a
b	c	b	a
c	a	a	c

21.

*	a	b	c	d
a	a	b	c	d
b	b	a	d	c
c	c	d	a	b
d	d	c	b	a

23. (a) n^{n^2} (b) $n^{[n(n+1)/2]}$

Section 6.2

1. (a) Not a semigroup (b) Monoid (c) Monoid
(d) Monoid (e) Semigroup (f) Not a semigroup
3. Commutative monoid; identity $= 1$. **5.** Semigroup; not commutative
7. Commutative monoid; identity $= S$.
9. Commutative monoid; identity $= 12$. **11.** Commutative monoid; identity $= 0$.

13.

*	a	b	c
a	c	a	b
b	a	b	c
c	b	c	a

15.

$\cup$	$\varnothing$	$\{a\}$	$\{b\}$	$\{a, b\}$
$\varnothing$	$\varnothing$	$\{a\}$	$\{b\}$	$\{a, b\}$
$\{a\}$	$\{a\}$	$\{a\}$	$\{a, b\}$	$\{a, b\}$
$\{b\}$	$\{b\}$	$\{a, b\}$	$\{b\}$	$\{a, b\}$
$\{a, b\}$	$\{a, b\}$	$\{a, b\}$	$\{a, b\}$	$\{a, b\}$

19. No

Section 6.3

5. No **7.** Yes **9.** Yes **11.** Yes **13.** Yes

17. (a)

⊛	$[a]$	$[c]$
$[a]$	$[a]$	$[c]$
$[c]$	$[c]$	$[a]$

(b) $f_R(a) = [a], f_R(b) = [a], f_R(c) = [c], f_R(d) = [c]$

Section 6.4

1. No **3.** Yes, abelian, identity $= 0, a^{-1} = -a$. **5.** No
7. No **9.** No **11.** Yes, abelian, identity $= \emptyset, A^{-1} = A$.
15. (a) $\frac{8}{3}$ (b) $-\frac{4}{5}$

17.

	f_1	f_2	f_3	f_4	f_5	f_6	f_7	f_8
f_1	f_1	f_2	f_3	f_4	f_5	f_6	f_7	f_8
f_2	f_2	f_3	f_4	f_1	f_8	f_7	f_5	f_6
f_3	f_3	f_4	f_1	f_2	f_6	f_5	f_8	f_7
f_4	f_4	f_1	f_2	f_3	f_7	f_8	f_6	f_5
f_5	f_5	f_7	f_6	f_8	f_1	f_3	f_2	f_4
f_6	f_6	f_8	f_5	f_7	f_3	f_1	f_4	f_2
f_7	f_7	f_6	f_8	f_5	f_4	f_2	f_1	f_3
f_8	f_8	f_5	f_7	f_6	f_2	f_4	f_3	f_1

21. No **23.** Yes
29. $D_4, \{f_1, f_3, f_5, f_6\}, \{f_1, f_2, f_3, f_4\}, \{f_1, f_3, f_7, f_8\}, \{f_1, f_5\}, (f_1, f_6), \{f_1, f_3\}, \{f_1, f_7\}, \{f_1, f_8\}, \{f_1\}$

Section 6.5

1.

	$(\bar{0}, \bar{0})$	$(\bar{0}, \bar{1})$	$(\bar{0}, \bar{2})$	$(\bar{1}, \bar{0})$	$(\bar{1}, \bar{1})$	$(\bar{1}, \bar{2})$
$(\bar{0}, \bar{0})$	$(\bar{0}, \bar{0})$	$(\bar{0}, \bar{1})$	$(\bar{0}, \bar{2})$	$(\bar{1}, \bar{0})$	$(\bar{1}, \bar{1})$	$(\bar{1}, \bar{2})$
$(\bar{0}, \bar{1})$	$(0, \bar{1})$	$(0, \bar{2})$	$(\bar{0}, \bar{0})$	$(\bar{1}, \bar{1})$	$(\bar{1}, \bar{2})$	$(\bar{1}, \bar{0})$
$(\bar{0}, \bar{2})$	$(0, \bar{2})$	$(\bar{0}, \bar{0})$	$(\bar{0}, \bar{1})$	$(\bar{1}, \bar{2})$	$(\bar{1}, \bar{0})$	$(\bar{1}, \bar{1})$
$(\bar{1}, \bar{0})$	$(\bar{1}, \bar{0})$	$(\bar{1}, \bar{1})$	$(\bar{1}, \bar{2})$	$(\bar{0}, \bar{0})$	$(\bar{0}, \bar{1})$	$(\bar{0}, \bar{2})$
$(\bar{1}, \bar{1})$	$(\bar{1}, \bar{1})$	$(\bar{1}, \bar{2})$	$(\bar{1}, \bar{0})$	$(\bar{0}, \bar{1})$	$(\bar{0}, \bar{2})$	$(\bar{0}, \bar{0})$
$(\bar{1}, \bar{2})$	$(\bar{1}, \bar{2})$	$(\bar{1}, \bar{0})$	$(\bar{1}, \bar{1})$	$(\bar{0}, \bar{2})$	$(\bar{0}, \bar{0})$	$(\bar{0}, \bar{1})$

5.

⊛	$[0]$	$[1]$	$[2]$
$[0]$	$[0]$	$[1]$	$[2]$
$[1]$	$[1]$	$[2]$	$[0]$
$[2]$	$[2]$	$[0]$	$[1]$

7. (a) H, $[1] + H = \{[1]\}$, $[2] + H = \{[2]\}$, $[3] + H = \{[3]\}$ (b) H, $[1] + H = \{[1], [3]\}$ (c) H
9. (a) H, $[1] + H = \{[1], [5]\}$, $[2] + H = \{[2], [6]\}$, $[3] + H = \{[3], [7]\}$
   (b) H, $[1] + H = \{[1], [3], [5], [7]\}$
11. A coset of H in G is the set $(m, n) + H = \{(m + x, n + x) \mid x \in G\}$, where (m, n) is a fixed element of G. Thus this coset consists of all the points in the plane that have integer coefficients and lie on the line that passes through the point (m, n) and is parallel to the line $y = x$.
15. D_4, $\{f_1, f_3, f_5, f_6\}$, $\{f_1, f_2, f_3, f_4\}$, $\{f_1, f_3, f_7, f_8\}$, $\{f_1, f_3\}$, $\{f_1\}$

Section 7.1

1. 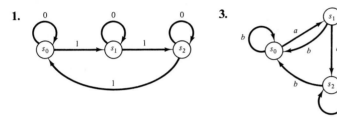   3.

5.

	a	b
s_0	s_1	s_1
s_1	s_1	s_2
s_2	s_0	s_2

7.

	T	F
s_0	s_1	s_0
s_1	s_1	s_1
s_2	s_1	s_2

11.

	0	1
a	a	a
b	a	a

	0	1
a	a	a
b	a	b

	0	1
a	a	a
b	b	a

	0	1
a	a	a
b	b	b

	0	1
a	a	b
b	a	a

	0	1
a	a	b
b	a	b

	0	1
a	a	b
b	b	a

	0	1
a	a	b
b	b	b

	0	1
a	b	a
b	a	b

	0	1
a	b	a
b	a	a

	0	1
a	b	a
b	b	a

	0	1
a	b	a
b	b	b

	0	1
a	b	b
b	a	a

	0	1
a	b	b
b	a	b

	0	1
a	b	b
b	b	a

	0	1
a	b	b
b	b	b

13.

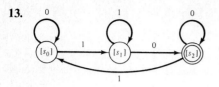

Section 7.2

1. $f_w(s_0) = s_2, f_w(s_1) = s_3, f_w(s_2) = s_0, f_w(s_3) = s_1$

3. The number of 1's is divisible by 4.   **5.** The number of 1's is of the form $2 + 4k, k \geq 0$.

7. $f_w(s_0) = s_0, f_w(s_1) = s_0, f_w(s_2) = s_0$ **9.** w must end in a "b."

11. Strings of 0's and 1's with $3 + 5n$ 1's, $n \geq 0$.

Section 7.3

1. All strings of 0's and 1's with the number of 1's equal to $3 + 4n, n \geq 0$.

3. All strings of 0's and 1's that contain the substring 010.

5. All strings of 0's and 1's that end with the string 01.

7. $G = (V, I, s_0, \longmapsto), I = \{0, 1\}, V = \{s_0, s_1, s_2, s_3, 0, 1\}$

$\longmapsto$:

$s_0 \longmapsto 0s_0$

$s_0 \longmapsto 1s_1$

$s_1 \longmapsto 0s_1$

$s_1 \longmapsto 1s_2$

$s_2 \longmapsto 0s_2$

$s_2 \longmapsto 1s_3$

$s_2 \longmapsto 1$

$s_3 \longmapsto 0s_3$

$s_3 \longmapsto 0$

$s_3 \longmapsto 1s_0$

9. $(0 \vee 1)*1$

11. $G = (V, I, s_0, \longmapsto), I = \{a, b\}, V = \{s_0, s_1, s_2, a, b\}, \longmapsto$ in BNF is

$\langle s_0 \rangle ::= a \langle s_0 \rangle \mid b \langle s_1 \rangle \mid a \mid b$

$\langle s_1 \rangle ::= a \langle s_0 \rangle \mid b \langle s_2 \rangle \mid a$

$\langle s_2 \rangle ::= a \langle s_2 \rangle \mid b \langle s_2 \rangle$

13.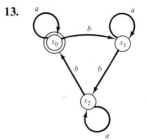

15.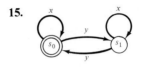

17.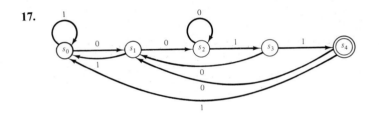

19.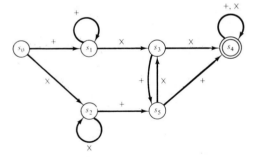

21.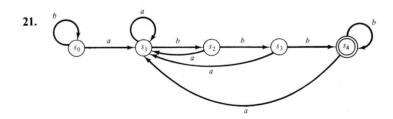

23.

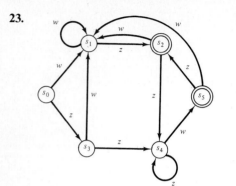
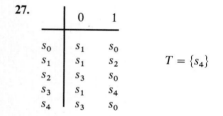
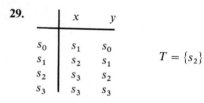

25.

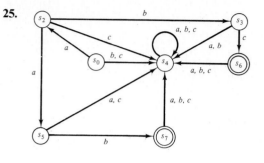

27.

	0	1
s_0	s_1	s_0
s_1	s_1	s_2
s_2	s_3	s_0
s_3	s_1	s_4
s_4	s_3	s_0

$T = \{s_4\}$

29.

	x	y
s_0	s_1	s_0
s_1	s_2	s_1
s_2	s_3	s_2
s_3	s_3	s_3

$T = \{s_2\}$

33. SUBROUTINE PAIRSOFTWOS (RESULT)
1. RESULT ← T
2. STATE ← 0
3. UNTIL (EOI)
    a. CALL INPUT (X, EOI)
       1. IF (EOI = F) THEN
          a. IF (STATE = 0) THEN
             1. IF (X = b) THEN
                a. STATE ← 1
          b. ELSE
             1. IF (STATE = 1) THEN
                a. IF (X = a) THEN
                   1. STATE ← 0
                b. ELSE
                   1. STATE ← 2
                   2. RESULT ← F
4. RETURN
END OF SUBROUTINE PAIRSOFTWOS

Section 7.4

1. $R_0 = \{(s_0, s_0), (s_0, s_1), (s_1, s_0), (s_1, s_1), (s_2, s_2)\}$
3. $R_1 = \{(s_0, s_0), (s_0, s_3), (s_3, s_0), (s_3, s_3), (s_4, s_4), (s_1, s_1), (s_1, s_2), (s_2, s_1), (s_2, s_2)\}$
5. $R_{127} = R_1 = \{(s_0, s_0), (s_0, s_3), (s_3, s_0), (s_3, s_3), (s_4, s_4), (s_1, s_1), (s_1, s_2), (s_2, s_1), (s_2, s_2)\}$
7. $R_2 = \{(s_0, s_0), (s_0, s_1), (s_1, s_0), (s_1, s_1), (s_2, s_2), (s_2, s_3), (s_3, s_2), (s_3, s_3), (s_4, s_4), (s_4, s_5), (s_5, s_4), (s_5, s_5)\}$
9. $R = R_1 = \{(s_0, s_0), (s_0, s_3), (s_3, s_0), (s_3, s_3), (s_4, s_4), (s_1, s_1), (s_1, s_2), (s_2, s_1), (s_2, s_2)\}$
11.

	0	1
$[s_0]$	$[s_5]$	$[s_2]$
$[s_2]$	$[s_0]$	$[s_0]$
$[s_3]$	$[s_3]$	$[s_5]$
$[s_5]$	$[s_3]$	$[s_0]$

$P_R = \{\{s_0, s_1, s_4\}, \{s_2\}, \{s_3\}, \{s_5, s_6\}\}$

13.

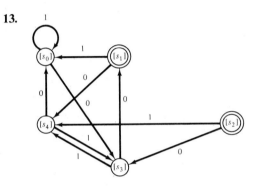

$P_R = \{\{s_0\}, \{s_1\}, \{s_2\}, \{s_3, s_6\}, \{s_4, s_5\}\}$

Section 8.1

1. (a) 3   (b) 2   (c) 3   3. (a) Yes   (b) No   (c) Yes   (d) No
5. (a) 4   (b) 2   9. 1   11. (a) 3   (b) 2 or fewer   15. 2
17. $\begin{bmatrix} 0 & 0 & 0 \\ 1 & 0 & 0 \\ 1 & 0 & 1 \\ 0 & 1 & 0 \end{bmatrix}$

19. $\begin{bmatrix} 0 & 1 & 1 \\ 1 & 1 & 0 \\ 0 & 1 & 1 \end{bmatrix}$

21. $e(000) = 000000$
    $e(001) = 001111$
    $e(010) = 010011$
    $e(011) = 011100$
    $e(100) = 100100$
    $e(101) = 101011$
    $e(110) = 110111$
    $e(111) = 111000$

Section 8.2

1. (a) 011   (b) 101   **3.** Zero   **5.** Zero   **7.** One
9. (a) 01   (b) 11   (c) 10
11. Possible answers: (a) 011   (b) 101   (c) 010
13. Possible answers: (a) 001   (b) 101   (c) 110

Appendix

1. (d) and (e)
3. (a) $3 + 1 < 5$ and $7 = 3 \times 6$; $3 + 1 < 5$ or $7 = 3 \times 6$
   (b) I am rich and I am happy; I am rich or I am happy.
   (c) I will drive my car and I will be late; I will drive my car or I will be late.
5. (a) T   (b) T   (c) T   (d) F   **7.** (d)
9. (a) If I am the King of England, then $2 + 2 \neq 4$.
   (b) If I will not walk to work, then I am President of the United States.
   (c) If I did take the train to work, then I am not late.
   (d) If I will not go to the store, then I do not have time or I am too tired.
   (e) If I will not buy a car or I will not buy a house, then I do not have enough money.
11. (a) $(\sim r) \to q$ (b) $(\sim q) \wedge p$ (c) $(\sim p) \to q$ (d) $(\sim p) \to (\sim r)$
13. (a) Absurdity   (b) Tautology   (c) Contingency   (d) Contingency
    (e) Contingency   (f) Tautology   (g) Contingency
15. Invalid   **17.** Valid, $((p \to q) \wedge p) \to q$ **19.** Valid, $((p \vee q) \wedge \sim q) \to p$

Index

$$G = (V, S, \text{identifier}, \mapsto)$$

$$N = \{\text{identifier, remaining, digit, letter}\}$$

$$S = \{a, b, c, \dots, z, 0, 1, 2, 3, \dots, 9\}, \qquad V = N \cup S$$

1. $\langle \text{identifier} \rangle :: = \langle \text{letter} \rangle \mid \langle \text{letter} \rangle \langle \text{remaining} \rangle$
2. $\langle \text{remaining} \rangle :: = \langle \text{letter} \rangle \mid \langle \text{digit} \rangle \mid \langle \text{letter} \rangle \langle \text{remaining} \rangle \mid \langle \text{digit} \rangle \langle \text{remaining} \rangle$
3. $\langle \text{letter} \rangle :: = a \mid b \mid c \mid \cdots \mid z$
4. $\langle \text{digit} \rangle :: = 0 \mid 1 \mid 2 \mid 3 \mid 4 \mid 5 \mid 6 \mid 7 \mid 8 \mid 9$

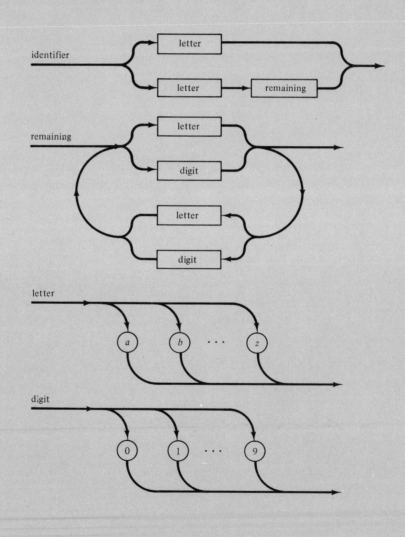